DOWN TO EARTH

THE UNIVERSITY OF CHICAGO PRESS, CHICAGO
THE BAKER & TAYLOR COMPANY, NEW YORK; THE CAMBRIDGE UNIVERSITY
PRESS, LONDON; THE MARUZEN-KABUSHIKI-KAISHA, TOKYO, OSAKA,
KYOTO, FUKUOKA, SENDAI; THE COMMERCIAL PRESS, LIMITED, SHANGHAI

DOWN TO EARTH

AN INTRODUCTION TO GEOLOGY

BY

**CAREY CRONEIS AND
WILLIAM C. KRUMBEIN**

DEPARTMENT OF GEOLOGY
THE UNIVERSITY OF CHICAGO

DECORATIVE DRAWINGS BY

CHICHI LASLEY

THE UNIVERSITY OF CHICAGO PRESS
CHICAGO · ILLINOIS

COPYRIGHT 1936 BY THE UNIVERSITY OF CHICAGO. ALL RIGHTS RESERVED. PUBLISHED MAY 1936. SECOND IMPRESSION SEPTEMBER 1936. COMPOSED AND PRINTED BY THE UNIVERSITY OF CHICAGO PRESS, CHICAGO, ILLINOIS, U.S.A.

DEDICATED WITH AFFECTION TO
THOSE WHO TRIED TO TEACH US
AND TO THOSE WHO TEACH US
WITHOUT TRYING

PREFACE

> Explain a thing till all men doubt it.
> —Pope
>
> Thanks, the exchequer of the poor.
> —Shakespeare

SAMUEL JOHNSON once said that "no man ever read a book of science from pure inclination." This unguarded statement was of course made long before the flare for popular science made "best sellers" possible in the fields of chemistry, physics, astronomy, biology, and geography. But what about geology? Since the turn of the century there have been many excellent textbooks issued on the subject, and a few outstanding popular accounts also have appeared. Yet none of these works had a very large circulation when contrasted with the many editions run through by Hugh Miller's general works on geological subjects and by Sir Charles Lyell's *Principles*, published about a century ago. Moreover, even during  the past generation physical geology (usually called "physiography") has declined from one of the most popular of the high-school sciences to a position of little consequence in the curriculum of the secondary schools.

To us these are situations which seem paradoxical in the extreme. We find it just a little hard to understand man's intelligent and commendable interest in Arcturus, and the time it takes light to come from that distant star, when he seems so abysmally ignorant of many facts concerning the planet from whose vantage point he scans the heavens. We are continually amazed at the student's familiarity with hydrogen, which he cannot see, and his apparent woeful ignorance of the hydrosphere, which he can see and feel and which is all about him. We are eternally surprised at the layman's relatively decent understanding of atoms, electrons, and genes, which he cannot directly observe, and his lack of knowledge concerning rocks, minerals, and fossils, which he cannot only see but can collect in his ordinary rambles in the out-of-doors.

WE DO not believe that these apparent anomalies can long continue with the ever growing interest in getting "back to nature" which is so clearly evidenced in various ways, such as by the ever increasing flow of tourists through the national parks. Similarly, modern trends in education are resulting in larger numbers of "liberal-arts" students attending orientation courses in the physical sciences. Several years of experience with such courses, and indeed the students themselves, have indicated the need for a brief text covering the principle features of the science of geology. Accordingly, we have prepared this volume for use in any general course in the physical sciences and also as a text for shorter courses in geology.

We have attempted to enliven the subject without in any sense writing it down. Contrary to most introductory treatments, which lean toward the descriptive aspects of the science, this text attempts to develop a more analytical attitude. In fact, we have included more rigorous treatments of many geological phenomena than commonly have been presented in texts, in order that we might bring out the true relationship of geology to the other sciences. We believe that this inclusion of certain astronomical, chemical, physical, and biological discussions makes for a more unified treatment of the entire subject of the earth sciences. For this reason we hope that the intelligent layman also may find in this volume an introduction to geology which will be relatively easy, or "popular," scientific reading without being at the same time spectacular or, at best, unauthoritative reading.

PREFACE

WITH Colonel Sir Ronald Ross we have wondered "in what witches' cauldron of folly the absurdity was brewed that poetry and science are enemies." And with Hooker we are also agreed that there is no surer sign of one's not appreciating the aim and scope of the science he cultivates than exhibiting a craving to load it with names. Consequently, we have embellished most chapters with what we hope is an apt quotation or two, and at the same time have denuded the pages of as much of the professional jargon of geology as possible. With a strong appreciation for all science, and with a real affection for our own subject, we nonetheless feel that many scientists take their own theories and beliefs too seriously. Consequently, although we have not approached our topic in any spirit of levity, we have made an attempt to avoid the ponderosity which marks some introductory works in science.

ALL science is intrinsically interesting; geology can be particularly so. We therefore have felt that the treatment of the subject at least should not be more dull than the science itself. We have not been impressed in the slightest by the widespread and absurd misconception that popularized science must necessarily be inaccurate science. Lyell said, "I must write what will be read"; and, studiously following his own dictum, he achieved the largest scientific audience of his day, without sacrificing accuracy of statement. Nor have we felt it necessary to give the reader the impression that all the questions in geology have been solved. The science of today is not the science of tomorrow, and we have a feeling that this is a point too little emphasized by scientific writers. The interest of the wide-awake reader is certainly more readily challenged by the reality of unsolved problems than by the myth of definite answers to all of Nature's riddles, however fascinating and plausible the solutions can be made to sound. Accordingly, we have made no attempt to conceal from the reader the fact that the science of geology is a living, changing, growing one.

Any attempt to discuss geology in all its ramifying aspects is confronted with the old problem of the molasses men who couldn't get hot until they ran, and, of course, couldn't run until they got hot. Naturally, historical geology cannot be well understood without a rather complete knowledge of physical geology, but the great subject of geological processes can be plunged into at almost any point with both advantages and disadvantages. We hold no particular brief, therefore, for the order of presentation followed in this volume. We do, however, feel that it is a thoroughly logical one and that it permits the development of a relatively continuous story.

THE sixty-four pages of rotogravure pictures represent an innovation in textbook illustration and make it possible to reproduce more pictures than can conveniently be inserted in a text by the usual procedure. Such excellence as the pictures may possess is due largely to the fine co-operation of various airplane and railroad companies, the Royal Ontario Museum, the Field Museum, the Smithsonian Institution, the Museum of Science and Industry, Chicago, the Illinois Geological Survey, and the National Park Service. All rotogravure photographs, except those taken by the writers, are credited directly to the organization which supplied them. The text illustrations have been culled chiefly from the ancient works in geology and other fields, or have been drawn especially for this volume by Chichi Lasley, to whom we owe a great debt of gratitude, not only for her interesting drawings, but for her many valuable suggestions. Raymond Hirsch also made a number of drawings for the historical section of the book. As a general rule the text illustrations have been selected with a view toward carrying out the same style throughout the entire volume. They have been designed, in most cases, to give general ideas rather than specific information.

WE ARE grateful to all our colleagues of the Physical Sciences General Course of the University of Chicago, especially to Drs. Harvey B. Lemon and Hermann I. Schlesinger for their encouragement and advice, and to Drs. Walter Bartky, Reginald Stephenson, and Eugene J. Rosenbaum for pertinent criticism particularly of those portions of the book which deal with physical geology. Professor Carl Eckhart, of the Department of Physics, has also given us the benefit of his experience in developing the plan of presentation of some of the problems in which the physical concepts are important.

PREFACE

Our colleagues in the Department of Geology have been extremely helpful, Drs. E. S. Bastin, A. Johannsen, R. T. Chamberlin, J H. Bretz, A. C. Noé, F. J. Pettijohn, and E. C. Olson each having read and made suggestions on that section of the book corresponding to his field of specialization. Dr. W. W. Rubey, of the United States Geological Survey, was kind enough to give us advice on the chapters dealing with the process of gradation; M. King Hubbert, of Columbia University, has given us his criticism on that part of the volume which concerns structural geology; and Professor J. B. Macelwane, S.J., of St. Louis, has critically examined the chapters on earthquakes and the interior of the earth. To all of these specialists we hereby tender our sincere thanks and appreciation, and, at the same time, following an ancient custom, acknowledge publicly that the sins of omission and commission with which this book may be charged are ours, not theirs.

We are deeply indebted to the following officials for supplying us with photographic material: Dr. S. C. Simms, director, Field Museum; Dr. Ray S. Bassler, curator of geology, United States National Museum; Mr. O. T. Kreusser, director, Museum of Science and Industry; Dr. W. A. Parks, director, Royal Ontario Museum of Paleontology; Messrs. H. C. Bryant and Earl Trager, of the Educational Division of the National Park Service; Mr. H. E. Stevens, vice-president of the Northern Pacific Railway Company; Mr. C. J. Hanratty, of the Canadian National Railways; and Dr. M. M. Leighton, director, Illinois Geological Survey. We are likewise under obligation to Messrs. Homer J. Smith and Dan Jones, of the Department of Geology, University of Chicago, for assistance in the preparation of the illustrative material.

Finally, we wish to express our appreciation of the kindly assistance given throughout the period of planning and writing this volume by Mr. Donald Bean, manager of the University of Chicago Press, and of the intelligent co-operation of Miss Mary D. Alexander, book editor of the same organization, in the arduous work of seeing the book through the press.

CAREY CRONEIS
WILLIAM C. KRUMBEIN

UNIVERSITY OF CHICAGO
March 1, 1936

CONCERNING ILLUSTRATIVE MATERIAL

THE rotogravure plates that accompany this book have been inserted in several groups throughout the volume. Each group contains the plates that illustrate the chapters immediately preceding or following it. The individual plates illustrate or elaborate on certain concepts presented in the chapters to which they are referred, but they have been designed to tell their own stories. They may be studied, therefore, as a group when the rotogravure sections are reached in the text, or they may be consulted in connection with each chapter.

NO ATTEMPT has been made to include every aspect of geology, but rather we have tried to show, by means of carefully chosen photographs, some of the more interesting or significant phases of the science. It is hoped that this type of organization will enhance the usefulness of the book by making available for comparative study, in appropriate groups, the photographs on related subjects that commonly are widely scattered throughout a text. Within the text itself many topics of geology are illustrated by reproductions of old cuts or by modern drawings and graphs.

TEACHERS of the sciences and scientific clubs also may be interested in the six reels of talking motion pictures on physical geology which are used to supplement this volume in the Geology Section of the Physical Sciences General Course at the University of Chicago. Prepared, under the direction of Dr. Croneis, by Erpi Picture Consultants, these films may be secured through the University of Chicago Press.

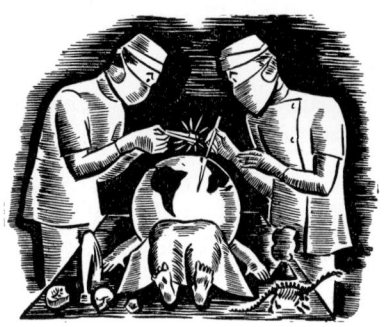

CONTENTS

CHAPTER	PAGE
1. INTRODUCTION, in Which We Come Down to Earth and Learn What Geology Is	1
2. SECOND-RATE PLANET, or an Investigation of the Size, Shape, and Motions of Our Terrestrial Sphere	5
3. SOLID, LIQUID, AND GAS, in Which Are Discussed the Major Divisions of the Earth, Something about Environments, and a Bit of Chemistry	13
4. ENERGY AUDIT, in Which We Examine the Earth's Sources of Energy, and Follow This Energy through the Economy of Our Terrestrial Globe	23
5. THE TOOTH OF TIME, Which Gnaws Away at the Solid Rocks of the Earth's Surface, Disintegrating Them to Loose Débris . . .	33
6. WINDS AND TURBULENT MOTION, in Which We Inquire into the Capriciousness of the Wind, and Its Transporting Power . .	44
7. STREAMS AND VALLEYS, Wherein We Learn the Transportation Secrets of the Babbling Brook	51
8. THE EVERLASTING HILLS, Which Are Not Everlasting . . .	60
9. UNDERGROUND WATER AND LAMINAR FLOW, Wherein We Delve beneath the Surface of the Earth and Observe the Behavior of the Water There	69
10. RIVERS OF ICE Resemble Rivers of Water in Some Respects, but Not in Their Manner of Transporting Rock Débris	83
11. END OF THE LINE, in Which Our Attention Is Turned to the Causes Controlling the Subsequent Deposition of Materials Carried by Wind, Streams, and Ice	90
12. THE BOUNDING MAIN, in Which We Visit the Seacoast and Observe the Geological Work of Waves and Currents	100

CHAPTER	PAGE
13. PLUS AND MINUS, Wherein Are Summarized Our Observations on Land Features in Terms of Their Origin	110
14. EXCURSIONS AMONG THE SAND GRAINS, or a Visit to a Geological Laboratory, Where We Strike Up an Acquaintance with Some of the Commoner Sedimentary Rocks	116
15. BRIEF INTERLUDE, during Which We Pause To Reflect on the Past and To Consider the Future	126
16. WHAT PRICE CONTINENTS? Wherein We Examine Evidences of a Disturbing Force within the Earth, and What It Has To Do with the Existence of Lands above the Sea	129
17. FOLDS AND FAULTS, in Which a Few Problems Are Raised concerning Structures among the Rocks, Some of Which Require a Bending of Those Brittle Substances	137
18. EARTHQUAKES, or Shivers and Spasms within the Earth's Body	145
19. JOURNEY TO THE CENTER OF THE EARTH, in Which We Use Seismologists as Our Guides	156
20. VULCAN'S CHIMNEYS, Those Fiery Mountains, and Why They Behave as They Do	166
21. IGNEOUS INTRUSIONS, or What Goes on beneath the Surface While Volcanoes Smoke above	177
22. FROZEN SEAS OF LAVA, Wherein We Introduce the Igneous Rocks, and Observe Chemical Reactions That Took Place Millions of Years Ago	183
23. HEREDITY VERSUS ENVIRONMENT, or What May Happen to Rocks That Become Involved with Certain Forces of Nature	196
24. THE EVOLUTION OF ROCKS, in Which We See That There Is No Beginning or End to the Subject	206
25. ANCIENT HISTORY, or How We Read the Oldest Story of Them All	215
26. UNIVERSAL CEMETERY, Wherein We See That, after All, Dead Things May Tell Tales	221
27. CORRELATIONS, or a Discussion of the Geologist's Methods of Determining the Age Equivalency of Beds in Widely Separated Areas	235
28. OLDER FLOODS than the Noachian Deluge Leave Their Records for the Geologist To Read	246

CONTENTS

CHAPTER | PAGE

29. MOUNTAINS AND CLIMATES, in Which Are Discussed Older Generations of Mountain Ranges, and Climatic Changes throughout the Past 257

30. VESTIGES OF CREATION, in Which We Examine the Geological, Astronomical, and Physical Evidences Which Throw Light on the Origin of the Earth 268

31. EARTH'S BIRTH, Wherein We Theorize as to Just How It Happened 275

32. GARGANTUAN CALENDAR, in Which We Discuss the Methods of Determining the Length of Geological Time 291

33. PRE-CAMBRIAN, or the Discussion of the Relatively Little-Known Oldest Periods of Geological History 301

34. LIFE BEGINS at a Very Early Date in the Earth's History and Starts To Record Its Complex History 315

35. EARLY PALEOZOIC EVENTS, or the Physical History of the Cambrian, Ordovician, and Silurian Periods 321

36. INVERTEBRATE HEYDAY, or the Time When Creatures without Backbones Ruled the Organic World 333

37. LATE PALEOZOIC EVENTS, in Which Are Described the Chief Episodes of Devonian, Mississippian, Pennsylvanian, and Permian Time 344

38. THE FOREST PRIMEVAL, or the Early History of the Plants . . 357

39. CROSSING THE STRAND, Wherein Animal Life Comes out of the Water onto the Land 366

40. AN EARLIER DEPRESSION Weeds Out the Unfit and Turns Up New and More Versatile Forms of Life 377

41. EARLY MESOZOIC EVENTS, in Which Is Recited the Main Episodes in the Physical History of the Triassic and Jurassic Periods . 385

42. END OF AN ERA, or the Outstanding Physical Events of the Cretaceous Period 392

43. MEDIEVAL LIFE, or the Animals and Plants of the Mesozoic Era 400

44. MEGALOMANIA, or the Curse of the Dinosaurs Who Placed Reliance on Brawn, Not Brains 408

45. CEPHALOPODS, or the Story of a Group of Animals Which Are Illustrative of Many Biological and Geological Principles 416

CHAPTER	PAGE
46. THE TERTIARY, or the Events of the Third Great Period of Geological Time Recognized by the Early Writers	424
47. THE QUATERNARY, or the Period of Recent Glaciation	432
48. THE WARM-BLOODED, in Which Is Discussed the Career of the Mammals	447
49. HOMO DILUVII TESTIS, Wherein We Discuss Ancient Man's Lineage as Influenced by Geological Changes	458
50. MONEY AND POLITICS, and the Influence of Geological Phenomena on These Two Much-Talked-About Subjects	467
51. PROSPECTS OF AN END, Which Is Both a Summary and a Conclusion	478
INDEX	485

CHAPTER 1

INTRODUCTION

I hate definitions.—DISRAELI

ACCORDING to Fontenelle, "people generally admire what they do not comprehend." If this were true, geology would indeed be regarded with wonder and delight. For geology is probably to be considered as the least commonly understood of the major branches of science. The fact is that people generally love the soil, but few know even the rudimentary facts concerning its formation; they thrill at the sight of majestic waterfalls, towering mountains, and breath-taking canyons, but only a handful understand much, if anything, about the origin of scenery; they sense romance in any story of famous gems, but not one in a thousand can explain how they were formed; they may even realize that the distribution of mineral wealth has caused wars and altered the course of history, but can you remember meeting anyone who knew how the metals were deposited and why they occur where they do? In short, although certain phases of the science of geology

According to Fontenelle .... (a geyser at Yellowstone)

have a real meaning in the life of almost every human being, the individual who knows very much about the subject itself is, outside of those in the mining camps and the oil and gas fields, a *rara avis* indeed.

SINCE many a "man in the street" has been instructed in the wonders of the heavens, has at least heard about Einstein and his theories, is familiar with "Creative Chemistry," and has a smattering of physics "From Galileo to Cosmic Rays," it is perhaps not unreasonable to suggest that we all come "Down to Earth," and also learn a modicum about the terrestrial sphere on which we live. In order to do so, we may well first briefly define our subject, suggest its importance, and discuss our plan of presentation.

GEOLOGY is the science which deals with the history of the earth and its inhabitants. More specifically, it is concerned with the earth's constitution and structure, with the various stages through which it has passed, with the living things it has nourished, with the agencies and processes which are continually altering it, and with the utilization of the earth's materials by man. Standing on the broad foundations of astronomy, mathematics, physics, chemistry, and biology, geology comprises in its superstructure a long list of divisions, such as physiography, meteorology, mineralogy, petrology, economic geology, paleontology, and too many other "ologies" to be mentioned.

Thus geology includes the study of such apparently diverse subjects as the earth's landscapes, the processes which actively shape its surface features, weather and climate, mineral resources, the origin and development of animals and plants, minerals, rocks and gems, and even the origin of man himself. From one point of view, therefore, geology may be regarded as a comprehensive study of ancient geography; conversely, modern geography makes up the last chapter in the great book of geology.

GEOLOGY has played an all-important rôle in the development of our modern industrialism; and, with physics and chemistry, it has sponsored the "technical ascent of man." In this technical rise, geology's main contribution has been its making available the raw materials upon which our present economic structure is based. One hundred years ago the United States produced practically no lead, zinc, or copper, only insignificant quantities of coal and iron, and so little gold that the total per capita amount was only about three cents. Today, owing to great advances in the basic sci-

INTRODUCTION

ence of geology and the profession of geological engineering, practically all of the mineral resources of the country have become known, at least in a general way, and the production of most of our raw materials now goes on at a rate which is almost incomprehensible. For example, this country alone is now producing approximately a billion barrels of oil a year.

"With mind and hammer"— the motto of the geologist

A billion barrels of oil is an almost inconceivable flood of petroleum; yet, the amount could be greatly increased. This fundamental liquid resource has been made readily available by geological engineering, which has advanced to its present high degree of efficiency solely through the scientific application of basic geological principles. Thus the location of many new oil fields is today determined by the application of physical methods of prospecting. Even the exact bed in which the oil is likely to be encountered, and consequently the depth to which the drill must penetrate, is commonly accurately determined in advance by examining the microscopic fossil animals which occur in the rock fragments bailed from the drilling well.

THE direct social consequence of the geologist's development of the petroleum and other mineral resources of the world are too far-reaching and commonplace to need elaboration. But equally important is the fact that, among the sciences, geology in particular develops an appreciation of the enormous duration of time involved in the gradual evolution of the earth and its inhabitants. It demonstrates the continuity of the past and the present, and furnishes the background for an understanding of man's origin and his present place in nature. Geology also deepens man's aesthetic appreciation of scenic wonders through imparting a knowledge of the processes through which the great physical features of the earth have originated.

IT MUST be obvious that, in a single volume, it is impossible to give anything more than a survey of such a broad subject. We shall therefore attempt merely to introduce the reader to some of the fundamental concepts of earth science. In order to do this, and finally reconstruct for him the most ancient of all histories, we must first describe the entire physical makeup of the earth, and build up a picture of the ebb and flow of energy across its face and within its interior, and demonstrate the profound changes which this energy-flow brings about. If we can absorb an understanding of what a dynamic thing this world is, despite the extreme slowness with which its major events take place, we shall be in a fair way to understand the majesty of the past.

Toward such an end, then, the volume begins with brief chapters on several fundamental earth characteristics. From astronomy comes a picture of the earth and its motions; from geography we obtain some knowledge of the present configuration of the earth and its major features; chemistry supplies us with a knowledge of earth materials; and from physics we learn of the energy relations of the earth. On this background we shall etch the principles of earth science proper. The interactions of air, water, and solid earth involve several great geological processes which pretty thoroughly shape the present events on the earth. Each of these will be developed in some detail; and when we are finished with them, we shall have some grasp of **dynamical geology.** The principles learned in this section will then be applied to the unraveling of the tangled past; and when it is demonstrated that these principles actually do apply, we shall finally sketch for the reader the major events of the past history of the earth, and carry the story of **historical geology** back to the very birth of this terrestrial sphere from the sun.

CHAPTER 2

SECOND-RATE PLANET

> The rounded world is fair to see,
> Nine times folded in mystery.
> —EMERSON

AMONG the nine planets that speed endlessly around the sun is this terrestrial sphere. It is neither the largest nor the most massive, and astronomers are apt to remind us not only that it is a second-rate planet but that our sun itself is only a third-rater among the other stars. But because the earth is our abode, to us it is of supreme importance.

One of the first things we learn about the earth is that it is round. Some of us may even remember that a simple proof is that large objects gradually disappear below the horizon as one moves away from them. This is at best only a qualitative argument, but there are several precise proofs. As an example of the scientists' method of attacking such problems, let us follow through the steps involved in determining both the shape and size of the earth.

Among the nine planets .... is this terrestrial sphere

A LITTLE study of a circle will show that the property of circularity involves certain relations between the length of an arc of the circle and the angle made by the radii limiting this arc. Note in the accompanying figure that the radii OA and OB establish the arc AB. This arc is so related to the angle a that if we choose another arc equal in length to AB, such as CD, then the angle b must be of the same size as angle a. In the same manner, an arc twice the length of AB will be subtended by an angle twice the size of a. We may now argue that if every section through the earth is a circle, then equal distances on its circumference will be subtended by equal angles made by the radii from the center of the earth to the ends of the arcs. In order to measure these internal angles, we must make use of astronomy. Let us, therefore, imagine two observatories about 2,100 mi. apart along a north-south line. In each observatory we shall suspend a plumb line, which is nothing more than a string with a weight on the end. This line will point to the center of the earth if the earth is round. Now at the first observatory a telescope is pointed at a star directly overhead. At the same instant, the telescope in the second observatory is pointed at the same star; but because the second observatory is some 2,100 mi. from the first, the star obviously will no longer be directly overhead. Thus there will be an angle between the plumb line and the direction of the telescope. This angle is measured, and it is found to be about 30°. We now make a diagram like the adjacent one. At A is the first observatory, with the plumb line and the telescope pointed directly overhead at the star (which is too distant to be included in the same diagram). At B is the second observatory, with its telescope also pointed at the same star, but with an angle between the telescope and the plumb line. Note, however, that the two telescopes themselves are parallel, because the distance to the star is so great that the rays of light reaching the earth from it are practically parallel at any point on the earth.

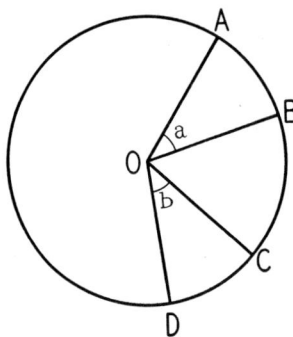

Angle a is the same as angle b, provided arc AB = arc CD

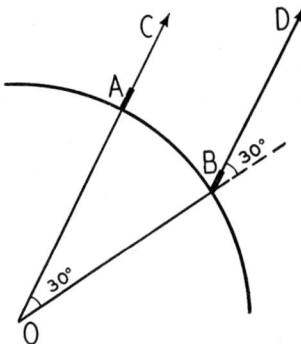

Two observatories 2,100 mi. apart simultaneously observe the same star

Remembering that the angle between the telescope and the plumb line of the second observatory is 30°, we next project the plumb lines into the interior of the earth, where they meet. The angle made by their junction is also 30°, because the lines AC and BD are parallel, and the line OB cutting these parallel lines makes an equal angle with each. Thus, it follows that an angle of 30° at the earth's center subtends a distance of about 2,100 mi. along the surface. If we next divide through by 30, we see that an angle of 1° must subtend a distance of some 70 mi.

Should the same experiment be repeated elsewhere on the earth, and the net result prove to be the same, then we might safely conclude that the earth is spherical. Such measurements have actually been made at many places over the surface of the earth, and everywhere the general relations hold, so that there is no doubt that the earth is spherical. Precisely speaking, of course, the earth is a spheroid, and flattened slightly at the poles. This flattening is so slight that between the polar and equatorial diameters there is a difference of only 27 mi., which is less than one-half of 1 per cent. It is, in fact, so insignificant that it amounts to only about $\frac{1}{2}$ in. on a 10-ft. globe, and consequently is not shown even on the largest of such representations of the earth.

WE HAVE by these experiments found not only that the earth is spherical but also that every degree of the interior angle subtends about 70 mi. on the surface. Since there are 360° in a complete circle, we need only multiply our 70 mi. by 360 to obtain the *circumference*, or distance around the earth. When this is done, the result is approximately 25,000 mi. Once we have the value of the earth's circumference, we can easily find its *diameter*, or the distance through the earth. We simply divide the circumference by 3.1416 (the value of the old familiar *pi* of school days) and the answer is about 8,000 mi., which is the approximate diameter of the earth.

AS LONG as we are on the subject of measurements, let us pursue it a bit farther. If we know the diameter of the earth, we can find its *volume*, because the volume of any ball is 3.1416/6 times the diameter cubed. If we put in our known values, we find that the volume of the earth is nearly 260,000,000,000 cu. mi. Similarly, since the surface of a sphere is equal to *pi* times the diameter squared, we may compute the *surface area* of the earth. Such a computation shows us that there are about 197,000,000 sq. mi. on its broad surface. But large as this area may seem, it is only about one twelve-thousandth the area of the sun's great surface. In other words, the earth's surface area is as much smaller than the sun's as tiny Switzerland is smaller than the earth's entire extent of land and sea.

ANOTHER important item concerns the *mean density* of the earth, or its mass per unit of volume. To find this, it is necessary first to determine the mass of the entire earth and then to divide this by its volume. The mass of the earth can be determined by several methods, but perhaps the simplest is that known as the **Cavendish experiment.*** By this procedure the attraction of two large lead balls for two small silver balls is measured, and then compared to the attraction between the earth and one of the smaller balls. From the exceedingly delicate measurements made in this experiment it is found that the mass of the earth is 6×10^{21} tons, which is the number 6 followed by 21 ciphers. To obtain the mean density of the earth, then, we divide the mass by its volume; whereupon the density is found to be approximately 5.5. In other words, a cubic foot of average earth-substance weighs about 5.5 times as much as a cubic foot of water. But right here we stumble onto an interesting thing. The average density of the familiar rocks of the earth's surface is only about 2.7. Thus the density of the earth's interior must be greater than 5.5, in order to counterbalance the lightness of the surface materials. This interesting problem of the earth's density will receive further attention in chapter 19.

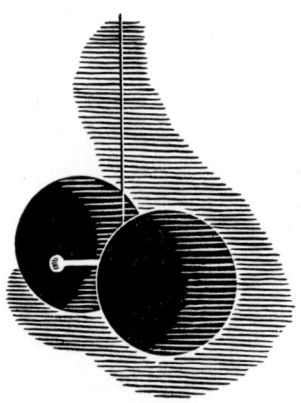

The Cavendish experiment

FOR the purposes of later discussions we also ought to have before us a fairly complete picture of the earth's motions in space.

Most proofs of the *earth's rotation* are difficult to grasp without some knowledge of physics. The one usually cited depends on the

---

* The interested reader is referred to H. B. Lemon, *From Galileo to Cosmic Rays* (Chicago: University of Chicago Press, 1934), chap. 5, for the details of this experiment.

fact that a heavy pendulum tends to swing in a fixed direction in space as long as no outside forces disturb it. Evidently, then, if such a pendulum is set into motion on a rotating earth, the earth should rotate beneath the pendulum without affecting the direction of swing of the pendulum. Such, in fact, is the result of the pendulum experiment. When first set into motion, it is directed so that the bob swings in a north-south plane. As time goes on, it is found that the pendulum continually departs from this direction, and that in the Northern Hemisphere the plane of swing tends to rotate in a northeast-southwest direction.* It must be borne in mind, however, that an observer on the earth has the same motion as the earth, and therefore he will feel himself at rest and will notice that the direction of swing of the pendulum appears to change. This departure would increase indefinitely were it not for friction stopping the swing of the pendulum. This experiment was first performed by Foucault in Paris, in 1851, and in his honor the device is called the **Foucault pendulum.** The theory of the Foucault pendulum is entirely beyond the scope of this book, but it is interesting to note that observation and theory agree so well that there is no doubt of the actual rotation of the earth upon its axis.

THE motion of the earth about the sun, or the *revolution* of the earth, also can be demonstrated in several ways. One of the simplest is by the periodic shift in the position of the nearby stars with

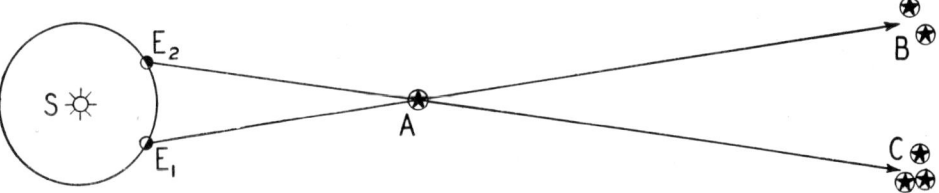

The apparent shift of the nearby star proves the earth's revolution. The sizes and distances are obviously not true to scale

respect to the more remote stars. In the adjoining figure the earth's orbit around the sun, S, is shown diagrammatically. The earth it-

---

* The direction of change depends on whether the observer is north or south of the Equator, and the rate of change depends on his latitude. In the latitude of Chicago the plane of swing changes about 10° an hour. At the Equator itself there is no change.

self appears in two successive positions, E_1 and E_2, several months apart in its journey around the sun. Star A is one of the nearer stars; and in the distance are two groups of stars, B and C, representing more remote constellations. Notice that when the earth is at E_1, the star A, as viewed from the earth, appears to be in group B, but that later, when the earth is at E_2, star A appears to have moved to a position among group C. Thus star A periodically appears to shift back and forth between groups B and C. Such shifts actually are noted among the nearer stars, and the phenomenon is cited as the **parallax proof** of the earth's revolution.

THE *inclination of the earth's axis* to the plane of its orbit is another feature of this second-rate planet into which we should briefly inquire. This inclination is demonstrated by the fact that at any point on the earth's surface the altitude of the sun above the horizon at noon on June 21 and December 21 differs by 47°. This apparent change in the altitude of the sun measures twice the angle of inclination of the earth's axis, which thus turns out to be $23\frac{1}{2}°$. In the adjacent figure the line OS, from the center of the earth to the center of the sun, lies in the plane of the orbit. Now at O let us erect a perpendicular OA, and then draw the earth's axis, OP, at an angle of $23\frac{1}{2}°$ to OA. This angle, POA, measures the inclination of the earth's axis.

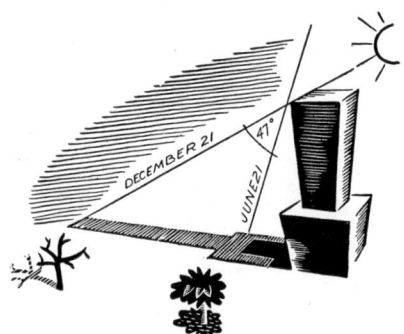

The angle of the sun's rays differs by 47° in June and December; a proof of the inclination of the earth's axis

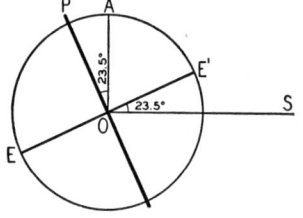

Diagram of the inclination of the earth's axis

Inasmuch as the Equator of the earth is at right angles to its axis, it follows that the equator is inclined $23\frac{1}{2}°$ to the plane of the earth's orbit. This is line EOE'. It is the inclination of the earth's axis that causes the seasonal changes on the earth, as the figure below shows. The earth, like any rotating body, tends to keep its axis always oriented in the same direction in space, regardless of the earth's position in its orbit. Notice, then, that when the earth is at the right in the diagram, the region receiving perpendicular rays from the sun lies $23\frac{1}{2}°$ *north* of the Equator; whereas when the earth is at the left, the perpendicular rays strike the surface $23\frac{1}{2}°$ *south* of the Equator. In the course of a year, then, the sun seems to swing through an angle of twice $23\frac{1}{2}°$, or $47°$, in its apparent motion above and below the Equator. When the earth is at the right, it is summer in the Northern Hemisphere, because at that time the northern half of the earth is receiving the more nearly perpendicular rays from the sun. In like manner, summer in the Southern Hemisphere and winter north of the Equator occur when the earth is on the left.

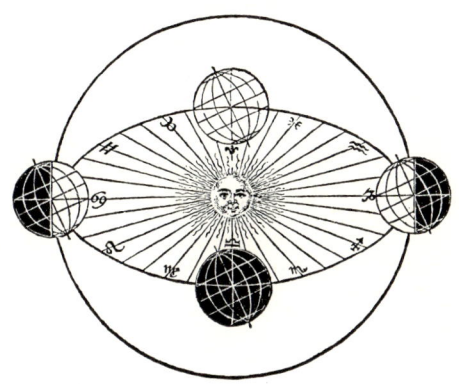

The earth in its journey around the sun. Old engraving from Blaev's *Institution Astronomique* (Amsterdam, 1669).

THE several points which have been discussed furnish us with the modicum of astronomical data that we need as a background for topics to be presented later. There are, of course, many other interesting features about the astronomical relationships of

Item	Value
Shape	Very nearly spherical
Average circumference	24,800 mi.
Average diameter	7,910 mi.
Volume	$2,599 \times 10^8$ cu. mi.
Surface area	197×10^6 sq. mi.
Mass (weight)	6.1×10^{21} tons
Average density	5.5 times that of water
Period of rotation	24 hr.
Period of revolution	$365\frac{1}{4}$ days
Inclination of axis	$23\frac{1}{2}°$

the earth, and the curious reader is urged to consult an elementary book on astronomy for details.*

Above, for convenience and future reference, we have included in tabular form a summary of the salient facts discussed in this chapter.

By this time the reader should have in his mind's eye a fairly clear picture of the earth as a sphere rotating daily on its axis, like a fowl on a spit, as it annually revolves about the warming fire of that unfailing solar hearth, the sun. Second-rate though it may be, this terrestrial sphere of ours has taken part in the stately procession of day and night, summer and winter, since time immemorial; and judging from its astronomical relations, it bids fair to continue into a future immeasurably long.

* See, for example, Walter Bartky, *Highlights of Astronomy* (Chicago: University of Chicago Press, 1935), chap. 1.

CHAPTER 3

SOLID, LIQUID, AND GAS

> The earth was made so various, that the mind
> Of desultory man, studious of change
> And pleased with novelty, might be indulged.
> —Cowper

THE scientist thinks of the earth as an insignificant spheroid reeling through space, but the majority of us are not inclined to think about it at all. This is not surprising. The earth usually is taken for granted as the stable part of our environment. Only during cataclysmic disturbances such as earthquakes, volcanic eruptions, or catastrophic floods are the unthinking ever reminded that this terrestrial sphere is indeed a mobile earth; is, in fact, an unimportant cosmic speck swayed by a myriad of external forces, wracked by a legion of internal stresses.

We all suffer from a provincial attitude, partly because of the limited part of the earth that we inhabit. Perhaps 90 per cent of the earth's population lives in a narrow zone between sea level and about 1,000 ft. above the sea. This is an extremely thin zone, when contrasted with the 4,000 mi. of the earth's radius. Thus it may be well to pause here a while and attempt to acquire a more cosmopolitan view of the world as a whole.

.... thinks of the earth as an insignificant spheroid reeling through space ....

IT IS difficult for us to visualize the earth in its entirety. Even extensive travel on its surface does not help. It is only through the study of maps or globes that its larger features may be understood at a glance. Let us, therefore, take an ordinary geographic globe—say 12 in. in diameter—and look at it from a distance of about 2½ ft. We will then be seeing the earth, on a proper scale, as if from the porthole of some superstratospheric rocket ship distant about 20,000 mi. from this terrestrial sphere. Let's not be disappointed at the smoothness of the globe; the world's highest mountains themselves are not very impressive from such a distance. In fact, on the scale involved in a 12-in. globe, the highest points above sea level, some 6 mi., would be about 1/100 of an inch high. Thus, Everest itself is scarcely a bleb in the thin film of varnish that covers our little man-made geographic globe.

As we watch the earth for 24 hr., it turns completely around, and we readily observe that most of the surface is covered with water. The continents occupy only about one-fourth of the earth's surface, or only 54 million of the globe's 197 million square miles. One of the features of the earth that we cannot see from our vantage point is its envelope of air. Shifting clouds, however, would suggest the presence of an atmosphere and blot out some of the surface features. Hence these armchair observations have some advantages over those made from a rocket ship.

B<small>UT</small> coming down to earth, and landing safely, we may recognize three states of matter comprising the globe as a whole. The main solid body of the earth is composed of rocks and soil, and is called the **lithosphere.** Partly surrounding this solid body is the **hydrosphere,** composed largely of the oceans, which keep some three-fourths of the solid earth from contact with the **atmosphere,** the layer of gas that surrounds this entire terrestrial sphere.

Approaching the earth . . . .

The lithosphere itself is known only to a very shallow depth. No wells or mines penetrate much more than 2 mi. below the surface, nor were any of the rocks now exposed on the earth ever more than a relatively few miles below the surface. In fact, on the scale of our geographic globe, we have direct knowledge of scarcely more than the layer of varnish on the surface.

SOLID, LIQUID, AND GAS

Although we are inclined to think of the oceans as being very deep, and the atmosphere around us as extending for considerable distances upward, **they are mere films wrapped around the solid body of the earth.** The average depth of the oceans is about 2 mi.; but this is an insignificant part of 4,000 mi., the earth's radius. In similar fashion, even if the true atmosphere extends out for as much as 150 mi. from the earth's surface, nevertheless this value is only about 4 per cent of the earth's radius. Thus we may safely speak of the air and water *films* about the earth, because on a world-scale they are little more.

BUT insignificant as these features might be to an astral being, they are quite impressive to us. Viewed from our limited zone on the earth, we justly admire mountain-climbers or stratosphere-flyers who reach dizzy heights, or those intrepid explorers of the oceans who descend a half mile or more into the deeps. We shall accordingly return to our customary restricted environment and look at some of the relief features of the earth in more detail. We should, however, at least keep the enlarged viewpoint in the back of our minds for later reference.

The major features of the lithosphere are the continents and ocean basins. If we include the shallowly submerged continental shelves with the continents, then we may say that about one-third of the earth's surface is occupied by the continental elevations and the remainder by the ocean basins. An idea of the differences in altitude between the ocean bottoms and the land surfaces may be had from the fact that the greatest oceanic depths are over 30,000 ft., or about 6 mi., and the highest mountains are nearly that high above sea level. Thus there is a maximum relief of over 60,000 ft., or roughly 12 mi., on the lithosphere. Viewed provincially, these are large figures; but in the light of our new cosmopolitan knowledge, we readily understand that the highest mountain and the deepest abyss are truly insignificant irregularities on the relatively smooth surface of this terrestrial sphere.

To help us visualize the distribution of irregularities on the earth's surface, let us consider the adjoining chart, called a *hypso-*

graph. The vertical scale expresses elevations above or below sea level, and the horizontal line represents the total area of the earth. The curve itself traces the distribution of relief on the globe. Notice how relatively little area is occupied by either the high mountains or the greatest oceanic depths. By far the greatest portion of the continents is within a few thousand feet of sea level; and most of the wide expanse of ocean bottom is between 10,000 and 20,000 ft. below sea level. Some of the relief features of the earth are illustrated in Plate 1, to which the reader is referred.

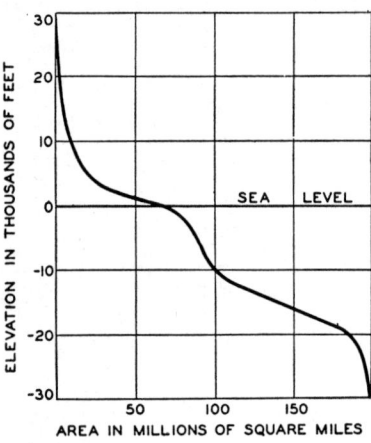

Hypsograph showing the distribution of relief on the earth's surface

Land features on the continents themselves may be divided into **mountains, plateaus,** and **plains.** Mountains are high lands of small summit area, prominent above their surroundings. Plateaus are high lands with considerable summit areas, rising abruptly above their surroundings on at least one side. Plains are low lands with little relief. In addition there are the smaller relief features, such as hills and valleys, familiar to all.

BY AND large, the major physical features of the earth are pretty well known to everyone. Not so, however, the chemical nature of the several parts of the earth. Most people have learned at one time or another that the atmosphere consists mainly of oxygen and nitrogen. It also contains, in addition to the rare gases, small amounts of carbon dioxide and water vapor, which play very important rôles in geological processes. Every high-school boy knows that water, and hence the hydrosphere, is a chemical compound of hydrogen and oxygen. When we consider the chemical composition of the lithosphere, however, we find that common knowledge is much more meager. And yet there are many important and interesting features of the solid part of the earth into which we should inquire.

SOLID, LIQUID, AND GAS

As a first approach to the subject, let us consider two rather broad types of environment that may be recognized in the lithosphere. At the surface the pressure and temperature are comparatively low. Beneath the surface of the earth, however, and particularly at depths of 10 or more miles, the temperatures and pressures are relatively high. The pressure effect can be understood easily when we remember that every cubic foot of rock must exert a pressure on the rock beneath it. A cubic foot of common surface rock weighs about 150 lb., and 10 mi. down the 52,800 cu. ft. of rock exert a pressure at the bottom of the column equivalent to thousands of pounds to the square inch, in contrast to the mere 14.7 lb. exerted by the atmosphere at the earth's surface. Likewise the temperature increases with increasing depth, and at a depth of 10 mi. it certainly is several hundred degrees higher than at the surface.

The point to this discussion of environments is that the kinds of chemical compounds that tend to occur in the lithosphere depend in large measure on the environment. We usually speak of the **surface environment** on the one hand, and **deep-seated environments** on the other, meaning by the latter an environment beneath the surface at a depth great enough to involve a considerable increase in temperature and pressure.

Now it happens that many rocks are formed deep beneath the surface of the earth, and as long as they are in their original environment of high temperatures and pressures they remain unchanged. Thus we speak of such rocks as being "in balance," or in *equilibrium*, with their environment. Broadly speaking, the deep-seated environments tend toward the production of complicated chemical compounds, whereas the surface environment is more suitable to simpler compounds. The rocks which are formed deep beneath the surface may be brought nearer to the surface through the operation of geological changes, and thus they pass from a deep-seated environment to the surface environment. Obviously, of course, our direct observation is confined pretty well to surface conditions, but in deep canyons we may find types of rock from far below. Likewise there are large areas (Eastern Canada is one of them), where rocks have been brought to the surface by geological changes and may still be found largely unaltered by surface conditions. It is from such rocks that we reconstruct and interpret the conditions of the deep-seated environments, those great laboratories in which the rocks of the earth are largely made. Furthermore, direct experiments in our chemical and physical laboratories, made at high temperatures and pressures, supplement and confirm this evidence from the rocks.

A GREAT many samples of all kinds of rock have been collected and chemically analyzed, so that we may confidently say something about the average composition of the outer part of the lithosphere. Tables have been prepared of the ultimate composition of the lithosphere in terms of the percentages of the elements present. When we look at such a table, we find to our surprise that out of the ninety-two elements, only eight are present to the extent of 1 per cent or more by weight.

Element	Percentage
Oxygen	46.46
Silicon	27.61
Aluminum	8.07
Iron	5.06
Calcium	3.64
Sodium	2.75
Potassium	2.58
Magnesium	2.07
	98.24

Let us look here at the part of such a table that includes the eight most abundant elements. An outstanding feature of this table is that oxygen is by far the most abundant element of all. We do not mean to say that the oxygen exists in its free state. Obviously, it is chemically bound up in the rocks. Similarly with the seven other elements: not one of them occurs in its elemental state in the rocks; all of them are combined in one way or another into chemical compounds.

.... hang on ....

AT THIS point some of us shall probably have to take a deep breath and hang on for a moment, because it may be easy to fall by the wayside. But if you are at all familiar with these ele-

SOLID, LIQUID, AND GAS

ments, you know that, with the exception of oxygen and silicon, the other six elements are true metals; silicon has some metallic attributes, but in its properties it lies between the true metals and the non-metals. All seven of the solid elements, however, do form definite oxides, and the suggestion is strong that the fundamental chemical unit with which we have to deal is the oxide. At any rate, if we talk about oxides we shall find that the discussion is much simplified.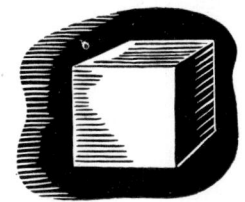

For example, metallic oxides tend to form *bases*, whereas the oxides of the non-metals tend to form *acids*. Silicon oxide shares this non-metallic attribute, at least in the presence of metallic oxides, and the result is that silicon oxide has a strong tendency to combine chemically with metallic oxides to form a **silicate.** For example, if magnesium oxide and silicon oxide combine chemically, a compound called *magnesium silicate* results. In chemical shorthand the reaction is expressed thus:

$$MgO + SiO_2 \rightarrow MgSiO_3.$$

The resulting compound is one of the kinds usually formed under deep-seated conditions. Commonly, more than one of the metallic oxides combines with the silicon oxide to form double or triple silicates, like *calcium aluminum silicate*, $CaAl_2(SiO_3)_4$, which may be written $CaO.Al_2O_3.(SiO_2)_4$. It is thus a combination of calcium oxide, aluminum oxide, and silicon oxide. Now these very silicates are the naturally occurring minerals that we find among the rocks of the lithosphere, and we shall become familiar with a few of them in this chapter.

A MINERAL is a naturally occurring chemical compound; and although we may not be aware of it, we already are familiar with several minerals. They commonly occur as **crystals**, that is, in definite geometric forms bounded by plane surfaces. The simpler types of crystals may be cubes or prisms; but complex forms occur, as the accompanying sketches show. **Rocks** are combinations of minerals in various proportions, or occasionally are made of a single mineral. Almost invariably the minerals making up a rock are present as a physical mixture of crystals or crystal fragments.

It is the crystalline, as opposed to the fragmental, nature of rocks that largely determines their classification. Rocks composed mainly of interlocking crystals are called **crystalline rocks;** whereas others, composed of crystal fragments, commonly with broken or rounded edges, are called **fragmental rocks.** The fragmental nature of this second type suggests that they were formed from the breaking-down or wearing-away of the crystalline rocks. This is so in fact, but the story is no simple one. Indeed, we shall spend the next dozen chapters describing the geological changes to which the crystalline rocks are subjected to form the fragmental rocks.

WE SHALL begin our story with the minerals contained in the crystalline rocks, because these rocks are associated with the deep-seated environment. Geologists recognize literally thousands of minerals among natural earth-products.

But do not despair. Kind Nature has decreed that if we are familiar with some half-dozen of them, we shall be able to talk intelligently about most rocks. In the present chapter we shall consider three groups of minerals, which we call **primary minerals** and later we shall introduce a few other important ones.

LET'S go back for a moment to the eight elements we mentioned. If we make a list of oxides that are formed from these elements, we have the following:

Silicon oxide, SiO_2
Iron oxide, FeO
Iron oxide, Fe_2O_3
Aluminum oxide, Al_2O_3
Calcium oxide, CaO
Magnesium oxide, MgO
Potassium oxide, K_2O
Sodium oxide, Na_2O

Iron forms two oxides (ferrous and ferric), while the other elements form one each. Some of the silicon oxide combines with other oxides to form silicates. Out of the welter of such silicates that are

SOLID, LIQUID, AND GAS

known, only two groups are just now important from our point of view. They are the **ferromagnesian minerals,** and the **feldspars.** In addition to these silicate minerals, silicon oxide itself commonly occurs as a mineral, in which case it is called **quartz.**

LOOK for a moment more closely at the mineral groups we have just listed. As the name of the first suggests, it is composed largely of iron and magnesium oxides. Actually, a number of individual mineral species are included in the ferromagnesians, but they have so much in common that we may well group them under this one term. In general the ferromagnesians are chemical combinations of magnesium oxide and iron oxide with silicon oxide, and in addition they may contain calcium oxide and aluminum oxide. If you are interested in actual chemical formulas, we may list one of them which is fairly typical of the group: $Ca(Mg, Fe)(SiO_3)_2$. This is a calcium-bearing iron-magnesium silicate, commonly called *pyroxene*. The ferromagnesians are usually dark green to black in color, owing to their high iron content.

The feldspars, in contrast to the ferromagnesians, are generally light in color. They range from white through cream to pink and pale green. The feldspars are combinations of the oxides of aluminum, calcium, sodium, and potassium with silicon oxide, and they form rather complex silicates. The chemically minded may be interested in knowing that one common feldspar has the formula $(K, Na)AlSi_3O_8$, and another is $CaAl_2Si_2O_8$. The first of these is a sodium-potassium aluminosilicate called *orthoclase*, and the second is a calcium aluminosilicate called *anorthite*. Both are naturally occurring minerals.

WE SHALL be quite chagrined if our readers feel it necessary to memorize any of these complicated chemical formulas; they are included merely as a shorthand method of indicating which of the eight most abundant elements they contain. Whenever we have occasion to mention these minerals in a chemical sense we shall point out the elements again, and thus refresh your mind. However, if you are one of those hardy folk who insist on remembering formulas, then remember the one that we shall introduce next.

The ancient Greeks believed that the mineral quartz was water frozen so firmly that it would never thaw again. This belief may have arisen from the finding of crystal-clear quartz in the Alps, among the perpetual snows that swathe those majestic peaks. We are more prosaic nowadays, and we recognize quartz as silicon oxide itself, the chemical formula for which is SiO_2. Quartz, as a matter of fact, is one of the most abundant minerals on earth. The sand of beaches and streams is practically all quartz, broken and ground into fragments by the forces of nature. The abundance of quartz suggests that in nature there is an excess of this oxide above what is needed to form silicates with the metallic oxides.

THE minerals we have been discussing occur most commonly in a particular group of the crystalline rocks, formed far beneath the surface of the earth, at high temperatures, and usually from liquid rock. Everyone is familiar with the phenomenon of **volcanism,** at least to the extent of knowing that there are vents in the earth's surface from which molten rocks pour. This **lava** comes from beneath the surface, obviously; but much molten rock never sees the light of day. It is such lava, buried beneath the surface, protected from rapid cooling by a layer of other rocks above, that cools and crystallizes into rocks which contain the primary minerals we have cited. Plate 2 illustrates these primary minerals and shows some of their more common characteristics.

.... the phenomenon of volcanism. .... Old woodcut of Barren Island, Bay of Bengal. From Lyell's *Principles of Geology,* 1842

BEFORE taking up in detail the changes that these crystalline rocks undergo to form the fragmental rocks, we must concern ourselves with one other important aspect of the physical set-up of the earth: the earth's sources of energy, and the passage of that energy through the economy of this terrestrial sphere.

CHAPTER 4

ENERGY AUDIT

> The energies of our system will decay.
> —Arthur James Balfour

BUSINESS corporations periodically issue financial reports to indicate their solvency, and to suggest, perhaps, the desirability of investing in their stocks and bonds. A complete report includes not only a balance sheet but also a statement of profit and loss which summarizes the income and expenditures of the firm. A study of the sources of income and the channels of expenditure generally yields a clear picture of the nature of the business and the efficiency with which it is operated.

Now this terrestrial sphere has sources of income and channels of expenditure of **energy**. Suppose we constitute ourselves a planetary commission to audit the earth's income and outgo of that extremely important commodity. To do so, we must first consider the sources of energy income and contrast with them the outgoing channels. We should also consider the manner in which the incoming supply of energy is distributed over the earth; and finally we should set up a statement of profit and loss, and note whether the balance is to be drawn in black ink or red.

. . . . issue financial reports . . . .

THE first and most obvious source of energy income is the sun. With the aid of suitable instruments and appropriate calculations, astronomers have been able to determine that the earth receives 1.94 cal. per min. per sq. cm. This is called the *solar constant* and, except for small variations, it appears to remain fairly uniform over long periods of time. The stars, too, send us radiant energy; but when we attempt to evaluate it in the same terms as that of the sun, it is found to be an extremely small amount—so small, indeed, that it is to be compared with the augmented income that would accrue to the Standard Oil Company for selling an additional gallon of gasoline a year. For the purposes of our audit, then, we may quite safely ignore the energy income from the stars, as far as its recognizable effects on the earth are concerned.

IN ADDITION to its one main external energy source, the earth has a supply of internal energy. When one descends into a mine, or lowers a thermometer into a deep well, he finds that the temperature rises at a fairly uniform rate toward the earth's interior. On the average, this increase of temperature is about 1° C. for every 100 ft. of descent. The name **geothermal gradient** is applied to this downward increase of temperature. Everybody knows that if we place a hot object in contact with a cold one, heat will flow from the hotter body to the colder. This flow of heat expresses itself to our senses as temperature decreasing away from the hot object; and, indeed, we can easily tell which way the heat is flowing by moving in the direction of decreasing temperature. If we apply this simple test to the earth, we find that the temperature decreases toward the surface, so that the heat itself must be flowing outward from its interior.

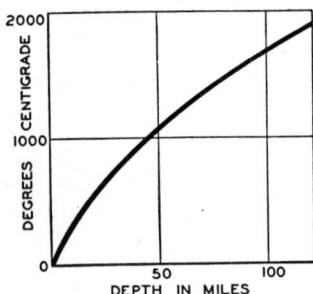

Graph of probable temperature increase beneath the earth's surface. The curve is hypothetical below a shallow observational zone

We on the earth's surface thus have two main sources of energy income; and of the two, that from the sun is by far the greater. But, although the heat radiated outward from the earth's interior is small in comparison with the energy received from the sun, it is nevertheless very much larger than the energy received from the stars; and it is of great importance in geological processes such as volcanism and the deposition of many of the ores.

Let us now consider how the energy income of the earth is distributed. If we confine our attention first to the external source of energy, we may see that the astronomical relations of the earth have much to do with it. The earth's spheroidal shape, the inclination of its axis, and its movements on its axis and in its orbit around the sun cause this external energy to be distributed over the earth's surface in varying quantities: a variation from place to place and from time to time. Consider for a moment some point on the earth's surface, such as Chicago, which is situated about 42° north latitude. During the day the sun warms the surface and raises the temperature. At night some of the heat is radiated off into space and the temperature drops. Thus does the rotation of the earth affect the energy relations.

In summer the sun is more nearly overhead at noon, and its rays pass almost vertically through the atmosphere. In winter the noon sun is lower down toward the southern horizon, and hence its rays must follow a longer route through the atmosphere, with a consequent greater absorption. The result is that in summer more energy is poured directly on Chicago than in winter. The change in position of the sun is, of course, caused by the inclination of the earth's axis, as we pointed out in chapter 2.

The spherical shape of the earth is brought into the picture when we contrast Chicago with some point nearer the Equator. There the rays are more nearly vertical the year round, so that annually a greater quantity of energy is poured down near the Equator than farther north at Chicago. It should be recalled, of course, that the number of hours per year that the sun is above the horizon is the same for all latitudes; but the actual distribution of heat depends partly on the angle at which the rays strike the earth.

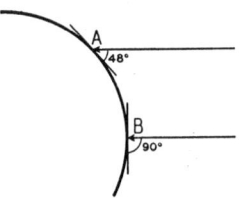

Effect of latitude on the angle of inclination of the sun's rays on the earth. The diagram represents the time of the equinoxes

In addition to the astronomical factors, there are certain terrestrial features that affect the distribution of the sun's energy over the earth. The amount of absorption by the atmosphere increases with its content of water vapor, carbon dioxide, dust, and several other minor constituents. Clouds reduce the amount of sun's rays directly striking the earth's surface, and differences in the nature of land and water determine largely the increase of temperature that follows the absorption of a given amount of sunlight on the surface.

The principal factor that controls temperature rise (assuming a given amount of absorbed heat and a given mass) is the *specific heat* of the substance.* Water requires about five times as much heat to raise its temperature a given amount than does an equal weight of rock. Thus with a given amount of absorbed sunlight, land areas will rise in temperature rapidly, while water areas will not. Of the heat which strikes the surface, some is directly reflected, of course, and so does not contribute to the temperature rise.

THERE are certain aspects of the sun's energy that we should consider. Much of the solar energy which penetrates the atmosphere is converted to heat, and it is almost entirely as heat that it is utilized on the earth. A number of important processes depend on this supply of heat and its distribution, so that it will be pertinent to pause here for a moment to examine them.

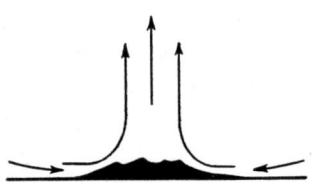

Cool air moves in and displaces the warmed air

The distribution of the sun's energy over the earth, and the wide fluctuations in its amount at any given time or place, result in a differential heating in the earth's air and water films. Since heated air expands and becomes lighter, colder and heavier air may flow into the area of heating to displace the warmed air. Thus are set up atmospheric circulations. In addition to the circulation of the winds, there is a constant passage from liquid water to gaseous water vapor, owing to evaporation from the surfaces of seas and lakes. This water vapor is swept along with the winds and eventually condenses back to water, falling as rain or snow. The rain feeds

* H. B. Lemon, *From Galileo to Cosmic Rays*, p. 139.

and swells the streams, which ultimately carry the water back to the ocean basins. Thus there are continual circulations within the films, and as the winds and waters flow over the earth they modify its surface. Everyone has seen examples of these air and water movements in wind storms that sweep clouds of dust with them, or in cloudbursts that swell small streams into torrents and load them with mud and débris.

ALL of these circulations require energy. As winds blow, they dissipate some of their energy in friction; and from other causes as well, they gradually lose their energy and die down. Similarly, evaporation requires energy, which is stored in the vapor as latent heat, only to be released again when the vapor condenses. We begin to see here some of the devious paths followed by the energy that reaches the earth, but it is too early yet to summarize the process.

THE circulations of the earth films cause a number of important phenomena known as **geological changes.** Air and water, or ice, acting on the solid part of the earth, modify its surface, by carrying away the transportable rock débris, in such manner that the irregularities of the surface, like hills and mountains, are slowly worn away and leveled down. The material carried off is eventually deposited in the seas, those great receiving-basins of the earth. Thus the two films serve as active **geological agents,** and the surface of the lithosphere is the object acted upon. These interactions of the films and lithosphere are summarized in the term **gradation,** the great leveling-down process of geology. *This is the first of three important geological processes.*

IN ADDITION to the inorganic world itself, we recognize a fourth sphere about the earth; this is the **biosphere.** It may also be likened to a thin film, because the life zone of the earth is confined within very narrow limits indeed. The biosphere extends from the ocean bottoms to a point several miles above the earth's surface, and that is all. Yet this limited zone is extremely important

Energy from plant to man

in the distribution and use of the earth's supply of energy. Plants absorb sunlight directly and convert it into potential chemical energy, which is stored in such substances as cellulose and starch. Animals may then eat the plants and use this energy for their own life-processes. Part of the plant material may, however, be buried beneath the surface of the earth as coal, and there the potential energy remains locked up until released by the magic wand of fire. In similar fashion organic matter may be buried and later give rise to valuable pools of petroleum, with their high stores of potential chemical energy. Eventually, however, most of the energy stored up by plants and animals is released again through the slow processes of decay, and thus contributes to the outgoing energy which no longer figures largely in the economy of the earth.

WE ARE now almost ready to prepare our statement of energy profit and loss, and we shall do so as soon as we have looked a bit more closely at the earth's interior sources of energy. Just as we do not know certainly what the ultimate sources of the sun's energy may be, so are we uncertain about the cause of the earth's interior heat. We may, however, reason intelligently about the matter. There is a strong likelihood that a considerable part of the energy is released by *radioactive decay*. In fact, chemical analyses of common rocks indicate that their radioactive content is sufficient to account for a good portion, if not all, of the heat escaping from the earth's interior. It seems likely, however, that other sources of heat are also present. For example, we know that within the earth there are tremendous pressures due to the weight of layers and layers of rock from the surface downward. The possible compaction of the earth's interior that may result from this could well generate heat in the struggle between the resistance of the earth body to shrinkage and the indomitable forces of compression. Whatever the sources (if, indeed, it is not merely some original amount of energy that is being dissipated), the heat is nevertheless conducted always to regions of lower temperature; and in the earth this means toward the surface.

WE HAVE ample evidence, however, that all the energy within the earth is not merely dissipated as heat. As a matter of fact, it is by no means certain that the heat being conducted outward from the interior of the earth represents even the major part of the earth's internal energy. The great compaction within the earth may there set up stresses and strains; and when the limits of adaptability are reached, the earth yields by deformation. We have ample evidences of earth deformation; *earthquakes* are an example. Such deformations or movements of the solid part of the earth, either rapid or slow, are called **diastrophism.** Usually the movements are very very slow, but occasionally there are rapid adjustments, and when they occur, the local area is shaken by earthquakes. Serious as they are from a human viewpoint, earthquakes are thus merely shivers and spasms within the earth body.

Earthquakes are an evidence of the earth's internal energy. Old woodcut of fissures formed during the Calabrian earthquake of 1783. From Lyell, 1842

Diastrophic events may express themselves in many ways, and later on we shall devote a series of chapters to the subject. For the present we may point out that in all its aspects diastrophism involves energy in some form or other, and thus belongs in our ledgers. So important is diastrophism in earth economy that it is called the *second great process of geology.*

THE energy of the earth's interior tends to be radiated outward into space, but of course it must first be conducted to the surface. During this conduction, large bodies of rock may become

Volcanoes are another evidence of the earth's internal energy. Old engraving of Ferdinand's Island, from Von Leonhard's *Geologie*, 1844

heated above their melting-points, despite the great pressures, and so liquefy. Movements of this liquid rock, or lava, within and upon the earth, give rise to **volcanism,** the *third great process of geology*. If the liquid rock finds access to the surface, we have the spectacular phenomena of volcanic eruptions. If it solidifies again underground, the results are no less far reaching, but of course not so spectacular, for, although the outpourings of lava on the earth cause havoc and destruction, the more quiet intrusions of lava among the rocks below the surface commonly give rise to valuable and extensive ore deposits.

WE NOW have all the data we need to complete our audit. We saw that the incoming energy arose from two sources—the sun and the earth's interior. In the transformation of the sun's radiation through the energy scheme of the earth, the short incoming light waves are largely converted to heat energy, and as such are used in the activities of wind, water, and animal life, or are stored

Profit or loss?

as potential chemical energy in plants and various fuels. Simultaneously, some of this heat energy is dissipated outward into space as radiations of long wave-length.* Likewise the energy from the

---

* The whole subject of the transformation of energy really belongs in the domain of physics. We suggest the reader refer to H. B. Lemon, *From Galileo to Cosmic Rays*, pp. 416 ff. for his discussion of radiation and energy.

ENERGY AUDIT

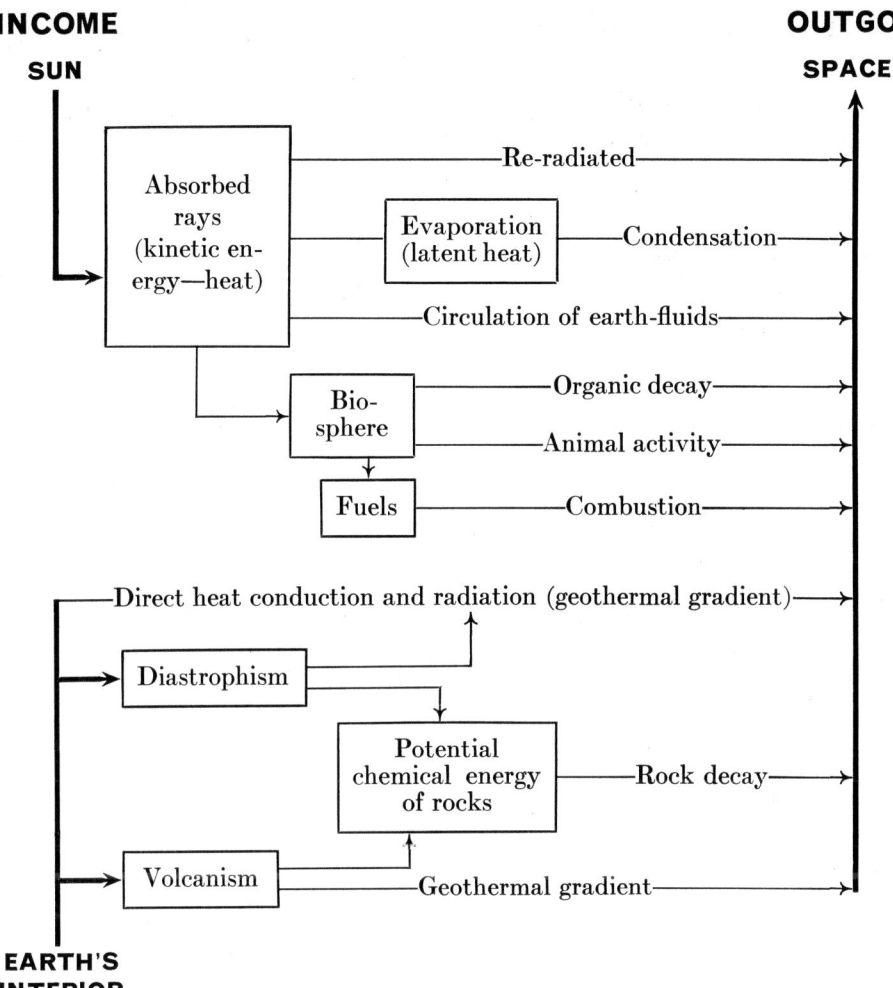

earth's interior is partly used in diastrophic adjustments and in volcanic activity; but eventually it, too, escapes as heat radiation. It thus seems that there is today essentially a balance between the incoming and outgoing energy, with perhaps a slight excess for the surplus account, stored up as potential energy in one form or another.

All these various aspects of the energy flow are shown in the appended diagram, which represents our condensed statement of profit and loss. It may not resemble an accounting record as much as an electrical circuit, but that is a minor matter. The essential thing is that it furnishes us with a picture of the ebb and flow of energy upon and through the earth, and demonstrates that, although the percentage of profit may not be very high, nevertheless this terrestrial sphere is still a going concern, and certainly is in a position to continue actively in business for many years to come.

CHAPTER 5

THE TOOTH OF TIME

> Against the wreckful siege of battering days
> .... rocks impregnable are not so stout.
> —SHAKESPEARE

HUMAN affairs and institutions are subject to decided changes. Styles, manners, traditions—even entire civilizations—wax and wane. But Good Old Mother Earth, one likes to remind himself, remains ever the same. But does she? The answer is relative. Just give the "old lady" enough time, and even the most unobservant may see that her visage is indeed one of great mobility. Her facial expression is changing today just as it has throughout the past. Everywhere about us is a ceaseless ebb and flow among earth-materials. These changes generally are slow, and the traces slight, so that it is not strange they were recognized relatively late in the development of human culture.

Styles, manners, traditions ....

Consider the rocks, for example. To the average layman, if anything is firm and stable, it is rock; and this almost universal belief has given rise to innumerable sayings to the effect that something is "hard as rock." And yet there is abundant evidence that even rocks are subject to changes: that under surface environments they tend to break down into débris. In fact, this gradual decay of rocks is common knowledge, when one really stops to think about it. Everyone knows that the writing on old tombstones is gradually obliterated by the tooth of time, and the blocks of most old stone buildings show signs of weakness and decay. Pitted surfaces develop, dark stains appear, or the blocks crack and crumble away. Few of us, however, have considered the implications of such slow changes extending over long periods of time. If marked changes occur within a few score of years, how much more apparent will they be in thousands of years.

GEOLOGISTS have accumulated abundant evidence that under surface conditions rocks are relatively easily altered. The process of alteration is called **rock-weathering.** The nature of this weathering process depends largely on the kind of rock and on the specific conditions of its environment. In some places the surface of the rock becomes dull and stained, and pitted and crumbly. In others the surface of the rock remains reasonably fresh, but thin shells break from its surface and joints and cracks appear in the body of the rock. The first kinds of change are chemical; the second are mainly physical. Both are included under the general term "weathering."

The net effect of most weathering changes is that hard and firm rock ultimately becomes a mass of small fragments and grains, which in most cases accumulate around the base or on the surface of the parent rock. This loose débris on the earth's surface is called **mantle rock.** Because it consists mainly of small particles, it is subject to the mechanical forces of wind and running water. These agents may sweep the particles away from their parent location and deposit them elsewhere to form a second generation of rocks. In this sense weathering is the first stage in the ever continuing processes that wear down the land surfaces.

DURING the discussion of shallow and deep-seated environments it was pointed out that quartz and the silicate minerals (ferromagnesians and feldspar) were in equilibrium with their environment as long as that environment remained one deeply buried beneath the earth's surface. But if the rocks containing these minerals are later exposed to the atmosphere by some geological change, the minerals which were formed under high temperatures and pressures are no longer permanent. In fact, the silicate minerals are relatively unstable and tend to break down to simpler compounds when exposed to the chemically active gases of the atmosphere, which are *oxygen, carbon dioxide,* and *water vapor.*

A complete analysis of rock-weathering deserves an entire volume, and so we must confine ourselves to certain fundamental notions. The principal points that we shall emphasize here are that (1) **weathering is an approach toward a state of equilibrium under surface conditions** and that (2) **this approach toward a stable state usually involves the breaking-down of the rock into finer débris.**

ROCK is a poor conductor of heat; and when the sun's rays pour down on exposed rocks, the surface layer is heated and expanded more than the interior. This expanding shell sets up strains within the rock and weakens it. At night the reverse process takes place, and the outermost layer contracts more than the inside. If these expansions and contractions continue for a long enough time, the rock may finally yield by shelling off its outer layer. This loosened shell then falls from the parent block and lies at its base. Thus is exposed a fresh surface, ready for a repetition of the process. In detail this phenomenon is probably much more complicated than we have indicated, and involves some chemical changes as well as a yielding to physical forces. Time is undoubtedly an important factor.

Rocks crack and crumble under the attack of the weather. Old cut of Skull Rocks, after Williams, 1879

Another aspect of temperature change is the disrupting effect of ice formed in rock crevices. All rocks, however dense in appearance, are penetrated by crevices and cracks, and into them rain water percolates. If the temperature drops sufficiently, the water freezes, and on freezing expands to about eleven-tenths its former volume. If there is no direction in which the expansion may take place freely, the ice exerts a pressure against the rock that may be great enough to disrupt it.

These are examples of physical weathering. Such physical

Talus slope at base of the Twin Sisters, Green River. After Williams, 1879

changes take place most rapidly when high altitude is combined with steep slopes, as on mountainsides. Here the extremes of temperature are more pronounced, and the broken fragments rapidly fall away from the parent rock. The blocks and fragments accumulate at the base of the mountains and build up **talus slopes,** which reach up the sides of the mountains and may ultimately completely bury them. The reader is referred to Plate 3 for several photographs of dominantly physical weathering phenomena.

SIMULTANEOUSLY with the mechanical or physical weathering of rocks, chemical agents attack them and produce various decomposition products. When rocks are weakened and broken by physical forces, the resulting fragments have a total surface area much greater than that of the original rock exposure. Inasmuch as the speed of chemical reaction depends in part on the intimacy of contact between the reacting bodies, the greater surface area becomes an important factor in decomposition. The active gases of the atmosphere are dissolved in rain water which percolates among the fragments of broken rock. Thus these agents are able more effectively to perform their work of chemical decomposition.

AN IMPORTANT distinction between *physical* and *chemical* weathering must be made clear. Although physical disintegration changes the size and shape of the parent rock masses, the minerals originally present remain intact. With chemical reaction, on the other hand, comes a change in the very composition of the minerals themselves, as well as a change in the physical state due to the formation of new compounds having their own physical properties. Chemical weathering, therefore, is a much more complicated process than physical weathering.

The chemical decay of minerals depends in large part on the fact that the surface environment contains the active chemical agents oxygen, carbon dioxide, and water vapor, which are rarely present in deep-seated environments. These three chemical agents tend to produce *oxides*, *carbonates*, and *hydrated compounds*, respectively, by combining with the primary minerals of the rocks that originated far below the earth's surface.

The low pressures and low temperatures of the surface environment, then, demand oxidized, carbonated, and hydrated minerals for stability. Among the primary minerals thus far mentioned, quartz is already an oxide, and it is indifferent to chemical combination with either carbon dioxide or water. Consequently it remains practically unchanged under surface conditions. Not so the minerals of the silicate groups. The ferromagnesians, for example, may contain only partially oxidized iron, with the result that decomposition of the molecules sets in during the further oxidation of the iron by atmospheric oxygen. During this decomposition the rest of the silicate molecule may combine chemically with water and thus become hydrated.

Just as the ferromagnesians decompose in the presence of atmospheric agents, so the feldspars may suffer chemical decay. The calcium oxide may be released from the silicate molecule and may unite with carbon dioxide to form a very common carbonate called **calcite**. Similarly the sodium and potassium may be released to form various soluble salts. Meanwhile, the silicate part of the feldspar molecule becomes hydrated by combination with water, much as the ferromagnesians do. The net result of this hydration and decomposition of the silicate molecules is a mixture of débris that we call **clay**. Since clay may consist of several minerals, depending on the composition of the original material, we apply the term **clay minerals** to the group as a whole.

WE HAVE introduced two new mineral names here. Let us pause a moment and look at them. We said the clay minerals are hydrated silicates. That means they still are silicates, but now water has become an essential part of the molecule. Here is the formula of one of them: $H_4Al_2Si_2O_9$, or $Al_2O_3 \cdot 2SiO_2 \cdot 2H_2O$, commonly called *kaolin*. Other clay minerals may contain magnesium. Clays are composed of very small particles; and, as everyone knows, they are adhesive and slippery when wet. **In the formation of the clay minerals, then, the hard and firm silicate minerals break down to finer débris.** This is the important point to understand.

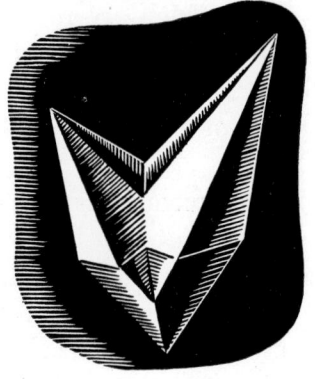

Dogtooth spar, a fanciful name ....

The other mineral, **calcite**, is calcium carbonate, $CaCO_3$. It almost invariably results from the decomposition of calcium-bearing primary minerals. In some cases calcite is found in crystals that resemble somewhat the canine teeth of dogs, from which the fanciful name *dogtooth spar* arose. In comparison with most other minerals, calcite is relatively soluble in water, and especially so in the presence of carbon dioxide. Hence, under many conditions of weathering the calcite may be carried away in solution by waters percolating through the ground. Quite similarly the soluble salts that are formed from the sodium and potassium in the silicates are carried away in solution and ultimately contribute to the salt supply of the oceans.

We may now summarize the process of chemical weathering something like this:

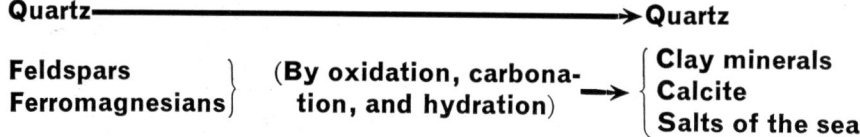

The whole process of weathering is thus an approach toward equilibrium or balance under surface conditions. With the exception of quartz, the primary minerals largely break down to new minerals (called **secondary minerals** because they are derived from the others) which are in adjustment with their environment of

THE TOOTH OF TIME

low pressures, low temperatures, and the chemically active gases of the atmosphere. Plate 4 illustrates some of the features of the secondary minerals.

THE careful reader may, at this point, recall that the chapter began with some emphasis on the slowness of the weathering process. If it is so slow, how do geologists know the exact stages through which the rocks pass during their decay? Simply by being able to observe rocks in all stages of weathering, and from this evidence building up a connected sequence of events. At the risk of becoming a bit technical, we shall let the reader look behind the curtains and see the methods by which the stages of weathering are studied.

Suppose we collect a sample of a partially weathered deep-seated rock from near the surface of a quarry, and then take a series of samples farther and farther down into the fresh, unaltered rock. We grind down chips of the rock with an abrasive until they are only a fraction of a hundredth of an inch in thickness. These *thin sections* are then examined under a microscope. Let us start at the bottom, and see what the fresh, original rock looks like. For simplicity's sake we shall concern ourselves only with the feldspar minerals, which quite adequately illustrate the process.

HERE we shall have to turn to Plate 4 and look at the photomicrographs toward the bottom. In the circle to the left, notice the large central crystal with its alternate white and dark bands, which are due to crystal properties. This is fresh, unaltered feldspar. Notice that it is crystal clear. Now look at the middle picture. See how the upper crystal of feldspar is blotched and mottled, although the light and dark bands (narrower in this case) may still be seen. The blotches are incipient changes, in which the feldspar is beginning to show the effects of surface conditions. When we ex-

. . . . mounted on a slide and examined under the microscope

amine the photograph on the right, we see a much later stage of weathering in a sample taken from nearer the surface. In this thin section only a "ghost" outline of the feldspar crystal remains. The original composition has almost wholly changed to clay minerals, although a few suggestions of the bands may still be seen. If we were to illustrate the fourth and final stage, it would be a handful of crumbly soil from the surface, where the originally crystalline rock has broken down to heterogeneous débris.

NOT alone are there physical and chemical agents of weathering, but many biological agencies as well. Lichens, for example, cling to the surface of rocks and gradually dissolve some of their constituents for plant food. These attacks commonly produce dull and pitted surfaces where the plant acids have leached the rock. In addition to this solvent action, plants may also exert mechanical forces by their growing roots. It is not uncommon to find trees growing from cracks and joints in rocks, where their enlarging roots have forced apart the rock by a type of wedge action. Animals also contribute to the ultimate decay of rocks in several ways. Burrowing animals dig up buried material from the mantle rock and expose it to the more concentrated action of the atmosphere. In this connection most persons are familiar with Darwin's classical researches on earthworms, wherein he showed that in many places annually the earthworms on each acre of land ingest and cast aside 10 tons of soil, constantly exposing fresh material at the surface. All of these factors contribute in some degree to the weathering process and afford one of the many examples of the close integration of geology with the biological as well as the physical sciences.

ONE of the most important factors in rock-weathering is the climate. In arid regions physical changes may predominate and chemical reactions be relatively inconsequential. Oxidation of the disintegrated rock, however, is fairly common, although hydration is relatively rare. Hence, in the formation of iron oxides the red ferric oxide may predominate. The absence of reducing agents such as organic matter (humus) tends to keep the iron fully oxidized, so that red and brown colors among ancient rocks (such rocks are called **red beds**) are regarded by many as an important feature in identifying desert conditions of the geological past.

In humid climates chemical weathering predominates, because the active atmospheric gases are dissolved in the rain water that enters and seeps through the rocks. Here hydration is pronounced, and the net result is a thoroughly decayed rock. Temperature also plays a rôle in this case, because the speed of chemical reactions increases rapidly with slight rises of temperature. Hence in humid tropical regions weathering proceeds most rapidly, produces the most highly weathered débris, and consequently very thick soils.

ONCE the surface becomes covered with a mantle of decayed rock, however, the percolating rain water has to circulate farther to reach fresh surfaces, and hence its dissolved gases will be less concentrated. Furthermore, the cover of mantle rock protects the fresh rock below from sudden or extreme temperature changes, so that the physical forces become less effective. Thus arises the commonly observed situation of a zone of mantle rock, with its surficial covering of *soil*, blanketing the fresh rock below, and with the degree of weathering becoming less pronounced downward. Some of the factors involved in the formations of soils and mantle rock are illustrated in Plate 5.

Bedrock grades upward into mantle rock and soil

If the weathered products accumulate indefinitely, weathering of the underlying rock virtually ceases. Other forces, however, are continually at work preventing this situation from developing in all but a few favored localities. Usually there is a fairly constant removal of the surface materials by gravitative pull down slopes, by winds, and by running water. The surface of the mantle rock is thus gradually removed; and as the mantle thins, weathering processes are again able to penetrate more deeply. In this way there tends to develop a balance between the rate of weathering and the rate at which surface agents carry away the débris.

IN THE final analysis the removal of the weathered rock débris depends on the force of gravity. It is the component of gravity along the slope that determines the flow of water in streams, and winds are generated by relative gravitative effects between warm and cool bodies of air. There is one mode of transportation, however, that depends directly on gravitative pull and which involves no moving medium with its frictional grip on the débris. We have already touched upon it: it is the downward tumble of loose blocks of rock from steep slopes. When rocks crack and break during weathering, fragments result. If this disruption occurs on cliffs or steep slopes, the loosened débris immediately falls or rolls down, and collects as talus at the bottom. The slope of the talus is adjusted to the *angle of rest* of the blocks; it is the steepest angle at which the blocks may lie. Further breaking of the blocks in the talus pile causes shifting of the material, so that over long periods of time there is a gradual movement downslope.

A somewhat similar phenomenon takes place along river valleys, where the mantle rock on the valley walls gradually moves downward under gravitative pull. This slow movement is called *slope creep*. It is aided by the lubricating effect of water which percolates among the soil particles. In some situations, such as in steep mountainous regions, the entire mountainside may give way and relatively rapidly slide downslope. Such **landslides** are often destructive. In these movements underground water also may loosen or lubricate the material, so that the force of gravity may more readily set it in motion.

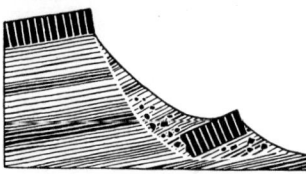
Diagram of a landslide, showing displaced massive layer

THESE last two cases of movement are somewhat transitional between gravity alone as the operating force and the movements of surficial agents, like wind and water, which transport the débris directly. In fact, most movement of rock débris is carried on by the **geological agents,** which are **wind, running water** (rivers and streams), **glacial ice, shore agents** (waves and currents along shores), and **ground water,** which seeps and percolates through the body of the rocks.

The current chapter has shown how weathering breaks down the massive rocks into smaller débris, which is thus subject to movement by the geological agents. We may accordingly turn our attention to these agents, considering first some of the underlying principles of transportation that are common to several of them.

CHAPTER 6

WINDS AND TURBULENT MOTION

The way of the Wind is a strange, wild way.
—Ingram Crockett

A MODERN, open-type freight car is about 50 ft. long and has a capacity of some 50 tons. To transport 400,000,000 tons of gravel, sand, and mud requires 8,000,000 such cars, enough to make a train about 80,000 mi. long. Such a train would girdle the earth more than three times. And think of the freight bill!

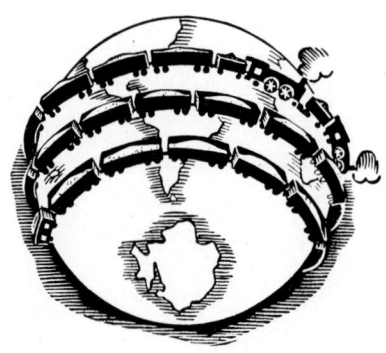

.... a train of cars 80,000 mi. long

It is estimated on good grounds that annually the Mississippi River carries enough sediment to fill such a train of cars. Yet the Mississippi is only one of multitudinous rivers on the earth; how much more, then, must be their total load! Add to that total the dust carried annually by the winds, and the great quantities of sediment shifted along the seashores, and the imagination reels under the resulting figures. There is no escaping the fact that this old globe is in the transportation business in a big way.

All of this transported rock débris must come from somewhere on the earth's surface. Unfortunately, too much of it is derived from our plowed fields, where wind and water are stripping off good topsoil at an alarming rate. The widespread interest in the prevention of destructive erosion of our farm lands attests the importance of the subject in national economy. It is pertinent, therefore, to consider this geologic process somewhat closely. Accordingly, we propose to develop in some detail the principles underlying the movement of rock débris, in order to see just how it enters into the complete geological picture.

WINDS AND TURBULENT MOTION

WINDS usually are not smooth flows of air over the surface, but rather they act as alternations of gusts, eddies, and whirls. The velocity of the wind also varies considerably from place to place and from time to time. In similar fashion, when water flows over a stream bed, the irregularities along the bottom, such as boulders and stones, give rise to diversions of the current. The result is that intermingling threads of water, as it were, wind and whirl about in many directions, with, however, a net movement downstream. Such irregular movement in a fluid is called **turbulent flow.**

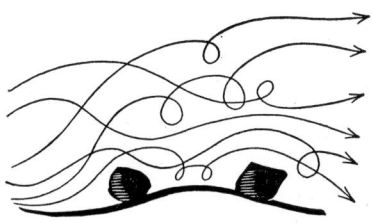

Eddies and intermingling threads of liquid in turbulent flow

The irregular surfaces over which fluids commonly must flow in nature, and their average ranges of velocity, account for the fact that, except under rather quiet conditions, most movements of air and water are turbulent. But in contrast to this irregular motion or turbulent flow, there is a quiet type called **laminar motion** which is also of some importance in geology. We shall devote some space to it in chapter 9.

TURBULENT motion in air and water, then, involves certain upward and sideward currents, as well as the forward mass movement. The irregularities of movement, especially those with upward components, may lift objects of small size up into the main stream, and so carry them along. The greater the turbulence, broadly speaking, the greater is the lifting effect on the particles. The moment the particle is lifted off the ground into the liquid, however, it tends to settle again. The settling rate, or terminal velocity,* of the particle depends on its size, shape, density, and the nature of the fluid. To consider all these factors in detail would require most of the remainder of this book. Accordingly, we shall content ourselves with those whose effects may readily be seen; the more complicated ones involve technicalities beyond the scope of this volume.

* H. B. Lemon, *From Galileo to Cosmic Rays*, chap. 5.

Most everyone would predict that if a sphere and a cube of the same material and volume are placed on a smooth level table, the sphere could more easily be set in motion by blowing on it than the cube. Furthermore, it takes no physicist to see that this result depends partly on the fact that the cube rests on the table with a greater part of its area than the sphere. Applying the same principle to a mixture of rock particles *of a given volume* but of different shapes, we should expect that a gentle breeze will roll away the more spherical grains and leave the others, unless, indeed, the wind is strong enough to move them all. This action of choosing certain grains on the basis of size or shape is called **selective transportation,** and it is an important factor in the work of wind and water. In fact, when a gentle wind acts on the dried sand of the seashore, it actually does roll the more spherical grains along with it, and leaves the flatter ones behind. This accounts for a phenomenon that you yourself may observe with a low-power magnifying glass, namely, that in general the sand grains in sand dunes are more rounded than those on the beach.

.... easier to move a sphere than a cube ....

FOR a still better understanding of this situation, let us distribute a layer of sand grains of various sizes and shapes on a table-top and turn an electric fan on them. Most commonly the immediate effect will be a whirling cloud and a layer of débris over all the furniture; but if the fan is adjusted to move all but the largest grains, the selective action of the air current may be observed. In the somewhat complex shifting of grains that results, it will be possible at least to distinguish that the very small grains of all shapes are whisked up into the air, the larger spherical grains are rolled along the surface of the table, and the largest particles remain behind. The grains that are rolled along constitute the **traction load,** the particles that are carried into the air constitute the **suspension load,** and the material left behind is the **lag sediment.**

The traction load, being confined to the surface of the ground,

WINDS AND TURBULENT MOTION

may be stopped by any obstacle in its path; whereas the suspension load, carried higher up, may move over the obstruction. When later we analyze the factors causing deposition, we shall see that the deposits resulting from the action of wind or water are pretty intimately bound up with this difference in the traction and suspension loads.

BUT for the moment let us look more closely at the suspension load. We have seen that it is made up of a mixture of very small particles of all shapes, ranging from flat to spherical. As soon as these rise above the surface, they are acted upon by gravity and tend to settle. Now there is a property of spheres that is unique among solids, whatever their shape, namely, a sphere has the minimum surface area per unit of volume. Of particles having the same volume, then, the flat ones will encounter more resistance from the air because of their relatively greater areas, and so they will tend to settle more slowly than the spherical grains. As long as the lifting effects of the air currents are greater than the settling velocities of the grains, the grains will rise. But there are so many intermingling currents in turbulent air that, as the grains rise, they are shifted about in a most complex manner. If they are shifted from an upward current to one having a horizontal movement, the grains will immediately start to fall. In general they will not fall back to the starting-point, however, because as they settle the wind carries them forward. The path they follow, therefore, is a line that trends forward and downward simultaneously.

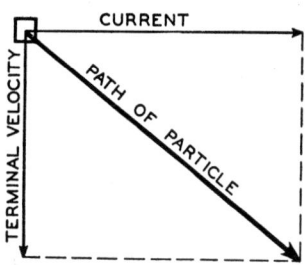

Path of a uniformly settling particle acted on by forward motion of a fluid

IF WE now consider the entire picture, we see that the action of the wind has been to sort out and classify the miscellaneous mixture of débris that results from rock-weathering. The various loads that are carried away, as well as the material left behind, have certain characteristics in terms of size, shape, and density. The density, in turn, depends on the chemical or mineralogical composition of the grains; so it is apparent that many factors are involved. Because of the ways in which rock débris is sorted during transportation, we may expect that, if the resulting deposits are studied in detail, we should be able to learn something about the transporting agent. We shall see in chapter 14 that the detailed examination of a sediment actually does afford a number of clues to its origin and to the manner in which it was carried from place to place.

WHAT we have been saying about wind applies equally well to running water, for in general the principles of transportation are the same for the two agents. We may note, however, that the differences in the densities of the two fluids partly determine their carrying-capacities. That water can move larger objects than wind is not surprising when we remember that the density of air is about one one-thousandth that of water, or, in other words, it is only one one-thousandth as heavy, volume for volume. Consequently, if equal volumes of air and water are moving with the same velocity, the kinetic energy, on which the effective work depends, is a thousand times as great in the water as in the air. In fact, the average range of wind velocities, combined with its low density, pretty well confines wind action to the smaller particles like dust and sand. On the other hand, water suffers no such restrictions; and indeed it is capable of moving very large boulders.

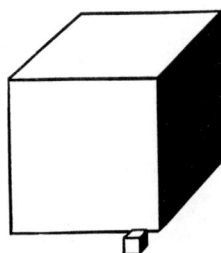

Comparative kinetic energies of equal volumes of wind and water moving with the same velocity. The energy of the water is 1,000 times as great

Winds and running water are the most obvious transporting agents that display turbulence, but they are not the only ones. Ocean waves, as a matter of fact, also are turbulent as they dash against the shore within the line of white-capped breakers. Waves are peculiar in one respect, however, in that they display a *pulsating effect*, in a degree much more pronounced than in the general forward movement of wind and running water. We shall accordingly defer our treatment of ocean waves to a later point, especially since certain aspects of wave work also involve laminar motion, and thus fit into our story better in a later chapter.

LET us now look somewhat more closely at the movements of wind and see what sort of work it performs on weathered débris. It requires no great knowledge of geology to understand that wind work is most effective in arid regions. There the decayed rock at the surface is usually quite dry and powdery, and it has no protective cover of vegetation. In desert regions, then, wind may be one of the most effective agents of transportation, and during windstorms may carry loads of sand and dust far and wide.

Dust storm on the western plains

Not only deserts, but even our western plains, have witnessed the havoc and destruction of wind action. The plowed fields of those western states, where rainfall is scarcely 10 in. a year, are peculiarly susceptible to wind work. The years 1934 and 1935 saw dust storms of such magnitude that practically complete destruction of large areas of farm land followed in their wake. The great clouds of dust were borne upward into the atmosphere and carried as far as the Atlantic seaboard, where they constituted dust storms of no mean magnitude.

THERE is more to wind work than we have mentioned. Not only is the sand and dust swirled about by the wind, but as it moves it may be blown against other rocks and act on them like a sand blast. In fact, **wind abrasion,** as this scouring effect is called, is of considerable importance in the arid Southwest; many of the prominent cliffs of varicolored sandstones there have been shaped and polished largely by wind abrasion. Commonly the traction load of sand, blown along near the ground, may also undermine the rocks and develop curious pedestal effects.

Among the various rocks that may readily be acted on by wind abrasion are sandstones. Composed, as they are, of sand grains loosely cemented together, individual grains are easily torn off by the blast of wind-borne débris, and they themselves join the load carried on by the wind. Their erosional effectiveness is found, of course, in the fact that they lie within the range of sizes that can be most easily swept along by winds. As wind abrasion goes on, the rocks affected are worn away, so that the landscape is gradually changed. Plate 6 illustrates some of the aspects of wind as a geological agent.

NATURALLY enough, the material carried by wind tends ultimately to be deposited somewhere as the winds die down or as obstacles are met in the path of the traction load. The deposits thus built up often assume characteristic shapes, and are composed of typical sizes of débris. Ordinary sand dunes, familiar to everyone, are wind deposits. In their formation, once again the selective effect of turbulent agents enters the picture, and we shall have to devote some space to these depositional effects after we have considered a few other transportational agents.

The similarities between wind work and the work of running water are due mainly to similarities in their movement. In detail there are wide differences between them, because, whereas winds sweep over large areas simultaneously, rivers are confined to their valleys and concentrate their work along narrow sinuous paths. We may with profit, then, turn our attention to running water as a geological agent and see how and why it acts as it does. In this connection, Plate 7 should prove instructive.

CHAPTER 7

VALLEYS AND STREAMS

> Large streams from little fountains flow.
> —David Everett

EXCEPT in the most arid parts of the country, it is impossible to drive a car many miles without crossing bridges. The observant thus tend to recognize that there are several kinds of streams, some rushing through narrow rock gorges, others winding across wide flats with gentle valley walls in the distance; all of them, however, are confined to some kind of **valley** or trough cut below the general level of the land. Furthermore, everyone knows that main trunk rivers have tributaries, and these in turn have smaller branches, so that the whole drainage pattern, as spread out on a map, bears some resemblance to the branches of a tree. Finally, we all know that major rivers eventually flow into the oceans.

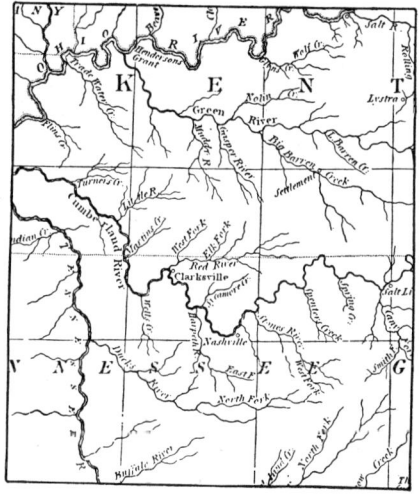

.... like the branches of a tree..... Part of an old map of the southern states, by J. Russell, 1795

If we start at any point along a river and follow it upstream, we note as we pass successive tributary-mouths that the main river becomes a little smaller. Finally, near the headwaters the river itself has become a small stream, in some cases beginning in a swampy tract, in others tumbling down a mountain side as a brook, or merely heading in a dry ravine among the hills. There are other features to be noted too. The valley flat, or **flood plain,** becomes narrower as it is followed upstream, until eventually it disappears and the stream occupies all of the valley bottom. The valley walls then become steeper, and the slope of the valley bottom itself increases toward the head.

NOW let us see what happens at the very head of the valley during a rainstorm. Let's choose a simple case, where the terminating ravine is cut into mantle rock. When rain falls on land, it first soaks down through the spaces between the soil particles until the surface is thoroughly wetted. Then, as more rain falls, the water tends to collect in irregularities and to form tiny streamlets which flow downslope. As the volume of water increases, it gains energy and sweeps along the finer soil particles in its path. By this process it carves a little trough, which may assume the proportions of a fair-sized gully if the land slopes away abruptly. Nearly everyone has seen such gullies along the sides of hills, especially where the soil is reasonably free from a protective covering of vegetation.

A geologist gets down to earth

The rain water and its load of débris are led by the slope into the ravine at the head of our valley. Here, during rainy weather, the concentration of water may become considerable; and as it tumbles along it carries mud, sand, and even pebbles along with it. When the rain has ceased, the ravine again becomes dry; but in the meantime several important things have happened. In the first place, the rain water has washed rock débris down the slopes into the ravine, and the water there has carried much of this débris along, to supply it to the permanent stream farther down the valley. As a consequence of the removal of this material, the ravine gnaws a little farther into the hillside. Thus the valley is lengthened, **and this lengthening always takes place in a headward direction.** Now, during any one rainstorm the amount of material swept from a square yard of land surface may not be great; but, given enough time, it becomes obvious that the valley will increase both in length and depth. Furthermore, the gully that we saw develop will itself increase by headward growth, and the rainfall that enters it will flow into the main ravine. Thus the gully becomes a lengthened tributary of the main stream and will in time develop branches of its own.

VALLEYS AND STREAMS

THE reader at this point may well ask where this process will end. In brief, we may reply that the typical stream pattern is gradually enlarged by this headward growth until it comes into direct competition with another drainage system working in the opposite direction. Each of these headward-cutting stream systems will then control a series of slopes leading to itself, with the result that a **divide** will lie between the opposing ravines. At this point a state of equilibrium is reached between the two drainage basins, and successive rainstorms will lower the divide, but, except under unusual conditions, the two drainage basins will remain separate.

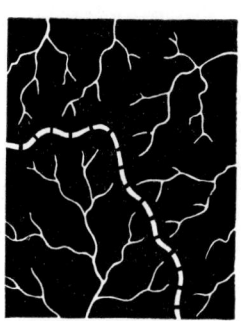

The line of dashes locates the divide between opposing drainage basins

In the foregoing discussion we assumed a simple case in which a supply of weathered rock was always available for the stream. This situation, however, is by no means universal. When rain falls on a bare slope of solid rock the clear rain water merely flows over the surface and does not develop a gully. For water must have tools with which to work before it can attack solid rocks. In the first situation there was a layer of soil or decayed rock débris which was fine enough to be picked up by the running water. The carrying-away of this débris must, of necessity, leave a hollow in the soil; and since water in rivulets flows as a continuous stream, the hollows must be shaped like the path of the water, which is sinuous and winds down the slope. If water which carries sand and pebbles with it now flows over solid rock, the particles strike against the rock surface and abrade it like sandpaper. Thus are other grains torn out; and these in turn act on more rock, gradually carving from even the most resistant rock a path for the water to follow.

THIS wearing-away of rock masses by running water, or by air or ice, for that matter, equipped with tools for abrasion, is an important part of **erosion,** or the wearing-away of land masses by transporting agents. The carrying-along of weathered rock débris (and, in fact, the process of weathering itself) is also a part of erosion, since it tends to wear away the land surface; and the tools thus furnished to the stream are in turn able to cut away the solid, and as yet unweathered, rocks that they encounter.

Observations of streams and their activities have been carried on in many parts of the world, and there is no doubt that the network of rivers everywhere is the result of just such headward encroachments of streams, spread out over great lengths of time. We are able, then, to state with certainty that streams serve the function of delivering the rain water back to the oceans, and at the same time they transfer tremendous loads of débris from place to place over the earth. Eventually, all débris carried by the streams is dumped into the sea; and, given enough time under stable conditions, land areas may be worn down nearly to sea-level by this process of erosion.

THESE almost boringly simple principles of stream development have not always been recognized. Indeed, not much more than a century ago a controversy raged over even the manner in which valleys originate. It seemed unthinkable to many that streams could themselves have carved their valleys from the land. A series of catastrophic events was called upon, during which great clefts opened in the earth, which later became occupied by their present streams. Even the source of water from rainfall was not recognized. Writers spoke of great subterranean chambers which fed the streams and which were tapped by the very catastrophes that formed the valleys.

THERE were, however, some observers who noted that in practically every case the size of a stream bore a definite relation to the size of the valley it occupied, that tributaries entered the main stream with an acute upstream angle, and that at the juncture of two streams the water generally met at the same level. To them it seemed that, if valleys were purely accidental features due to earthquakes, for instance, there should be no reason why this systematic relation should hold in practically every case. Hence in

VALLEYS AND STREAMS

opposition to the catastrophic notion there arose the present, almost self-evident, idea that streams themselves accomplish the work of cutting their own valleys.

ONE of the first geologists who recognized that valleys and other land features are due to the slow work of everyday forces, was James Hutton of Edinburgh. In 1795 he supported his views in a two-volume work on a *Theory of the Earth*, but unfortunately his style of writing was involved and his ideas did not gain the audience they deserved. Consequently, his friend, John Playfair, became his disciple and set out to explain Hutton's ideas "in a manner more popular and perspicuous than is done in his own writings." But Playfair was no mere Boswell to his Dr. Johnson; rather he was a Huxley to his Darwin. He himself gave numerous examples of the principles announced by Hutton and attracted a respectable group of followers. Opposition remained strong, but with the accumulation of more and more data the fundamental truth of the new doctrines became clear. Hutton's observations were admirably summed up by himself in his now classic quotation, that "in the economy of the world I find no traces of a beginning,—no prospect of an end."

James Hutton of Edinburgh

WE ARE at present much farther along than Hutton or Playfair in our knowledge of the details of stream action; and among the several factors that can be evaluated, there are some that deserve our present attention.

ALL the factors that control the velocity of a stream along its channel are not easy to evaluate, because of the complexity of some of them. But we may at least discuss several of them that will help us to understand how a stream transports its load. Certainly, the *slope* of the bed is involved; and it is easy to see that the greater the slope the larger the component of gravity that operates on the water. The *discharge* of the stream, or the volume of water flowing through a given cross section per unit of time, is also involved: the greater the discharge, the higher the velocity. Likewise the *form* of the channel, whether it is broad and shallow or narrow and deep, is an important factor because of the frictional effects of a large area of stream bed in contact with the water. The frictional effect is also dependent partly on *channel roughness*, for a smooth bed offers less friction to the flow of water than a rough one. We may for our purposes, however, summarize these complexities as three main factors—**the slope, the volume or discharge, and the frictional effects.**

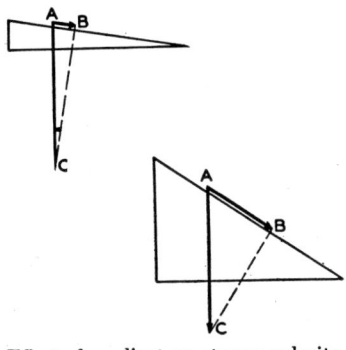

Effect of gradient on stream velocity. With an increase of gradient, the component of gravity along the slope (*AB*) increases notably

Whatever factors may predominate in a given case, the velocity of the water determines its kinetic energy, which in turn controls its carrying-power; and consequently the carrying-power is related to all the factors we touched upon. As soon as material is picked up by the stream, some of the stream's energy is used in the transportation; and as a result the velocity of the stream is diminished.

THE gradient of a stream is determined by the difference in altitude of the head and the mouth and by the distance between these two points. If a valley is incised into a plateau, where the relief is great, the gradient may be steeper than where the valley is cut into the walls of a rounded hill. The lowest level to which running water commonly may flow is sea level; and since gravitative forces seek to drain water to the lowest possible point, the waters of most streams ultimately find their way into the oceans.

The depth to which a stream may cut is determined in part by the original configuration of the land and by the difference in elevation between the head and mouth of the stream. There must inevitably come a time, then, when the stream has excavated its

VALLEYS AND STREAMS

valley about as deeply as local conditions will permit. When this point is reached, the excess energy of the stream is expended in sideward cutting against its banks. This sideward cutting may be due to original sinuosities of the channel, which divert the water from side to side as it flows along. The effect of the sideward swing is to widen the valley and to develop a flood plain along the bottom of the valley. Thus the change in direction of cutting results in the stream valley changing from a V-shaped notch, with relatively steep walls and a stoop gradient, to a flat U-shaped valley, with more gentle valley walls and a gentle gradient. With continued widening of the valley the stream is unable to cover the entire width of the flood plain, and hence it winds about the valley bottom in a series of loops or *meanders*.

NOW, if there is any critical time in the life-history of a stream, it is this one, where the downward cutting becomes less pronounced and the sideward cutting predominates. The change in direction of cutting at this stage results in the stream valley changing its cross-sectional shape, as we have seen. The point at which the flood plain begins to develop is easily recognized. The shift in the direction of dominant cutting is expressed by saying that the valley has passed from **youth** to **maturity.**

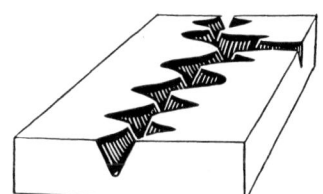

Block diagram of a youthful valley, showing its typical V-shaped cross section

The downward-cutting stage, with its steep gradients, steep walls, rapid movement, and small volume, marks the youthful stage of the stream's life-cycle. The flood-plain stage, characteristic of its maturity, is marked by sideward cutting, a lessened gradient, and a greater volume of water. Finally, with a still greater width of the flood plain, an even lesser gradient, and slow movement, the **old-age** period of the stream is entered. The distinction between maturity and old age is one of degree rather than

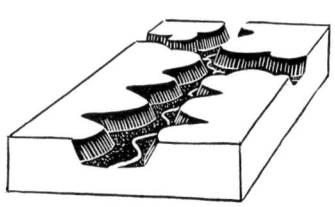

Block diagram of a mature valley, with its broad U-shaped cross section, due to the presence of a flood plain

of kind, and it is not as sharply marked off as the change from youth to maturity. Refer, in this connection, to the photographs of streams and valleys in Plate 8.

We must remember that we have here a dynamic process—that, while the stream is continually cutting headward in its youthful, lengthening ravines, the older parts of the valley are approaching the lower level to which they may cut, and that near the mouth the river may even be in the old-age stage. This accounts for the fact that, as a river is followed downstream, the configuration of its valley changes markedly.

IT IS important to recognize that this age classification of valleys is not based on time in years, but rather on the characteristics of the valley itself in terms of the stages through which it passes. In fact, the length of time required for such a set of events depends upon the nature of the rocks, on the climate, the relief of the region, and so on, so that some streams may pass through their entire cycles while others are slowly passing through maturity.

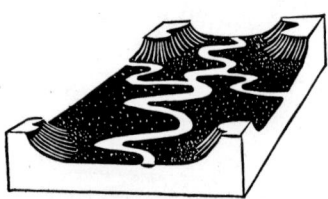

Block diagram of a valley in old age. The stream meanders at will over the broad flood plain

THE great importance of streams from an earth-science point of view is that they act as agents in wearing down the land and in carrying rock débris from higher to lower elevations. Ultimately, if enough time is afforded them, this débris is carried to the seas and oceans, and the general level of the land is lowered until only flat featureless plains remain.

RUNNING water is but one of the several geological agents that wear away the land, and we may here recall from chapter 5 that the others are wind, glacial ice, shore agents, and ground water. Now, in some regions one or the other of these agents may predominate in the transportation of rock débris: thus, in arid regions the work of the wind may be of great importance, and along the coasts ocean waves and currents may be most effective. By and large, however, it is the streams and rivers that are the most important erosive agents. Given sufficient time, whole regions—nay, entire continents even—may be worn down essentially to sea level by the relentless gradational activities of streams. Before we consider some of the other agents, therefore, we may complete our picture of running water by turning now to the *regional*, as opposed to the *single-valley*, aspects of running water and its geological work.

CHAPTER 8

"THE EVERLASTING HILLS"

> The hills are shadows and they flow
>   From form to form, and nothing stands;
>   They melt like mist, the solid lands,
> Like clouds they shape themselves and go.
> —Tennyson

IT IS difficult to conceive of natural processes so slow that during the entire course of written history they have effected no outstanding changes; and yet our terrestrial globe knows many such processes slowly but inexorably altering its surface configuration through the ages. As far back as reliable records go, the general configuration of land and sea has been about the same, and the ruins of the most ancient civilizations lie in geographic settings not markedly different from those when pomp and ceremony graced their age-old halls.

....ancient civilizations..... Old woodcut of the Nimroud Mound, from Buckingham's *Nineveh* (1851)

True, everyone realizes that some changes have taken place. Earthquakes, volcanic outbursts, tidal waves—all have left their marks here and there on the earth's surface. By and large, however, the same valleys are still present, the same mountain ranges, the same seas and oceans. We today use the same passes over the Alps that were traversed by Hannibal, and the same majestic glaciers wind their frozen way along the same rock-ribbed clefts. Little wonder, then, that as far as average human experience goes, the hills are considered everlasting.

NOTWITHSTANDING this apparent permanence of hills and mountains, Tennyson was right. The hills *are* shadows. Every trickling rivulet, every gust of wind, carries with it some rock débris.

Therefore, given time enough, whole land areas will be carried away by wind and water, each agent constantly at work reducing the continents toward sea level. True, it may require millions of years; but after all, a million years is not so very long in comparison with the awe-inspiring length of geological time. For definite evidence, which we shall examine later, indicates that the age of the earth is to be reckoned in billions of years. Even the most case-hardened skeptic will grant that lots of things can happen in a billion years, though it is doubtful if any but the most imaginative can grasp the real meaning of such a great lapse of time.

If the process of wearing away the land is so slow from a human viewpoint, how do we know that any vital changes do take place? The answer is a long one, and will require much of the remainder of the book. In this chapter we shall consider only part of the answer: that part which deals with the gradual erosion of land masses to low-lying lands nearly at sea level. In short, we are about to follow our chapter on the transportation of rock débris by running water with one in which the effects of such transportation and erosion upon the surface features of the earth are systematized into a complete picture.

OBSERVATIONS by geologists in many lands have shown that, from an erosional point of view, there are areas in all stages of development; and by classifying these areas we are able to discern the underlying pattern of regional erosion. Thus it is possible to work out a sequence of events that not only explains the land forms themselves but actually permits the development of a logical theory of the origin of land forms, relating each feature to the geological processes that developed it.

YOU may remember that we anticipated the present discussion when we described how streams develop first a small valley in which downcutting is dominant, followed later by a dominance of sideward cutting which widens out the valley and forms a flood plain. Furthermore, as the valley increases in length, it develops tributaries, until an entire area is covered by a network of streams.

We saw how all these features fitted hand in glove with the idea of a gradual development of stream valleys through several stages.

In our first consideration of stream valleys we confined our attention to a single valley system; but of course there are many valley systems in any given area of a continent, and all of them are going through similar stages of development. Consequently, there should also be cycles of development through which entire areas go as a result of normal stream development. The study of numerous regions does, in fact, permit us to pick out a sequence of regional forms. For example, there are large tracts of land in Utah and Arizona, called the Colorado Plateau, where the elevated land surface is nearly flat between the gorges and canyons that here and there deeply incise it. In other regions, as in much of West Virginia, the land surface is mainly in slope, with very little level land, and with streams cutting into every hill and mountain. Finally, we have examples of regions, such as the lower Mississippi Valley, in which the land surface is nearly flat. In such areas a winding stream, which is not confined within a deep gorge as in the first example, flows on a wide, shallow plain with very gentle valley walls.

BETWEEN these several types of surface are all gradations, which a detailed analysis shows result from a continuous process. We may sketch the broad outlines of regional development by considering some large plateau area, high above sea level, ignoring for the present the question of how the plateau originated. The elevation of the plateau gives steep gradients to streams; and as time goes on, we may visualize the slow but unceasing excavation that produces valleys in the plateau, starting at the edges and gradually lengthening toward the interior. The valleys will be deep and steepsided because of the relative height of the plateau; and the tributaries will be similar to the main streams, but smaller. In the course of events, then, we may picture to ourselves an area in which the flat top of the plateau still stretches out in unbroken monotony for many miles, but with here

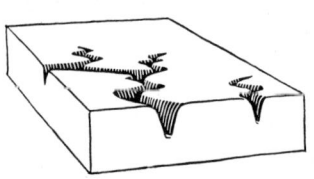

Block diagram of a youthful region. Note the dominance of upland

and there a localized drainage pattern made up of a main gorge and its tributaries, the gorges perhaps several thousand feet deep. If we allow time enough for the main gorge to be cut down as far as the base of the plateau, it may even have a narrow strip of floodplain adjacent to the stream. Between the flat top of the plateau and the flat stretch of the flood plain there is considerable relief; and to distinguish the several levels we call the plateau top the **upland,** the valley walls the **slope,** and the floodplains the **bottom land.**

LET'S stop a moment and see whether we can describe the area in this stage of development a little more precisely. We see before us a region in which most of the surface as seen on a map is still an unbroken upland. Here and there, where the streams have penetrated the plateau, there are steep slopes leading down to the valley bottoms. The total amount of slope, again considered as an area on a map, may be only a few per cent of the total. Finally, along the bottoms of the larger canyons there are narrow flood plains here and there, making up the very small remainder of the total map area. Since we started out with 100 per cent upland, it is obvious that the present stage of development cannot be very far along, because there still is a great preponderance of upland left. Consequently, we call this a **youthful stage,** on the basis of the ratios of upland to slope and upland to bottom land.

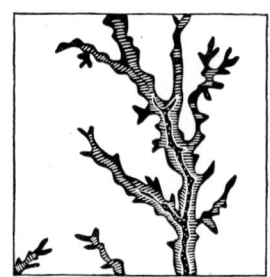

Map of youthful region. The white area is upland. Note the narrow flood plain at the bottom of the valley.

WHILE we have been considering this first stage, the process of erosion has continued; and, coming back now a million or so years later, we find that the streams have grown headward farther and farther into the plateau and have spread a network of tributaries over most of the region. We may still find a few small areas, however, where the original flat top of the plateau remains; but, if so, these are now merely narrow, flat-topped divides between drainage basins. With the increased amount of stream-cutting, the amount of land in slope has greatly increased, although the slopes themselves may not be quite as steep as they were in the earlier stages. The major streams, too, have developed wider flood plains, and as a consequence we find that the ratios of upland, slope, and bottom land have changed considerably. The upland has been decreased to a small percentage of the total map area or has been obliterated entirely. Meanwhile the area in slope has increased to dominance in the region, and the bottom land itself has grown to some extent in area. Here the slopes clearly predominate, and the characteristic youthful aspects are gone. We accordingly describe the region as **mature** on the basis of the high ratio of slope to bottom land and upland. Somewhere in the transition from youth to maturity the last remnants of the original upland entirely disappear, so that slopes occupy all of the region except for the still relatively small amount of bottom land. When the land is mainly in slopes like this, we naturally expect that the number of streams and tributaries will be at a maximum, because it is the streams that are responsible for the slopes. This is actually the case, and during the stage called *middle maturity* the streams are as closely crowded as they can be, and the drainage pattern has reached its greatest development. In this stage also, the amount of débris that is being carried away is at a maximum.

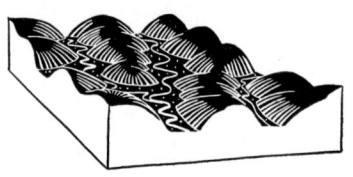

Block diagram of a mature region. Note the predominance of slope

Map of mature region. The upland has entirely disappeared, and the area in flood plain (dotted) has increased

WE MAY with advantage pause here to remind the reader that all during this erosion of the land, the streams are carrying away rock débris. The process we are describing is thus *unidirectional:* all the factors involved tend in one direction, the direction of *degradation* or downcutting. During regional youth, when streams are few, the amount of débris carried off is relatively small; but as more streams incise the land, and as the proportion of slope increases, the total load carried away also increases, until in middle maturity the rate of removing material reaches a maximum, as we have said. Beyond middle maturity, we shall find that the area in slope decreases; and as it does, the amount of rock débris carried away also decreases.

WHILE we are pausing, we may also draw an interesting contrast between the maturity of land areas and the maturity of individual valleys. A mature valley, we recall, is one in which downcutting has become negligible and sideward cutting dominant. Does it follow that a mature region will have mainly mature streams? Not at all, because if the number of streams has been increasing all the time, it is obvious that there must have been considerable high land available so that new tributaries could continually cut headward. These new tributaries will be in their dominantly downcutting stage, so that, as a matter of fact, most of the smaller streams in a mature region are themselves youthful. This fact ought not to cause confusion, however, as long as we remember that we are talking about two different things when we contrast stream age and regional age. Furthermore, as emphasized earlier, the ages are not measured in years, but in stages of development. According to the natural laws that control the development of both streams and of regions, it follows that during maturity in a regional sense the conditions are most suited to youthfulness in the majority of tributary streams.

NOW, as we follow our regional development still further, the reader will be able to anticipate that, as the numerous streams of middle maturity continue their work, they will gradually deepen their valleys, and finally more and more will reach the lower level of downcutting. From this point on, sideward cutting will increase relatively, and the total amount of bottom land in the area will correspondingly increase. Consequently, a few large streams will take the place

Block diagram of a region in old age. Note the predominance of bottom land

of many small ones. In a stage somewhat later than middle maturity, then, we should expect to find that many of these streams have relatively wide flood plains and that the slopes are no longer nearly as steep as they were. In fact, the divides between adjacent streams tend to become rounded off into gentle hills, and much of the ruggedness of the earlier stages is lost. The upland, we saw, disappeared long ago; and with the widening of the flood plains the ratio between slope and bottom land is increasing in favor of the latter. The stage will inevitably be reached, given enough time, when the total area occupied by bottom land will be greater than that occupied by slopes. When this situation develops, we say that the region has reached **old age.** These three principle stages of regional age are shown in the photographs of Plate 9.

Map of region in old age. The bottom land predominates

HERE, then, are three distinct stages through which entire regions may pass; and as we should expect, there are all possible gradations between the stages, because this is a continuous process and moves along as smoothly as time itself. It will be interesting to venture a bit farther and inquire what the very final product of this whole process is. Well, as the streams continue carrying away the débris, the slopes become still more gentle, the divides are lowered still more in height, and the main streams become more and more sluggish as they reach a state of complete equilibrium between their valley bottoms and sea level. No stream can cut much below sea level, of course; and in addition, there must be a slight gradient from the ocean upward into the continent, to allow the water in the streams to flow into the sea. Can you picture to yourself a flat, or nearly featureless, plain, with sluggish streams that slowly wind toward the seas? All about your point of vantage stretches this monotonous surface, with the horizon broken only by a few low hills;

Block diagram of a peneplain; the ultimate stage of stream erosion

it is, indeed, a nearly horizontal line. If you can picture such a plain, then you are visualizing the final **peneplain stage** of the cycle of erosion, the lowest level to which streams can reduce a land area if they are given unlimited time in which to work.

WE REALIZE that it is difficult to obtain such a long-time point of view on short notice, but we believe that the logical sequence of the events described should impress the reader with the essential simplicity of the process when it is shorn of details. There are complexities galore in actual practice. The rocks of the area may not be all of the same kinds, so that some will yield more readily than others to weathering and erosion; the original land may not have been a plateau at all; and finally, various other types of events may interrupt the process so that not enough time is afforded for a completion of all the stages. Changes in the levels of land and sea may, and usually do, interrupt the procedure, so that one cycle of erosion becomes superimposed on another until such a degree of complexity is reached that even a full-fledged geologist finds great difficulty in deciphering the record.

Fortunately, there are many regions available for study in which the complexities are not insurmountable. It is from these that the underlying principles have been drawn; and, armed with the laws that control the situation, geologists are pretty well able to attack even the more special and complex situations. We may see, then, why we can state with reasonable assurance that land areas undergo systematic changes with time, because in nature we find regions in all stages of erosional development. It is a monument to human thinking that, from the apparently confusing and chaotic abundance of land forms, it has been possible to set up a logical system of events that both explains and systematizes the many details of land sculpture. Furthermore, the possibility of relating these details to the underlying erosional processes places a finishing touch, as it were, on our ability to understand the devious ways in which nature does her work.

THE study of land forms, and the classification of them in accordance with the processes that formed them, is called **physiography,** an important branch of earth science. In the discussion of this subject we have been concerned with the land forms that result from the erosive work of streams, but of course streams are only one of several geological agents. Each of the agents—and here are included wind, glacial ice, the waves and currents of the oceans, and ground water—have their own sequences of land forms, due both to erosion and deposition. For we must not forget that while erosion takes place at one point, deposition takes place somewhere else. Thus far we have not said much about these depositional types of topography, but their time and place will come. Finally, in addition to the particular work that each agent performs, usually more than one act simultaneously on a region, thus giving rise to a whole series of related cycles moving along side by side. As an example of such a situation we may think of most any high mountain region where wind, streams, and glaciers are all contributing to the downcutting process.

THE past history of the earth is marked by numerous cases in which whole masses of land have been cut down, from lofty mountains to actual peneplains. The process we have been sketching is thus not purely hypothetical, or one that is just starting for the first time; but rather it is a process which has seen fruition on many occasions in the past. So important are these ancient peneplains in marking off stages of the earth's history that later we shall have occasion to mention them time and again.

CHAPTER 9

UNDERGROUND WATER AND LAMINAR FLOW

> The thirsty earth soaks up the rain,
> And drinks, and gapes for drink again; . . . .
> —ABRAHAM COWLEY

ANASTHASIUS KIRCHER, one of the greatest writers on natural science in the seventeenth century, thought that all rivers and streams arose from underground lakes or reservoirs, which in turn were supplied by conduits leading to them from the oceans. When we remember that this was back in 1665, before many scientific observations had been made, we see that Kircher's idea was rather ingenious. He at least recognized that the lithosphere contained abundant water, and that this water somehow escaped to the surface, where it contributed to the supply in rivers and streams. In short, he sensed what we know quite well today, namely, that water is one of the most ubiquitous things on earth. The hydrosphere, which covers some three-fourths of the globe's surface, is named after it; and the atmosphere contains on an average about 1 per cent of water vapor. The lithosphere also has its own important share of water.

Old engraving from Anasthasius Kircher's *Mundus Subterraneous* (Amsterdam, 1665), showing his notion of the origin of streams. Water from the seas is drawn into whirlpools and from there feeds underground lakes, which in turn supply surface streams

Whenever rain falls on the ground, some of the water inevitably sinks down among the interstices of the rock and thus enters the body of the lithosphere. This is called **ground water,** and locally it may completely saturate the pores of the rocks. Ground water circulates beneath the surface, and during its travels it performs work which is geologically important. Suppose we accompany it for a part of its journey and see what happens.

WE MENTIONED the pores in rocks. To the average person the very phrase "porous rock" seems a contradiction, and it may be well to explain further. We already know that rocks are not the impregnable, everlasting things we may have thought they were; so it is not difficult to understand also that most rocks are not perfectly dense. They are made up of solid particles like crystals and grains, among which are open spaces, so that a varying amount of pore space is present in almost every rock, at least near the surface of the earth. To make the notion concrete, imagine a rock made up of small spheres, packed together as closely as possible. Then there

Spheres packed in two manners. Even under the greatest possible compaction there is some 26 per cent of pore space left

would still be about 26 per cent of open spaces among the spheres. Here is an interesting thing, however: no matter what size the spheres are, any given volume of them packed in the same manner has exactly the same amount of pore space. That means that a cubic foot of rock made up of spheres one-hundredth of an inch in diameter would have as much water-holding capacity as a cubic foot of rock made up of spheres 1 in. or more in diameter. With more irregular grains, and also with mixtures of various sizes, this **porosity** of course varies considerably.

Rock porosity is obviously important in the general subject of ground water, but even more significant is the rate at which the water is able to circulate among the rock pores. Consider our 1-in.

UNDERGROUND WATER AND LAMINAR FLOW

spheres, for example, in which the open spaces will be rather large. We can easily see that water would have little difficulty in moving about between such spheres. **Permeability** is a measure of the ease with which fluids move among the pores of rocks. Hence a rock made up of these 1-in. spheres would be quite permeable. Now if the spheres are gradually decreased in size, the spaces between them also become smaller, and consequently the water must pass through smaller and smaller openings in the rock. Eventually a stage is reached in which the total surface area becomes so great that the spheres begin to attract an appreciable part of the water by molecular forces. When this happens, it requires a greater pressure or a longer time, or both, for the water to move through the pore spaces. Permeability, then, plays an important rôle in the movement of ground water. Near the surface, of course, the rocks may be shot through with cracks and crevices, so that the water has many lines of descent; but with increasing depth, such openings are less common, and the water in large part must pass through the pores of the rocks. The latter may be likened to a sort of continuous sieve, with openings of various sizes.

As a general rule, ground water moves very slowly, because of the variations in the permeability of the rocks it encounters, and also because of the character of the forces which set it in motion. We shall discuss these forces later, but for the present perhaps we had better get back to the surface and see what is going on there.

THREE things may happen to rain that falls on the land. It may evaporate and return to the atmosphere, it may run off down the slope, or it may sink into the ground. If it goes back into the atmosphere, it ultimately contributes to the rainfall somewhere else; if it runs downslope, it feeds into streams and so contributes to the work of running water. If it sinks down into the earth, it becomes ground water, and so enters our present bailiwick. Now the relative proportions of rain water that follow one or another of these three paths clearly depend on the conditions of the locality. In a region of steep slopes, most of the rain flows directly to streams; but in a region dominantly flat, a fairly large amount of water sinks downward into the ground. Atmospheric conditions, climate, and a number of other factors control the rate of evaporation.

WHATEVER proportion of rain water enters the soil and rocks beneath the surface tends to sink straight down under the influence of gravity until it reaches a zone where the rocks are already completely saturated. Everybody knows that the soil is not usually sopping wet to the surface, but that, when wells are dug, it is necessary to penetrate some distance downward before a supply of water is found. The level below which the rocks are saturated is called the **ground-water table,** and in any given region it bears a certain relation to the configuration of the land surface above it. A considerable amount of fairly exact data has been collected on the subject of the ground-water table, particularly by observing the height of water in wells. From such data it is easy to construct a map or surface showing the relationship of the ground-water table to the land surface above.

The adjacent diagram shows the general relation of these two surfaces. Note that the ground-water table is at the same level as the stream in the valley, and that it rises beneath the divides, so that literally the hills are mounds of water as well as of land. The fact that the ground-water level coincides with the stream level is important, because it explains why streams continue to flow even after the rain has ceased. The water that entered the ground during rains is slowly fed to the streams, and so maintains them in the dry weather. This is not universally true, of course, because ravines and gullies usually are quite dry between rains. The reason, obviously, is that these little fellows have not yet cut down far enough to intersect the permanent ground-water table. As the ravine is followed downslope, however, a place is reached where it carries a small permanent stream. The head of this stream marks the level of the ground water.

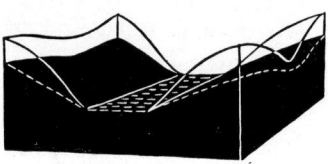

The surface of the ground-water table approximately reflects the surface of the land

UNDERGROUND WATER AND LAMINAR FLOW

WE SEE from the foregoing discussion that where the ground-water table intersects the surface, water issues from the earth. Thus *springs* are nothing but seepages of ground water where the land surface is cut by the water table. Naturally enough, the water that seeps out usually comes from a higher level. There are many complexities among individual springs, and we shall not pursue them all. In some cases the water issues from more permeable beds; at others there may be hydrostatic pressures involved; and so on. In the main, however, it is reasonably accurate to say that wherever the land surface dips down beneath the ground-water table, there water will issue from the earth. Lakes and swamps, then, are merely depressions in the land where the ground-water table comes out of concealment into the open.

THE upper level of ground water can be clearly established. Let us see now whether there is also a lower level, or whether, conceivably, the ground water fills all rocks far down into the interior of the earth. Observation again comes to our aid, for we know that in many deep mines and deep wells the rocks are quite dry. Surprisingly enough, this is only a matter of some thousands of feet. In other words, it is not unusual to find that deep wells, such as are drilled for petroleum, penetrate through a zone of wet rocks, where water is abundant, down to a zone of dry rocks where little or no water is encountered. It seems to be pretty generally true, then, that a completely saturated zone of rocks occurs only relatively near the surface—say within a mile or two at most. And yet there is some water at greater depths, because steam is given off abundantly by most volcanoes, and it is generally believed that this steam may come from considerable depths. Whether it is water that originally arose as ground water percolating farther and farther down, or whether it is some original source of water deep within the earth, we are not certain.

It is likely that the increasing pressures on the rocks far beneath the surface may have an effect on the depth to which the ground water penetrates. Eventually, if one considers a point deep enough beneath the surface, the pressures of the overlying rocks are so great that none of the rocks buried there has open cavities or pores; but such a depth is a matter of miles rather than of feet. One reason why the problem is not completely solved is that the number of deep wells that go below the ground water is limited. After all, most wells are drilled for water itself, and so they stop when they penetrate some short distance into the ground water.

OCCASIONALLY, a rather curious situation is met when wells are drilled. The water, instead of remaining at about the level of the ground-water table, rises up into the well, and in some cases flows out on the surface. Such wells are called **artesian.** The explanation of these wells is not difficult in the light of what we already know. Consider a layer of highly permeable rock between two that are relatively non-permeable or impervious. If these rock layers are inclined downward from the surface, the ground water within the permeable bed is confined by the impervious layers and thus exerts a pressure against its surroundings. If man drills a well through the overlying impervious layer, he affords the confined water an opportunity of relieving its pressure, which is promptly done by some of the water flowing up through the well.

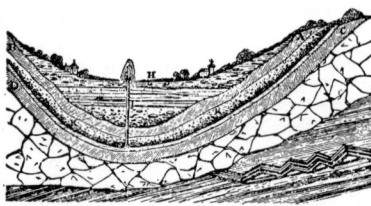

Conditions necessary for an artesian well. Impervious layers lie above and below the permeable bed, which is tapped by the well in the center. Old woodcut from Silliman's *Principles of Physics* (1860)

Artesian wells are very important, economically speaking. Many cities and towns depend entirely on artesian waters; and even in Chicago, with a huge lake at its doorstep, many industrial firms have used artesian water. One of the important sources of this water is a sandstone which is exposed at the surface in Wisconsin at altitudes higher than Chicago. These beds dip downward beneath the surface toward the south, affording a means of circulation of rain water from the more elevated exposed catchment area to the metropolis distant several hundred miles.

UNDERGROUND WATER AND LAMINAR FLOW

BUT after all, artesian conditions demand special relations among the rocks. More usually there are alternate layers of rock of varying degrees of porosity and permeability through which the ground water as a whole tends to circulate. When we come to consider this general circulation of ground water, we find ourselves faced with some problems, for it is difficult to make direct observations on the movement of water beneath the surface of the earth.

WE DO know definitely, however, that ground water moves very slowly. The water percolates among the rocks at a rate that may be expressed as a few hundred feet per year. This sluggishness is a result of the great frictional resistance the water encounters as it migrates among the small interstices in the rocks. Ground water displays no turbulence in its slow movement among the pores. Here is a case where water flows with another type of movement, which is called **laminar flow**.* In laminar flow, any given thread of the liquid retains its identity and flows smoothly alongside its neighbors. It is a sort of streamline flow which adjusts itself to irregularities in its path by curving around them, rather than by setting up whirls and eddies in the current.

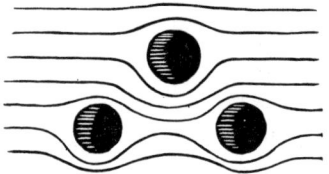

Streamlines in laminar flow

Laminar flow involves the concept of *viscosity*, which is an internal friction in the liquid, so that there is a drag of the liquid layers one on the other. Now, water actually is a viscous fluid; but the running water of surface streams, because of its relatively high velocity and the irregularities in its path, seldom displays laminar motion. Beneath the surface, however, laboratory experiments have indicated that the slow movement of water through rock pores due to gravity is definitely laminar.

Some readers may be interested in the experiments that study the flow of water and other fluids through rocks. In the field of petroleum geology it is important that the nature of the movement of oil and its associated water be known. Consequently, cores of actual rock are mounted in suitable apparatus and water

---

* We are using *laminar flow* here as an antonym of *turbulent motion*, although we realize that the physicist may place more rigid restrictions on the term.

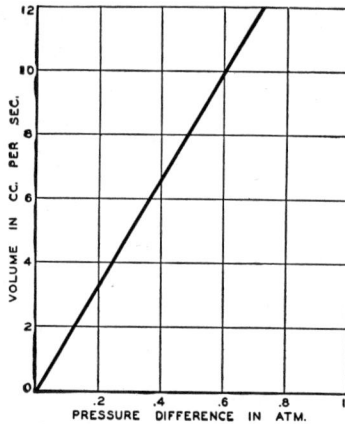

Graph showing flow of water through Cisco sandstone (after Plummer, Harris, and Pedigo). The straight line shows that the flow is laminar, because the volume is directly proportional to the pressure difference

or other liquids are forced through under pressure. The amount of water flowing through per unit time and the pressures that cause this flow are measured. These observations constitute the factual data. From theoretical grounds it is known that if the volume of water flowing through the rock is directly proportional to the difference in pressure at the two ends of the rock core, then the flow is laminar. In other words, the graph of the experiment is a straight line in laminar flow; whereas, if the curve departs from a straight line, then turbulence is present, for the energy dissipated in eddies and whirls reduces the volume below its expected theoretical amount. Graphs, such as the accompanying one, are based on experiments of this sort. Covering a wide range of pressures and velocities, they prove that the ordinary gravity movement of water in rocks is laminar.

TO GET some idea of the causes of ground-water circulation, we shall have to assume a rather simple situation in nature. Imagine a valley between two divides, in which it has been raining "for forty days and forty nights," as it were, until all the pore space in the rocks of the divides is filled with water. Now imagine the rain to cease. The excess water on the surface will flow down to the stream and be carried away. Meanwhile the surface of the ground water in the rock coincides with the surface of the land. This underground water also has a gradient, then, down which it may flow. Gravity is able to operate on the mass of water, and a circulation results. Some of the water, perhaps a large proportion of it, tends to move down parallel to the land surface toward the stream. The slope of the ground-water surface will tend to remain about parallel to the land surface, but with an ever increasing departure in its higher portions due to the greater circulation from those higher points. Meanwhile the stream itself is exerting an effect on the circulation of the ground water. But here we must pause for a bit of theory.

UNDERGROUND WATER AND LAMINAR FLOW

IF TWO ships pass each other at sea, their captains have to be careful so that their ships do not approach each other too closely. If they do, then the streamlines of water movement between the boats are constricted, and a powerful suction pressure is generated between the ships. As a result they may be drawn together and collide. Now a somewhat analogous thing is going on in our picture. As the water in the stream flows along its valley, it generates a suction pressure in its immediate vicinity which tends to draw the ground water into the stream itself.

Ships that pass in streamlines

Thus are set up slow and almost imperceptible circulations which may theoretically extend to considerable depths below the gradient that is established by gravity alone. For practical purposes, however, effective circulation probably does not extend much below the level of the main stream in the region. Below some such level the slight mass movement that may be present is probably negligible.

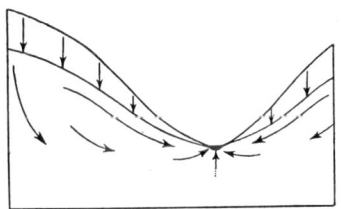

Ground water descends vertically from the surface of the land to the ground-water table. Within the zone of saturated rocks the circulation is in response to gravitative effects and suction pressure from the stream

The very slow circulation of ground water, combined with the fact that it is passing through what may be considered a continuous sieve, means that most ground water does not transport solid matter. By far the greatest transportation work of ground water is done by solution, and consequently we should expect its action to be limited to the more soluble minerals. In fact, *limestone*, among all the common rocks, is most susceptible to ground-water attack, because the calcite of which it is composed is the most soluble of the commoner minerals.

WHEN we talk about the solubility of calcite, we do not want our readers to think that it compares with the solubility of sugar, for example. As far as ordinary things go, calcite is relatively insoluble; but in contrast with most of the rocks that occur in nature, it is readily subject to solvent action by percolating waters. To make the thing more specific, we may say that only about 0.001 gm. of calcite can be dissolved in 100 gm. of carbon-dioxide-free water. That isn't much in comparison with sugar, say, of which about 200 gm. will dissolve in 100 gm. of water. But if we compare calcite with feldspar, we find that calcite is many times as soluble. After all, even a slightly soluble material, acted on by water for a long enough time and in great enough volume, will be completely dissolved. This is the case in nature.

There is, however, an important factor that tends to make calcite even more subject to attack by solution than the figures just given would indicate. When there is any carbon dioxide in the water, then the solubility of the calcite goes up by leaps and bounds. That is because the calcite is converted into *calcium bicarbonate*, $Ca(HCO_3)_2$, as expressed in the reaction:

$$CaCO_3 + H_2O + CO_2 \rightleftarrows Ca(HCO_3)_2 \,.$$

Calcium bicarbonate has a solubility of about 0.04 gm. per 100 gm. of water, which makes it about forty times as soluble as calcite itself. It takes no expert to see where the carbon dioxide comes from. As the rain falls through the atmosphere, it dissolves some of that gas from the air; and as it percolates through the rocks, any calcite in its path soon goes into solution.

SUPPOSE rain falls on an area in which limestone lies just beneath a thin layer of soil. Near the surface, where the loose soil affords an easy path for the water, there will be a percolation of carbon-dioxide-charged water coming in contact with the limestone. As the water runs down the cracks in the rock, some of the limestone is dissolved, the cracks are widened, and there is an even easier path for later rain water. The limestone also varies in permeability here and there, as all rocks do, so that in the course of time the percolating waters will dissolve out a regular series of passages. When they do, **caves** and **caverns** result. Surprisingly enough, these scenic underground features, which annually attract so many tourists, are due chiefly to the unobtrusive work of ground water operating on soluble rocks.

As a general thing, caves are in those zones where the circulation of the ground water is most pronounced. That is, the lower levels of a series of caverns tend to be adjusted to the level of the major stream in the area. If the caverns become large, and are dissolved out near enough to the surface, their roofs may lack support and consequently fall in. The steep-walled circular or oval openings that result

.... the unobtrusive work of ground water.
.... Old woodcut of Sundwig's Cave in Westphalia, from Ludwig's *Buch der Geologie* (1861)

are called **sinkholes.** They are relatively common in some cave areas, as we should expect. Sinkholes may also be formed by the enlarging of crevices leading down into the limestone; but whatever their origin, they act as funnels to lead water down into the rock layers, and thus help to carry on the work of solution. In regions having numerous sinkholes most of the rain that falls on the surface may run down into the sinks, with the result that surface streams may be largely or entirely absent.

ALMOST everyone who has visited a cave or who has seen pictures of them seems to be more impressed by the curious formations in them than he is by the cave itself. From the roof of the cavern may depend long slender "icicles" of stone; and upward from the floor spring similar, but generally more squatty, grotesque deposits. These formations are only indirectly due to the solvent action of ground water; they are the direct results of its depositional activity. The "stonecicles" that hang from the ceiling are called **stalactites,** and those that rise from the floor are **stalagmites.** In some cases they join; others are sheetlike instead of cylindrical; but one and all are due to a redeposition of calcite dissolved elsewhere by the ground water. See Plate 10 for photographs of these striking cave features.

Most caves have entrances, and so there usually is air circulating through them. Thus there is a chance for evaporation. If rain water, circulating downward through the rocks above a cavern, dissolves some calcite along its journey, it may drip from the roof of the cave. As it reaches the air it evaporates in part; and when it does, calcite is deposited. Thus by a slow process stalactites are formed and elongated. Meanwhile the water that drips to the floor may evaporate there, and thus build up a stalagmite. As usual, this simple picture is complicated in nature, and we find that relative concentrations of carbon dioxide, pressures, and especially temperature changes enter into the interplay of factors that decide whether the ground water is to be in a dissolving or depositing mood. Again we must forego technicalities, but we can explain that if the environment entered by the water is one of relatively high carbon dioxide concentration, solution will be the tendency, whereas a lower concentration of carbon dioxide will cause deposition. There really is a nice balance here in any given case, and it affords one more example of the approaches toward equilibrium that Nature seems so anxious to achieve.

Those among our readers who like to study chemical equations will be interested to note that the following reaction really expresses the relationship between ground-water solution and deposition:

$$CaCO_3 + H_2CO_3 \rightleftarrows Ca(HCO_3)_2$$
$$\updownarrow$$
$$H_2O + CO_2$$

For example, if conditions tend toward an increase of CO_2, the lower reaction is driven in an upward direction, and this drives the upper reaction toward the right.

GROUND-WATER work is not confined solely to the solution and deposition of calcite, for it plays an important rôle in many geological phenomena which can scarce be mentioned here. For example, ground-water deposition of minerals about some nucleus may form *concretions* which range in size from microscopic objects to symmetrical structures a dozen feet across. The deposition of minerals in rock pores is responsible for cementing together sediments deposited by other agents. Ground-water deposits in larger cavities or fissures may result in the formation of *mineral veins*. Furthermore, deep beneath the zones of ground-water circulation, there may still be a diffusion of dissolved materials; and when such diffusing solutions meet, chemical reactions may take place. Thus are formed some types of ore deposits. In fact, very commonly ground water acts to concentrate otherwise thinly distributed ores, and so contributes definitely to human economy.

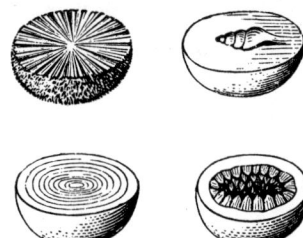

Concretions are formed when ground water deposits material about a nucleus. Woodcut after Dana, 1879

GEYSERS, those openings in the earth from which steaming hot water is ejected at intervals, are more spectacular evidences of ground-water activity. Yellowstone Park is the region par excellence for these phenomena, and annually thousands of tourists become acquainted with this relatively rare aspect of ground-water work. But ground water alone is not sufficient for geyser action: there must be a source of heat, as well as peculiar conditions among the fissures in the rocks. Since the source of heat is volcanic, the subject of geysers is obviously transitional between the current topic and volcanism, and we shall defer our detailed discussion of them to chapter 20. Meanwhile, Plate 11 illustrates some of these rather unusual aspects of hot ground water.

Beehive Geyser, Yellowstone. Old woodcut from Hayden's *Report* (1871)

With our treatment of ground water we have finished two important aspects of the geological work of water. Our consideration thus far has been largely with regard to *rain* as the ultimate source of the water. But what if it *snows* instead? The snow may melt or evaporate, as most of it does; but here and there may be an excess of accumulation over melting. The story of this excess forms the topic of our next chapter.

CHAPTER 10

RIVERS OF ICE

> Torrents, methinks, that heard a mighty voice,
> And stopped at once amid their maddened plunge!
> —Coleridge

NATURE is infinite in her moods. Consider water, for example. Under moderate temperatures it tumbles along the surface, constantly adjusting itself to irregularities in its path, seeking ever the lowest level it can find. Raise the temperature somewhat and the water vaporizes, disappearing into a tenuous gas which finds itself at the mercy of even the most capricious winds. But now reverse the conditions, allow winter to hold the water in its frozen grip, and the turbulent spirit of liquid or gas is stilled. But is all movement necessarily lost? By no means. Almost everyone has read about the movement of those great rivers of ice called **glaciers.** Yes, given suitable conditions, even the apparently solid ice may flow; and on this curious planet of ours it is not so unusual to find just such favorable conditions for glacial movement.

.... rivers of ice. .... Old woodcut of Zermatt Glacier, after Agassiz, 1840

WHEN precipitation takes the form of snow instead of rain, the blanket of snow does not run off down the slope, but accumulates until it evaporates or melts; in the latter case the *meltwater* at once runs off. Now under certain favorable situations, such as high latitudes and high altitudes, the heat of the summer sun is not great enough to melt or evaporate all the snow of the preceding winter. Hence there is an accumulation of snow from year to year, which finally assumes considerable proportions. Large snow fields are built up in this manner. Their lower limits form the permanent **snow line** of the mountains. Near the Equator this snow line occurs only on the tops of the loftiest mountains; in the middle latitudes it is found at much lower elevations; and in frigid zones it may descend to sea level. Thus, from Equator to poles there is a descending surface above which snow may remain all the year around.

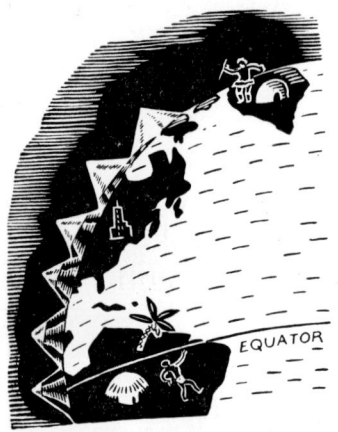

The snow line descends from the equator toward the poles

Although all the snow does not melt, some of it does, and part of the meltwater, percolating down into the snowbank, recongeals to form granular particles of ice. Similarly, the pressure of the overlying snow aids in converting the deeper layers into ice granules, which pass downward into porous ice; and toward the base, where the pressure is greatest, the ice may be densely compacted. The pressures developed by the overlying snow and ice tend to seek relief, and this relief is found by the lowermost marginal parts of the ice mass moving down the slope of the mountain. Thus are glaciers born, and where movement begins is commonly regarded as the beginning of the glacier.

GLACIERS do not move with a velocity anything like that of water. The ice moves slowly and ponderously: it is a matter of feet per year rather than feet per second as in streams. The movement of glacial ice, indeed, is no simple matter, as we shall shortly see; at present we are concerned mainly with the fact that it does move downslope.

As the glacier moves downward, it tends to follow any irregularities along the surface. Commonly, glaciers follow pre-existing stream valleys down the mountain side. These we call **valley glaciers**. The distance downslope that the glacier can extend depends in part on the increase of temperature in lower altitudes and in part on the rate at which snow is accumulating in the snow field above. There tends to be a balance reached in any given situation; and consequently glaciers may be large or small, and long or short.

RIVERS OF ICE

In high latitudes the glaciers may persist right down to the sea, and there break up into *icebergs*. Elsewhere the snout of the glacier reclines along the valley, and equilibrium is reached when the rate of arrival of new ice to the margin is just balanced by the rate of melting. From the terminus of the glacier issues a river which carries off the meltwater.

VALLEY glaciers are fairly common in the mountainous areas of the world, especially in middle and high latitudes. The Alps are famous for their glaciers; it was there that they were first studied, and for that reason valley glaciers are often called *Alpine glaciers*. In the United States valley glaciers are rather rare, being confined largely to the higher peaks of the Cascade and Wind River Mountains. Farther north, in British Columbia and Alaska, there are thousands of them, many of which persist down to sea level.

THE similarities of valley glaciers to rivers may be seen by considering the snow fields above, which constitute their catchment basins; their confinement to valleys; and their movement downslope. In addition, valley glaciers may have tributaries, so that two or more glaciers may merge to follow along a single valley from their point of juncture. In contrast to these similarities, of course, there are striking differences.

IN ADDITION to valley glaciers there are much larger masses of glacial ice called **continental glaciers** or *icecaps*, which cover all elevations of the region. Antarctica is covered with such an icecap, as is the greater part of Greenland. In fact, an area about twice the size of the United States, or nearly 6,000,000 sq. mi. of the earth's surface, is today covered by such ice. These icecaps are much thicker at the center than at the edges, so that there is a surface gradient from the center outward.

Greenland with its cap of ice

This surface gradient causes a slow movement in such ice masses also, and the position of the outermost edge again represents an equilibrium between the rate of movement and the rate of melting, or the rate of breaking into bergs at the seaward margin of the ice field. Present-day icecaps are largely confined to the high latitudes, but in the geological past they covered much larger areas of the globe.

GLACIERS, like rivers and wind, wear away the surface over which they move. The débris they pick up is largely carried to the terminus of the glacier, and there some of it is delivered to the streams of meltwater that flow on down the valley. Much of the rest of it is literally dumped near the margin of the ice as a heterogeneous pile of débris. In many respects, however, the carrying of material by glacial ice strikingly differs from transportation by turbulent agents. This difference depends largely on the motion and physical properties of glacial ice as compared to those of other agents.

In the first place, glaciers move ponderously downslope at a slow and snail-like pace. Furthermore, glaciers are composed of solid ice granules, and thus are not liquid. As a matter of fact, as far as direct observation goes, ice behaves as a solid; its outer portions yield suddenly and discontinuously, like any other brittle substance. Great slicing or shearing movements transect the ice at an angle, so that masses of ice are literally shoved over the lower portions, as the glacier yields to the forces operating on it.

IN DETAIL the structure of the ice often shows thin laminae arranged essentially parallel to the bed of the glacier. Relative movement takes place along these laminae, which may be zones of weakness in the ice. (See Plate 12 in this connection.) There is a suggestion that these laminae may also reflect a more continuous plastic yielding within the body of the ice, similar to laminar motion. The net effect of all factors involved in glacial movement is a complex yielding to gravitative forces; and whatever else may be involved, turbulence, in the sense that we have used it, plays no part in the story. Furthermore, there is no sorting effect in glaciers.

RIVERS OF ICE

In essence the ice simply freezes over, and carries with it, every movable thing, including rock masses literally torn from the glacier bed. Obviously, the shapes, sizes, or densities of the débris are not the determining factors in such transportation.

The absence of selectivity means that there is no traction or suspension load in the ice, in the sense that we earlier discussed. Ice, in fact, carries a heterogeneous load of débris of all sizes and shapes, intermingled haphazardly. The load ranges from immense blocks of rock down to the finest clay particles; and although much of it is concentrated at the base, yet in valley glaciers part of the load results from blocks falling on the surface. Similarly, the slicing or shearing movements within the ice carry some of the basal portions of the load up into the main body of the glacier, and even to its surface. In contrast to the heavy loading in some zones, there may be layers of ice almost entirely free from débris, but this is rarely or never true of the basal portions of the ice.

If the frictional resistance of the basal load against the glacier bed becomes very great, the ice may be unable to move the load along. Then that basal portion of the ice becomes stagnant; and the overlying ice, less heavily loaded, may slide over it, using the basal layer as its new bed. Such complications naturally render an exact evaluation of all the factors involved in ice movement a bit difficult.

SINCE glacial ice carries its load in a frozen grip, the fragments at the base of the ice are held rigidly; and as they are shoved along the glacier bed, they literally gouge out great scratches and grooves in the underlying rock. It is easy to see that a really good-sized glacier may account for a tremendous amount of erosion under these conditions. A glacier is thus like a huge plane scraping off the bedrock over which it moves. Even large rock hills are profoundly modified or entirely worn away by the overriding ice. If a part of the hill remains, the side which faced the ice movement is, in most cases, decidedly rounded off.

Glaciated pebbles are marked by parallel scratches

IT MIGHT be predicted that the erosive power of ice should give rise to typical land forms, and detailed investigations have shown that it does. In the first place, valley glaciers scour out the valleys they occupy, and change the cross sections of water-carved valleys considerably. Instead of being confined in their erosive action to the bottom of the valley, as water is, the glacier may occupy large parts of the valley. Thus the ice has an opportunity of cutting the bottoms and sides simultaneously, and in this manner it produces a typical valley form of its own. We say that **glaciated valleys** are like broad round U's in cross section.

At the heads of valleys occupied by glaciers are huge amphitheaters called **cirques**. They are due to erosion, occasioned apparently by a plucking action at the head of the glacier. Alternate thawing and freezing, combined with an accompanying movement of compacted snow and ice, result in the excavation of a notch in the mountain side. As this increases in size, it develops steep walls and ultimately becomes a conspicuous feature in the landscape.

Bridal Veil Falls plunges from a hanging valley. Old woodcut of Yosemite Valley, after Williams, 1879

Since some glaciers are larger than others, cirques vary considerably in size. More than that, glaciated valleys also differ in the degree of glacial scour. Thus it happens that where two glaciers merge, or where a glaciated valley passes an unglaciated tributary, the larger valley may be carved more deeply into the rock, so that the bottoms of the valleys do not coincide in level. When the ice later disappears (as even glaciers do when the climate becomes warmer), a **hanging valley** results. Present-day streams, flowing through formerly glaciated valleys, commonly develop waterfalls at the juncture of the main valley and the hanging tributary. Bridal Veil Falls in Yosemite is a striking example. Plate 13 includes several of the more common erosive features of glacial ice.

IT IS to be expected that when ice, charged with rock fragments, moves over the glacier bed, the fragments, as well as the floor, should be scratched and ground. The exact results depend on the relative hardness of the rock masses; but in general we do find within ice-transported débris a large number of pebbles with somewhat flat or soled sides, bearing sets of parallel scratches on them. These *soled pebbles* are evidence that the ice held them rigidly while it moved along, otherwise the sides would not be flat, nor would the scratches be parallel. In some cases there are two or more intersecting sets of parallel scratches on the same face, indicating that as the pebble was moved along it struck one or more snags and was slightly rotated.

WE MAY see now that glacial ice is individualistic in its movement, in its mode of transporting rock débris, and in the land forms that result from its erosive action. In fact, each geological agent has associated with it certain land forms and deposits; and from the geological point of view, the important thing is to recognize and understand how these features arise. Thus far we have considered four of the five geological agents, largely in terms of their transporting work and the erosive land forms that accompany the removal of rock débris. We have yet to consider the opposite phase of the subject, namely, the *depositional* features of these geological agents. In the long run, erosion is balanced by deposition in nature; and it will be convenient for us now to turn our attention to this opposite aspect of our subject.

CHAPTER 11

END OF THE LINE

> What goes up must come down.
> —*Old Saying*

"WHAT goes up must come down" is a homely old saying, but it applies quite well to the transportation of rock débris. The material that is picked up at any point by wind, water, or ice must be deposited somewhere else. This is inevitable because the energy of the transporting agent is limited, and eventually this energy is dissipated in the very act of transporting or by the many sources of friction that the agent encounters.

The transporting-capacity of a stream depends in large part on its velocity. As long as the velocity is increasing, the load may also be increased. If the velocity remains constant, the stream may continue to pick up material until it is utilizing its maximum carrying-capacity, when it is said to be *loaded*. If the velocity decreases after this loaded state is reached, some of the débris must be dropped. Consequently deposition also depends on the velocity, but in quite an opposite sense from transportation.

Up ....

.... and down

IN THE upper reaches of a stream valley the gradient is steeper than it is farther along its course. Now, in a perfectly frictionless fluid the velocity would continue to increase as long as any slope remained, because there always would be some component of the force of gravity operating on the fluid. In nature, however, the sources of friction are legion, and a stream's velocity tends to decrease as the slope becomes more gentle. There is an inner friction within the water itself, turbulence consumes some energy, and there

END OF THE LINE

is the friction of the moving mass of water against the banks and bottom of the stream.

The tendency toward decreased velocity due to energy losses through friction is in part counteracted by an increase in the volume or *discharge* of the stream as it flows along its course. An increase in discharge tends toward increased velocity, as we saw in chapter 7. It may also act in the opposite sense, however, because a decrease in the amount of water will decrease the velocity of the stream. Evaporation is constantly taking place along rivers, and in some cases they may lose part of their water because of seepage through the bed. There is thus a balance of forces in any given case which tends to determine the conditions of transportation or deposition that will apply.

NOW, although the velocity determines whether transportation or deposition will take place, the velocity itself depends largely on changes in the gradient, changes in the volume or discharge of the water, and in the friction it encounters from its channel. We should accordingly expect sites of deposition to be those very places where the stream gradient decreases appreciably, where decreases in volume take place, or where the frictional effect increases. The best examples of gradient decrease are to be found at the bases of mountains. A mountain stream, tumbling down a rather steep gradient with a high velocity, carries a heavy load of débris, as it actively erodes its channel. At the base of the mountain the land surface flattens out rather abruptly into a plain. This relatively abrupt change in the gradient results in quite as sudden a decrease in the stream velocity, and hence there is deposition at this critical point. In fact, large aprons of deposited material, which spread out in fanlike forms, are found at the foot of many steep slopes where streams issue from the mountains. These deposits are called **alluvial fans** and are due to conditions precisely like those outlined above.

Block diagram of an alluvial fan at the mouth of a mountain valley

ANOTHER example of deposition, involving a change in volume (and hence in discharge) of the stream, may be mentioned. Some streams arise in regions of appreciable rainfall, and during their course flow across lands that are more arid. Such a stream is the Humboldt River of Nevada, which rises among the Sierra Nevadas and flows over the arid plains of the Great Basin. Evaporation and seepage losses are high in the arid part of its course, and the result is that the amount of water continually diminishes as the river flows along. Ultimately the river dries up completely, about 200 mi. from its source. With a decrease of volume comes a decrease in the kinetic energy, and hence in the carrying-capacity of the stream. Finally, of course, when all the water has been dissipated there can be no more transportation at all.

IN THE lower reaches of valleys the flood plains are usually very wide and flat, so that during spring floods the streams commonly overflow their banks and overrun their flood plains. Since the velocity of the stream in its channel is greatly increased during such high-water stages, the floodwaters may be carrying a considerable load of débris. Now the moment the flooding begins, the sheet of water that spreads sideward out from the main channel meets with considerable friction along the bottom, and thus is greatly retarded. This is due to the large area of contact which such a shallow sheet of water has with the flood plain. What happens is simply explained: As soon as the velocity is checked, some of the load is dropped; and since the greatest checking of the velocity takes place right at the main channel edge, it is there that the bulk of sediment accumulates. Thus are built up **natural levees,** or low, broad ridges of débris along the channels of these old streams. When the volume of water in the channel again subsides, these levees stand above the new level of the stream and serve to confine the water within their bounds. People living along many old valleys build up these natural levees to greater heights, in an attempt to prevent the floodwaters

Block diagram of a natural levee

from flowing over onto the flood plains. The lower reaches of the Mississippi Valley afford good examples of such levees, and everyone is familiar with the havoc raised when destructive floods do tear away or overflow these ridges of earth.

THE building of natural levees results from a velocity decrease due to frictional forces. Similar frictional situations also occur at the mouths of streams, where rivers empty themselves into lakes or seas. What happens is that the velocity of the river is checked by the inertia of the relatively quiet waters into which it flows. Thus there is a dissipation of energy due to friction. The Mississippi River will again serve here as a good example. The river empties into the Gulf of Mexico; and as it does so, the velocity of the water is checked to such an extent that most of its load is dropped. In the absence of strong currents along the shore at that point, the deposited material accumulates near the mouth of the river. In the course of time a **delta** has been formed, which grew out into the bay as a fan-like deposit spread out widely from the river's mouth. Such deposits usually assume the form of a rough triangle, which is the shape of the Greek letter delta (Δ); and from this symbol they derive their name. In the formation of deltas an equilibrium is maintained between the deposition of the load that is continually arriving and the material already deposited. As the delta grows in length, its upper surface, being essentially at sea level, must be adjusted to allow the débris to be carried out to its edge. Thus it is necessary that a slope develop in the direction of the river's flow. This slope is formed by the deposition of débris toward the landward edge of the delta, and in this manner the upstream portions of the deposit gradually emerge above sea level. Plate 14 includes some of the more common features formed by stream deposition.

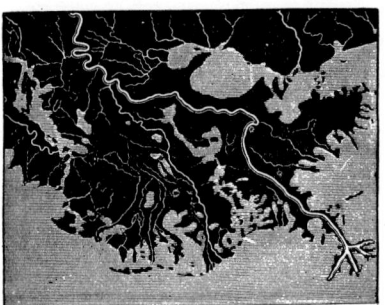

Old map of the Mississippi Delta, after Dana, 1879

WHEN the differences between traction and suspension loads are taken into consideration, we find further interesting details arising concerning sedimentary deposits. The suspension load is generally finer than the traction load, which we saw consisted of particles too large to be picked up and yet small enough to be rolled along the bottom. In any given stream—or air current, as a matter of fact—there is usually a gradation between the two types of loads. The traction load, however, is the marginal portion, and it will tend to be dropped at the slightest decrease in velocity. This effect is brought out most strikingly in the case of wind. A current of air carrying dust in suspension and sand as a traction load may move over an area where vegetation tends to slow down the velocity of the wind. As soon as the velocity is decreased, the traction load is deposited, but the suspension load may be carried right over the obstacles and so escape deposition. The fact that the traction load of sand is stopped by obstacles means that the sand will be deposited in mounds or hills; and as soon as some sand is dropped, it acts as an additional barrier. In this manner **sand dunes** are built; and such deposits are definitely localized as mounds, hills, or ridges, depending on the supply of sand, the direction of the winds, the obstacles, and other factors.

SAND dunes are fairly common land forms, and they are generally found near sources of loose sand. Hence we find them along coasts and also in arid regions where windwork has an opportunity to manifest itself. Dunes have typical forms which shed light on their mode of formation and their slow migration over the surface. For dunes do move; and cases are not unknown where even towns have been engulfed by the sand, only to appear again when the sand moves beyond.

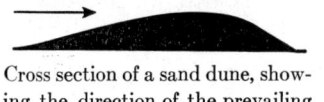

Cross section of a sand dune, showing the direction of the prevailing wind

In profile the dunes present a gentle slope to the wind, and a steeper slope to the lee. The grains of sand are blown up the gentle slope and tumble over the lee edge, where they lie at the angle of rest of loose sand, about 30°. The continual movement of sand from the front to the rear slope of the dunes results in a slow movement of the dune as a whole. Many dunes are rather low features, but hills up to 200 ft. high are not uncommon.

ON A much smaller scale are the tiny *ripples* that commonly form on dry sand; and if they are watched, it is seen that they move with the wind. Here also is a migration of the sand over a gentle windward slope and down a steeper lee slope. The development of the ripples appears to be a response to conditions demanding minimum friction between the moving air and the semi-mobile grains of sand. Between the ripples are tiny horizontal eddies, which allow the wind to blow over the crests and over these horizontal eddies with less friction than would be encountered over a smooth field of sand. Plate 15 illustrates these and other depositional features of the wind.

THE suspension load of dust in the air is carried beyond the localized obstacles and remains in suspension as long as the wind has enough motion to keep the particles off the ground. When the wind dies down, as winds inevitably do, owing to energy dissipation, the suspension load also tends to settle, not locally in mounds, but over a wide area. The net effect is that the suspension load forms a mantle of fine material over the landscape, which may be fairly uniform in thickness over large regions. There is a common wind deposit called **loess**, with exactly these characteristics. It is formed from the suspension load of winds, and it is disseminated wide-

Loess stands in steep walls during its subsequent erosion. Old woodcut from Dana, 1879

ly as a blanket which, unlike other deposits, clothes hills and valley slopes indiscriminately. Loess has the interesting characteristic that it is capable of standing in very steep walls when it is itself subjected to later erosion.

SUCH simple analyses as the foregoing indicate the importance of distinctions between traction and suspension loads, because in the interpretation of deposits they afford a means of determining the conditions that prevailed during the formation of ancient rocks on the earth. We must not gain the impression from our earlier discussion of traction and suspension loads that the materials deposited from the traction load are always quite spherical or that those from suspension loads are always flat. During transport even the particles in suspension may become more rounded from mutual impacts and other causes, and new loads are often picked up in the course of travel. The net result is that the final deposit may have quite an assortment of various shapes and sizes. Later on, in chapter 14, we shall see how both ancient and modern deposits are analyzed to determine their mode of origin.

THERE is an extremely important characteristic of deposits laid by turbulent agents, and that is the **sorting** of the deposits. From the miscellaneous débris made available for transport by weathering and erosion, the turbulent transporting agents sort out, or select, particular sizes, shapes, or densities by virtue of the differences between the traction and suspension loads, as well as by the carrying-power of the agent itself. When later these loads are dropped, the deposition is also somewhat selective, as we have already seen. The net result is that there is a fairly definite sequence of deposits sorted according to size and composition in any depositional environment. Thus arise the materials that we call *gravel*, *sand*, and *mud*. It is important to see **that only through sorting action is it possible to separate the miscellaneous débris of weathering into a series of sediments having restricted ranges of sizes, shapes, or densities.** Further, the particular agent that does the sorting often leaves other marks of its action on the material, and these furnish us with additional clues regarding ancient sediments.

ANOTHER striking feature of deposits left by turbulent agents is **bedding.** We saw that decreases in velocity cause selective deposition, so that a layer of débris is laid down. Later a slight shift in conditions or a difference in the load may bring about the deposition of another layer. These layers of material may then be separated by a distinct juncture plane, known as a bedding plane. The layers between these planes are called *beds*, or **strata;** and consolidated or cemented deposits showing such features are called **stratified rocks.** These rocks, composed, as they largely are, of fragments of weathered and broken deep-seated rocks, are part of the group of *fragmental rocks* that we mentioned back in chapter 3.

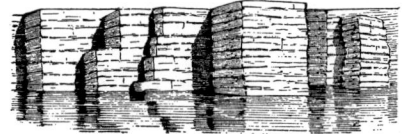

Horizontal bedding and vertical joints in sedimentary rock. Woodcut from Dana, 1879

Stratification of the sort just mentioned is due to selective transportation and deposition. But all stratified rocks are not necessarily fragmental rocks. Successive lava flows, poured out of volcanoes, may harden into layers of rock which also have a stratified appearance, although they lack the other characteristics of sediments. We should understand from this, however, that stratification is not confined to the sediments; indeed, not even all sediments are stratified.

IN THIS connection the deposits left by glacial ice are in striking contrast to the bedded deposits of wind and water. The fact that ice is incapable of sorting out the load that it picks up means that the deposits left by the ice must necessarily reflect the same heterogeneity as the original débris. This situation is, in fact, met in nature, **because the most outstanding feature of ice deposits is that they are unsorted.** In the deposits we find a heterogeneous mingling of all sizes and shapes, just as it was left when the glacier melted and dumped its load at the point of melting. This heterogeneous sediment laid by glacial ice is called **glacial till.**

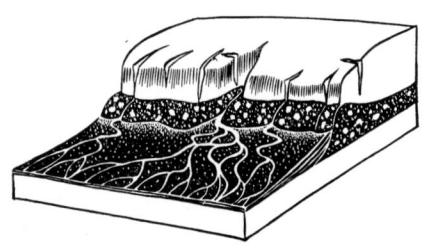

Block diagram of the terminal end of a continental glacier. The morainic ridge lies beneath the edge of the ice, and farther back under the ice is the till plain. The outwash plain, with its streams of meltwater, is in the foreground

Suppose a glacier moves down slope with its load until it reaches a level where the ice is completely wasted away. Here the bulk of the load is dropped, because with no more ice there can obviously be no more carrying. Consequently, at the ice edge there will be a great accumulation of débris, deposited in the same heterogeneous manner that it was carried in the ice. Of course, the meltwater will pick up some of this material and carry it along, but in this case the load is distributed in accordance with the laws of stream transportation and deposition. What we find then is a large ridge of unbedded and unsorted débris at the edge of the glacier, and stretching out from this ridge a series of bedded sands and gravels that have been carried away by the meltwater. Such ridges built at the edge of the ice are called **moraines,** from their wall-like aspects; and the gravels and sands beyond them are called **outwash plains.**

NOT ONLY is glacial till piled up in moraines along the ice edge, but it may also be deposited as a mantle beneath the ice behind the moraine. Such deposits are called **till plains** because they commonly occur as relatively smooth or gently rolling plains after the ice has disappeared. In fact, large areas of northern North America and Europe are covered with till plains, which are important evidences that these regions once were glaciated, although no remnants of the great icecaps now remain on either continent.

View of morainic hills near Amherst, Massachusetts. Old woodcut from Gray and Adams, 1853

Even in the case of the débris that is fed to the meltwater by the ice, there are other deposits in addition to outwash plains. For example, streams of meltwater flowing from the glacier's edge may carry material with them and, owing to changed gradients or pressures near the point of emergence, may deposit part of their load as hills of gravel or sand. These features, commonly found on moraines, we call **kames.** Plate 15 includes several of the more common features of ice deposition.

In fact, the work of glacial ice affords an interesting example of the great complexity of nature when examined in detail. The intermingling of ice-laid and water-laid deposits is so typical of glacial phenomena that we usually think of the associated outwash deposits as a part of glacial action itself. In a more advanced text we would call the outwash deposits *glaciofluvial* deposits, to indicate that both ice and water had a hand in their formation.

JUST as there is a close relation between the geological agents and their deposits in nature, so there may be a close relationship between erosion and deposition itself. Thus far we have kept transportation and deposition somewhat separate, but that has been largely to keep unity in our discussion. We shall combine the two, however, in turning to the next geological agent, and endeavor to show how both erosion and deposition may take place side by side in the development of certain land forms.

CHAPTER 12

THE BOUNDING MAIN

> I have seen the hungry ocean gain
>   Advantage on the kingdom of the shore,
> And the firm soil win of the watery main,
>   Increasing store with loss, and loss with store; . . . .
> —SHAKESPEARE

THE science of human geography concerns itself in large measure with the adjustments of human life and activity to present land forms. The very fact that this point of view may be assumed is concrete evidence that entire civilizations may rise and fall while the lands undergo changes so slight that not even one phase of the cycle of erosion is gone through during the entire expanse of written history.

Human adjustments to present land forms . . . .

Not only does the land surface appear to remain the same, but the configuration of coast lines also appears to be stable. However, coast lines are actually passing through slow cycles of development much like the cycle of erosion produced by streams. These changes are largely the result of wave and current action, and in the present chapter we shall see what some of these changes are.

PREVIOUSLY, we have only mentioned water waves in passing. We come now to the point where we ought to consider them in a little more detail. Suppose we start with waves of some reasonable size—say several feet from crest to crest. The general characteristics of such water waves are well described in most elementary textbooks on physics,* and we shall not dwell on them here in detail. Suffice it to say that during the passage of a wave the water particles describe circular orbits in a vertical plane, so adjusted that they are

---

* H. B. Lemon, *From Galileo to Cosmic Rays*, chap. 35.

THE BOUNDING MAIN

at the top of their orbits under the crest of the wave and at the bottom in the trough. As long as the waves move through deep water, they do not in themselves involve any mass movement of the water, and hence do not carry sediment with them. When the waves approach a shelving shore, however, an important change takes place in their characteristics. This change sets the stage for important geological effects.

IN DEEP water the speed of waves depends on the distance from crest to crest. In shallower water the speed is largely independent of the length of the wave, but it does decrease with decreasing depth of the water. Consequently, a point is finally reached near shore where the waves are crowded together and the towering crests overtake and fall into the preceding troughs. At this point the wave "breaks" and the energy of the wave motion is transformed into a horizontal movement of the water itself. In times of storm these nearly horizontal sheets of water, called *waves of translation*, strike against the shore like battering rams. The mass of water has considerable kinetic energy then, and on sloping shores sand and gravel may be swept along and piled up by the waves. Such deposits of gravel and sand constitute the **beach,** a depositional feature of the waves.

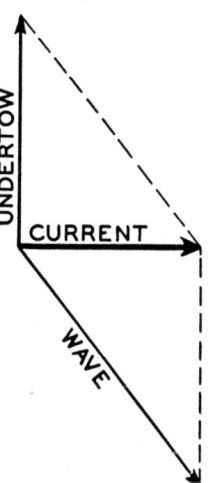

The longshore current is the resultant of wave and undertow

When the wave of translation has dissipated its energy, the water composing it moves back down the slope, beneath the oncoming waves. This reversed current, called the **undertow,** carries with it some of the finer material in suspension, as indeed it may also carry off the unwary bather. The undertow tends to be at right angles with the shore, whereas the waves may strike the shore at an acute angle. Thus the material which is borne up on the beach by the wave of translation, perhaps at an angle, is shifted nearly straight back by the undertow, only to be picked up by the next wave and carried forward again at an angle. In this manner material in transport may move along the shore in a zigzag path. Such movements result in **currents** which are essentially *longshore* in terms of their components parallel to the beach. These longshore currents are of considerable importance in shifting débris along the coasts.

THERE is a striking difference between the transport of material by these longshore currents and the movements effected by winds or streams. Though the movement of water in the wave of translation is largely turbulent, it is likewise a *pulsating action* due to the time interval between successive waves. Hence the transported material **comes entirely to rest** between its movement up on the beach and its subsequent journey downslope. It is this pulsating effect that causes the zigzag motion of the débris. In streams and winds the movement, although irregular in detail, is fairly uniform *en gross*.

The pulsating action of waves introduces an interesting feature about the shapes of pebbles observed on beaches. Many beach pebbles are flat disks rather than approximate spheres. This is due largely to the survival value of originally flat fragments in such environments. Suppose we start out with a mixture of somewhat flattened and spherical fragments on a beach. These original shapes depend on the kinds of rock present. As the wave breaks offshore, the turbulent translational wave moves up on the beach and sweeps with it all the pebbles, regardless of shape. As the water recedes in the form of undertow, however, it is less turbulent in motion, and the more nearly spherical pebbles are rolled along with it while the flat disks may slide along more slowly. Surface friction is one of the determining factors here, because a spherical pebble rolls under less force than is needed to slide a disk. The ultimate result is that the more spherical pebbles tend to be rolled away to deeper water, while the flat pebbles are slid up and down the beach, thus becoming relatively more numerous as time goes on.

THE prolonged action of winds on shallow water tends to drag the water along with it, and thus generates another type of current. Where streams empty into the sea, the added volume of water also sets up movements from the shore. In most cases these currents are movements along certain zones in the main body of water, and the great friction from adjacent masses of quiet water which they encounter insures their slowness of movement. Such currents probably involve little turbulence; they are apparently another example in which water displays typically laminar flow. Nevertheless, they may shift material already in suspension or pick up the finer débris along the bottom. These currents are not as important as the longshore currents in the transporting work of the oceans.

WE HAVE earlier pointed out that when waves of translation strike against the shores they drive water onto the land with great force. This incessant beating of the waves is bound to have pronounced effects on the continental margins. If we imagine a shelving coast descending to the sea, we may sketch some of the major features produced by this work of the waves. The pulsating impacts of tons of water against the shore, especially since there commonly is a supply of sand and gravel available, result in a wearing-away of the coast. Now, water in oceans performs its work of erosion in a manner different from that of running water in streams.

Sea cliff, wave-cut terrace, and associated stacks along the coast. Woodcut after Geikie

When rain falls over a region, it comes from clouds that are above the hills and valleys, so that running water may embrace almost the entire landscape in its sphere of erosional action. Not so the oceans. Their work is mainly confined to the zone between high and low tide, and hence wave erosion may be likened to a sharp saw cutting away at the land along a relatively narrow zone. Since in essence the waves of translation are horizontal forces, we should expect that they will be concentrated always on about the same level along the coasts. The result is that, as the sand and gravel are swept back and forth over the coast, they wear it down to a depth somewhat below the general water level. As this horizontal plane is cut farther and farther into the land, an originally shelving coast becomes terminated by a **sea cliff,** because, in cutting a notch always at the same level, the sea undermines the land above. The undermined rocks tumble into the surf, there to be used as tools for further cutting. Along actively eroded coast lines, then, we should expect to find sea cliffs—and, in fact, they are common features. Out beyond the sea cliff, beneath the water, the surface that has been cut by the waves displays a gentle slope, and is called a **wave-cut terrace.**

While the sea cliff is being cut back, the varying resistances of the rocks may result in greater erosion taking place along some stretches of the coast than along others. The sea cliff then will not present a smooth wall to the sea, but instead will have bays and promontories where the sea is able more or less effectively to do its work. Some masses of more resistant rock commonly are left behind as the adjacent sea cliff recedes, and form rocky islands or those curious coast features called **stacks** and **chimney rocks.** Plate 16 illustrates the erosive work of the waves.

AGAIN we could build up the concept of a cycle of erosion, from observations on the different stages to be found where the sea is encroaching on the land. Instead, we shall endeavor, from a single example, to show how a balance between erosion and deposition is reached in the work of the ocean. The ocean can be made to serve as an excellent example of the close manner in which erosion and deposition are related in some cyclical schemes, because, in this case, the material eroded is usually deposited close to its source. In the erosive work of streams, we may recall, the débris ultimately is transported to the sea, so that the erosive process can be observed somewhat apart from the depositional phases.

Block diagram of shore of submergence—youth in the cycle of shore erosion

As a starting-point for our scheme of shore-line development, suppose we consider a coast along which narrow promontories of land extend out into the sea. Between the promontories let us

THE BOUNDING MAIN

postulate estuaries of streams, so that our coast line will have the general appearance of a hand with somewhat blunt fingers stretching out into the ocean. Such a coast line is called a **shore of submergence.**

About the first erosive effect that may be noted is the cutting of sea cliffs along the more exposed headlands, with the result that these features soon become steep-sided. As these newly formed sea cliffs recede before the incessant attack of the waves, a wave-cut terrace is formed beneath the sea, extending out to the original line of the headlands. While this erosion is taking place, the débris will be carried out by the undertow and by currents to the deeper water, where their action does not penetrate to the bottom. Here it will be deposited, generally at the edge of the wave-cut terrace, because there the sea bottom drops rather abruptly. The net effect is that an extension of the terrace is continually built seaward. This is called the **depositional terrace.**

SOME of the material eroded by the waves is also carried by the longshore currents flowing parallel to the sea cliffs. It may thus be carried into the deeper waters at the mouths of bays and be deposited there. As such deposits accumulate, they form long and narrow extensions of sand and gravel across the mouths of the bays. Such features as these are called **spits** and **bars,** and they are an integral part of shore-line development. See Plate 17 for photographs of these depositional features of shore agents.

The cycle of shore erosion gets under way

At the same time that the depositional terrace and the spits are being built the streams that enter the bays from the land are carrying along their loads of sand and clay. The bays afford shelter from strong currents or waves, so that the material from the streams tends to fill up the bays. Such a process is called *silting;* and if the supply of débris is great enough, it may ultimately destroy the bay completely, leaving only a large swampy tract at sea level.

WE THUS have two processes taking place simultaneously: the waves are cutting back the headlands, and the river silts are filling up the bays. The net effect of these two processes is that the extreme irregularities of the coast tend to be reduced. The bays no longer reach as far inland as they did, and the promontories do not extend as far out to sea.

At about this point in our scheme we see that the originally complex shore line, with its deep bays and narrow headlands, has become a coast of much simpler outline. There still is some irregularity in the coast, but the bars and spits tend in part to reduce this irregularity. A state of equilibrium is approached as the headlands are eroded farther back and the filling of the bays advances outward, until finally an essentially smooth coast line results. This state of equilibrium is called **maturity** in the type of cycle being sketched. We could continue with the later stages of the cycle, but the data already presented furnish a sufficiently detailed picture of the gradual changes suffered by a shore line of submergence. But do not suppose that there are only shore lines of submergence. There are also *shore lines of emergence;* and, of course, they too have their own cycles of erosion and deposition.

Maturity in the cycle of shore erosion. Note the simplified coastline, as compared with that of the youthful stage

AN IMPORTANT point to be kept in mind is that, in the case of the oceans, not only are deposits formed by the material which it itself erodes from the land, but important contributions are also made by streams flowing into the sea. In some cases this sediment is distributed far and wide by waves and shore currents; and in others it is deposited locally as deltas or as fillings in bays. In either event, the material that comes to rest on the ocean floor, regardless of its ultimate source, is classified as an ocean-laid deposit. It must be obvious that such deposits are by far the largest and most extensive on the earth. Most of the present-day oceanic deposits, however, are laid on the submerged continental platforms, and not in the ocean basins themselves.

What we have said about the distribution of the deposits as terraces, spits, and in bays, deserves some amplification. It is already clear that the carrying of large rock fragments requires more energy than the transportation of small fragments. Turbulence is also a factor in the movement of débris. Now the only place in the seas where there is a high degree of turbulence is the zone of breaking waves near the shore. Here the most vigorous movement of débris takes place, and the retreating water of the undertow is usually incapable of carrying very coarse material beyond the line of breakers. The net effect is that there tends to be a *segregation* or *sorting* of the material on the basis of size, the coarser pebbles being confined close to the shore line and progressively finer materials being carried out and spread seaward from the land.

Marine sediments grade from coarse to fine

Thus we find coarse **gravel** near the shore, **sand** a bit farther out, and **mud** of various degrees of fineness out in the still deeper waters. Finally, out beyond even the finest fragmental material, deposits of *calcite mud* may be formed.

THERE is an interesting point in connection with the deposition of very fine material in the sea. As everybody knows, the smaller a particle is, the more slowly it settles in water. Now, when we examine the finer clays, we find they are composed of such small particles that they remain in suspension indefinitely. Particles as small as this are called *colloids*, and the reason they stay suspended so long is that they are kept agitated by the bombardment of the molecules of liquid surrounding them. Further, these tiny particles carry negative electrical charges, which render them mutually repellent. At that rate, one may ask, why do these fine particles ever settle at all? (They do, of course, because we find large areas of very fine mud deposits in the seas.) The answer is that the negative electrical charges tend to be neutralized by the sodium ions in the sea water, so that the clay particles, no longer mutually repellent, collect into aggregates which are large enough to sink to the bottom.

Thus clayey and turbid streams, carrying their load of materials, bring them to the oceans where dissolved salts are able to precipitate them. Naturally enough, the turbulence near shore is often great enough to keep even the coagulated colloids in suspension; but farther out, where the movements of the water are laminar, these small aggregates sink to the bottom as muds. In the quieter waters where deltas are being formed, this coagulation appears to be a factor of some importance in the precipitation of the stream-borne muds that partly make up the deposit.

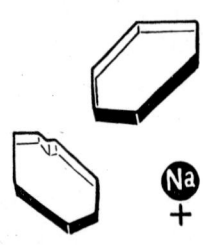

The positively-charged sodium ions attract the clay particles....

THE gradation of sizes out from the shore line is an extremely important thing, and it has deservedly been investigated thoroughly. One reason we are so interested in these size gradations of marine sediments is that, curiously enough, we find tremendous areas of stratified sedimentary rocks on the continents themselves. These sediments contain undoubted remains of marine life: fossils of clams, fish, and other creatures. We cannot doubt, then, that these sediments had their origin beneath the seas. Thus there must have been profound changes in the configuration of land and sea in the past. But the very idea of oceanic deposits standing high and dry on the continents involves another great geological process that we cannot stop now to discuss in detail.

However, when a size gradation of the type mentioned is found in marine sediments which have been raised above the sea, we infer that at one time there must have been a shore line along the zone involved. Thus, by a careful study of the deposits, we are able to map in the approximate coast lines of those ancient seas, which may be hundreds of miles inland from the nearest present seacoast. The significance of such ancient shore lines in the past history of the earth will become increasingly apparent as our story develops.

NOT only does the distribution of ancient ocean deposits give us a clue to the former extent of the seas, but it even tells us something about the cycle of erosion on the adjacent lands. During the early part of the erosion cycle on land the streams have steep gradients, and the material they carry to the sea may be relatively coarse. Further, in these early stages the erosive process moves forward as fast or faster than weathering, and the mineral content of the sediments reflects this fact in that undecomposed grains of primary silicates are found in the sediments. Feldspar is a common one. As time goes on, the land areas have gentler slopes, the gradients of the streams are less steep, and weathering proceeds further before erosion sweeps away the débris. Thus in the cycle of sediments that forms in the oceans, the lowermost beds associated with a given erosion cycle tend to be coarse. But as the beds are followed upward, we notice that they are composed of progressively finer débris, ranging through sand and finer and finer clay. Finally, when the land is worn so low that very little fragmental material is carried into the seas, we find that calcite mud, commonly called **lime mud,** is deposited.

THUS the seas, which have always been the great receiving-basins of the earth, have recorded perhaps our most important data concerning the earth's history. This is true not only of present seas but even more so for the ancient seas that covered so much of present-day continents, and thus left their rock record available for him who cares to "dig and discover."

CHAPTER 13

PLUS AND MINUS

> Hills peep o'er hills, and Alps on Alps arise!
> —Pope

EVERY cubic foot of rock débris that is picked up at one place and carried away by geological agents is deposited somewhere else. We have emphasized that ultimately this débris is carried to the seas, but in detail the process of transportation consists of numerous stopovers along the route. Material picked up near the heads of streams is carried along until local conditions change so that some of the load must be dropped. Later on, perhaps even in a later stage of the erosion cycle, this material is picked up again and carried farther along. If a single particle could be watched in its progress from source to sea, we would probably find it at rest for longer intervals than it is in motion.

Hills peep o'er hills . . . .

In many cases the débris dropped from transportation tends to develop recognizable land forms. We noted earlier that *alluvial fans* are formed at the foot of steep slopes along streams and that these alluvial fans constitute a definitely recognizable land feature. Similarly, farther downstream the river may be building a *natural levee*, which also involves an easily recognized topographic feature. Finally, when the stream flows into the sea, it may build up a *delta*, and thus make its last contribution with still another land form.

But we know that deposition requires erosion as an antecedent, so that the material built up into these land forms must have been removed from somewhere else. We know, too, that rivers erode canyons and valleys through the rocks they traverse and that the

PLATE 1
THIS TERRESTRIAL SPHERE

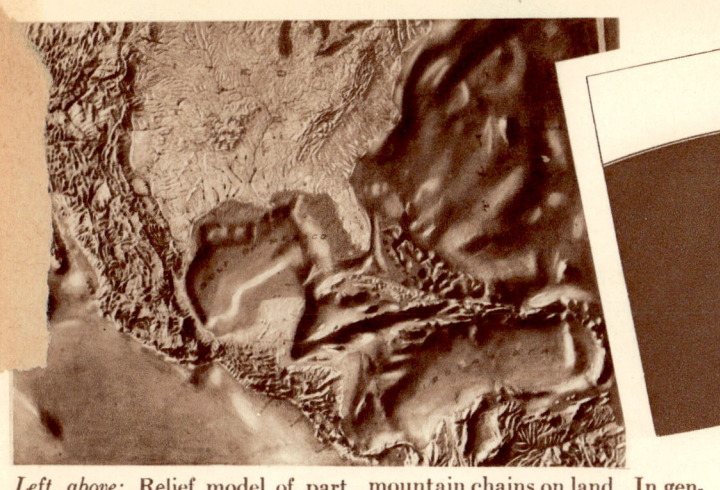

Above: Diagram illustrates the exceedingly shallow zone in which lie our direct contacts with the earth. The very fine outer line represents, to scale with the drawing, a zone deep enough to encompass the loftiest peaks, the deepest seas.

Left, above: Relief model of part of North America shows abrupt drop from continental platform to ocean basins. The ocean's smooth floor contrasts strikingly with mountain chains on land. In general mountain ranges parallel coasts, whereas continental interiors are often plains of low relief.

LAND AND SEA HERE BELOW

Small and insignificant as land features are when viewed from space, when we come down to earth they are majestic and awe-inspiring. We justly admire daring mountaineers who scale the heights, and adventurous explorers who wrest secrets from the seas.

Major relief features of the continents, mountains, pierce the clouds with their snow-capped peaks. Mountains are characterized by their small summit areas. (Monte Rosa, southern Alps; photo by R. T. Chamberlin.) But in contrast with such ruggedness is the boundless sea, stretching, as far as eye can reach, to a flat monotonous horizon. (Pacific Ocean viewed from American Samoa; photo by R. T. Chamberlin.) Between these extremes are such familiar land forms as hills and valleys, plateaus and plains. Details of such smaller land forms may be seen in succeeding plates.

PLATE 2 MINERAL KINGDOM

Minerals, scarcely noted in daily life form a fascinating field of study. *Upper left triangle:* Ferromagnesian minerals, named for high iron and magnesium content, characterized by dark color. Crystals, natural size, and a magnified rock section (*circle upper left*) illustrate their general appearance. *The diagonal band* shows three views of quartz, most common of all minerals. *Top:* crystals about natural size; *bottom,* magnified sand grains; *center,* crystal of quartz in rock, as seen under the microscope. *Lower right triangle* shows feldspars, third important group of minerals. Their light color is due to calcium, aluminum, silicon. The mounted specimen shows crystal form; below this an aggregate of crystals natural size, and magnified rock section, showing selective light absorption.

PLATE 3
MOUNTAINS CRUMBLE

Temperature changes, frost action are illustrated here in rock weathering. *Left, top:* Boulder cracks, Devils Lake, Wisconsin. *Right, below:* Broken fragments tumble down, accumulate as talus at bottom. (Lake Lace; photo courtesy Northern Pacific Railroad.) *Above, right:* Continued accumulation of talus buries mountains beneath own débris. (Photo Curtis and Miller; courtesy Northern Pacific Railroad.)

Above: Half Dome, majestic feature of Yosemite, shows strikingly the effects of weathering. Large slabs, paralleling the surface, crack, loosen, and slide off, exposing new surfaces to attack. As long as slopes remain, the process is a continuous and relentless attack by weathering agents. The dangers of traversing such smooth slopes is clearly implied by the railed path set up for hardy tourists. (Photo by Donald Martin; courtesy of National Park Service.)

PLATE 4
CLAY AND CALCITE

Above: Microscopic crystals of clay (inset highly magnified) are powdery when dry, turbid in suspension. *Left:* Crystals of calcite, natural size. Clear varieties highly prized in optical work. Calcite arises from calcium-bearing minerals; clay from the decomposition of silicates (see below).

DECOMPOSITION OF FELDSPAR

Above and left: photos by Marsh, courtesy A. Johannsen.

Three photomicrographs ($\times 30$) show progress of rock decay. *Left:* Crystal-clear feldspar (alternate light and dark bands) indicates freshness of rock. *Center:* Incipient alteration shown by spotted zones near top.

Above: A "ghost" crystal, all that remains after feldspar has altered to clay minerals. (Text, p. 39.)

PLATE 5
MANTLE ROCK AND SOIL

Left: Chemical weathering disintegrates limestone boulder, Palos Park, Ill.

Chemical weathering commonly accompanies physical breakdown of rocks. Biological agents as well contribute to rock decay (*above right*, lichen on rock, St. Cloud, Minn.). As weathering proceeds, massive rocks may crumble, débris be washed away, until only residual boulders remain. (*Above left:* Residual granite boulders, Matoppo Hills, Rhodesia, photo by R. T. Chamberlin. The scale of the photograph is indicated by the figures by the boulders.) In contrast to removal of débris is its accumulation as mantle rock. (*Right:* Road cut shows bedrock and mantle rock grading upward into soil.) As the mantle thickens, weathering slows down, essentially keeping pace with soil removal by geological agents.

PLATE 6
WINDS BLOW

Dust storms attest to the transporting power of wind. *Right:* Dust storm on western plains, 1934.

Left: Map showing part of the area affected by a single dust storm. The dust, carried high in the air, spread eastward to the Atlantic seaboard. Dust is carried as a suspension load, so that when the wind dies down, a mantle of fine material is deposited over the landscape. This photo and that above are from the motion picture, "The Work of the Atmosphere" (University of Chicago).

Wind-borne sand, carried near the ground, acts as abrasive on rocks in its path. Sandstone cliffs in the arid Southwest seldom develop talus because the débris is swept away by the wind. The photo to the right illustrates such a situation (*El Morro*, by H. Gleason; courtesy National Park Service).

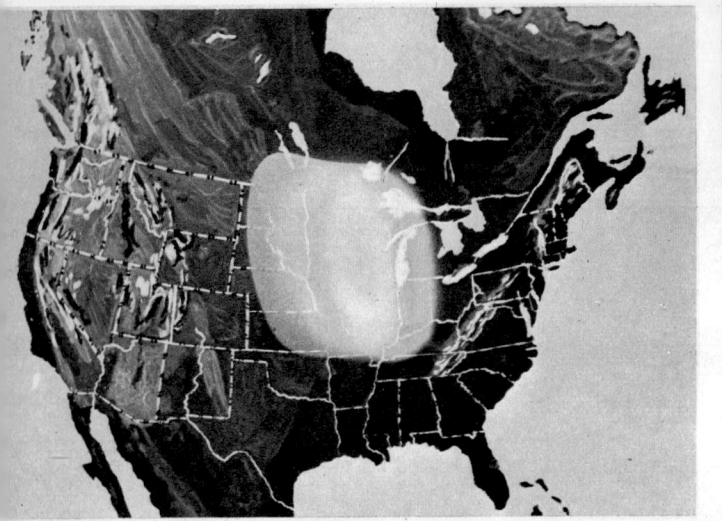

Pebbles in the path of wind-blown sand develop curious shapes. *Lower left:* Dreikanters, or angular pebbles, are found in deserts where the load of shifting sand abrades them as it sweeps along. Such pebbles mark sites of active wind scour. They are sometimes found associated with ancient sandstones, and thus afford evidence of the work of the wind in the geological past.

PLATE 7 RUNNING WATER

Ever restless, running water seeks the lowest level available. It tumbles among boulders, whirling in its turbulent flow. *Left top:* Short exposure photo of brook, showing irregular lines of flow. Such turbulent motion sweeps pebbles against the banks, often carving potholes from solid rock. *Right center:* Potholes in stream bed and, below this, descending series of potholes. (State of Minas Geraes, Brazil; photo by R. T. Chamberlin.) Waterfalls (*left center*, Skillet Falls near Baraboo, Wisconsin) are scenic features formed when water pours over ledges. Streams are erosive agents; the load of sand and cobbles they carry (*lower left*, dry channel, Waterpocket Fold, Utah; photo by F. J. Pettijohn) are tools with which they work.

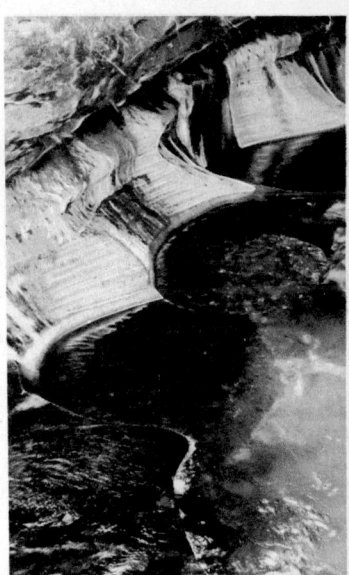

PLATE 8 VALLEYS

Right: Grand Canyon, Yellowstone (courtesy Northern Pacific Railroad), deep gorge carved in volcanic rock, displays youthful age by V-shaped cross section. *Below:* Gully shows same V-section, attesting to dominance of downcutting in both cases.

Flood plains develop along stream valleys by sideward cutting. (*Right center*, Zion Canyon, Utah; photo courtesy Union Pacific Railroad). Continuation of sideward cutting results in widened flood plains, eventually in monotonous expanses. *Right bottom:* Junction of Limestone and Nelson rivers, Manitoba; (photo courtesy Canadian National Railways), showing Limestone River meandering over its flat plain. Valleys are favorite haunts of geologists (*above, left*, small valley near Havana, Illinois; photo courtesy Illinois State Geological Survey) because the stream, cutting through overlying soil and mantle rock, exposes bedrock beneath the area. From such exposures is read the geological history of the region.

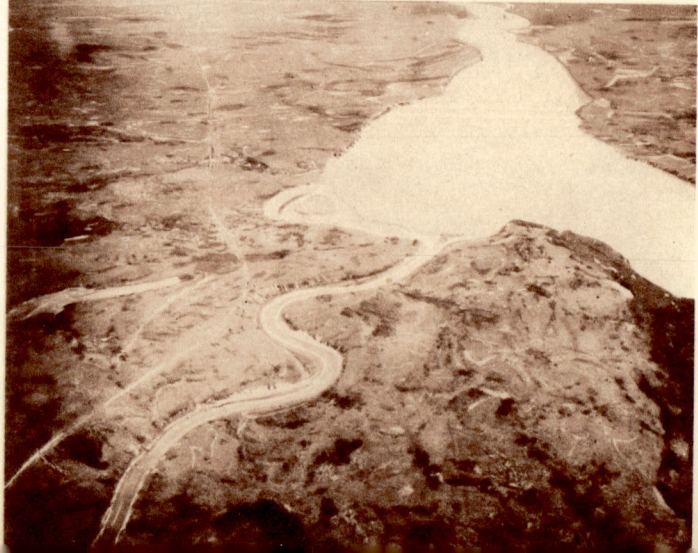

PLATE 9
REGIONAL AGE

In contrast to single valleys are regional aspects of stream erosion. Hardly a land form not carved in part by running water, from mountains (*left top*, Mount Samson, Jasper National Park; courtesy Canadian National Railways) to level plains (*left bottom*, peneplain, Baraboo, Wis.). Plateaus constitute upland (Plate 8, upper center and right photos) into which streams carve valleys. Elevation determines depth to which streams may cut; in the Grand Canyon, Arizona (*left center*, inner gorge; photo H. Bullen; courtesy National Park Service) erosion has cut downward a mile. As upland is reduced, slopes dominate and region becomes mature. *Right center:* Wasatch Mountains near Salt Lake City, Utah. Finally the region is reduced to a peneplain, ultimate stage of stream erosion (*left bottom;* also Plate 8, *right bottom.*)

Although peneplains constitute the last stage of stream erosion, they may themselves be uplifted, instituting a new *cycle of erosion.* Thus the bottom land of a previous cycle becomes upland in the succeeding series of events. Renewed cycles are not always easily recognized, but one criterion is the level summit of divides separating adjoining drainage basins. The view to the left shows such a level horizon.

PLATE 10 CAVERNS

← Entrance

Ground water dissolves limestone, forms caverns (*upper right*, Mammoth Cave). Balance between solution and deposition may swing to latter, giving rise to stalactites (*right center*, Luray Caverns) hanging pendant-like from ceiling, or to stalagmites (*left center*, Carlsbad Caverns), rising as squatty grotesque forms from the floor. Endless variety, intricate design of cave deposits shown at lower left (Totem Poles, Luray Caverns) and lower right (Helen's Shawl, Luray Caverns).

Photo by E. J. Hall, courtesy of Louisville and Nashville Railroad

Courtesy Luray Caverns Assn., Luray, Virginia

Photo by E. Kemp, courtesy of Santa Fe Railroad

Courtesy Luray Caverns Assn., Luray, Virginia

Courtesy Luray Caverns Assn., Luray, Virginia

PLATE 11
HOT SPRINGS AND GEYSERS

Above: Hot springs bear dissolved salts, form Mammoth Terraces, Yellowstone. (Photo by A. Curtis, courtesy of Northern Pacific Railroad.)

Above: Detail of hot spring, Mammoth. The terraces are formed of travertine (calcium carbonate).

Above: Other hot springs carry silica in solution. Note biscuit-like deposit about this spring, Yellowstone. (Photo courtesy of Northern Pacific Railroad.)

Above and right: Geysers eject hot water at intervals, puff steam between eruptions. Castle Geyser, Yellowstone. (Photo on right, courtesy of Northern Pacific Railroad.)

Above snow line of lofty peaks is an excess of snow accumulation over wastage. Thus arise snow fields, source of valley glaciers. *Left:* Glacier Natillions, French Alps (photo by R. T. Chamberlin), with view of snow field in the upper background. As glacial ice moves downslope, it yields as a rigid, elastic solid, by sudden fracture and shear. Thus arise crevasses (*below*, surface of Glacier d'Argentiere, French Alps; photo by R. T. Chamberlin), which render travel across surface so hazardous.

PLATE 12
GLACIAL ICE

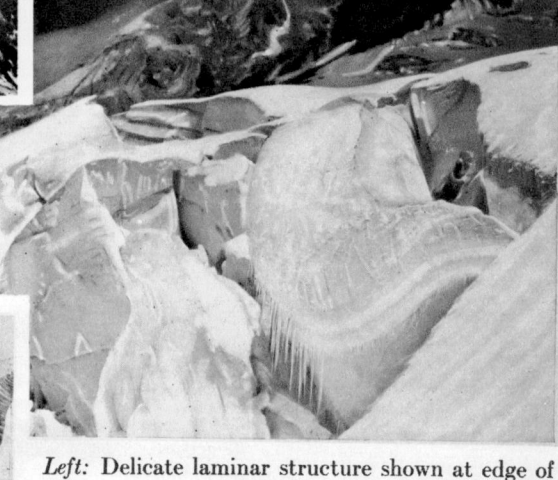

Left: Delicate laminar structure shown at edge of Glacier de la Brenva, Italian Alps (photo by R. T. Chamberlin). Such structures suggest a more continuous yielding movement within the ice, owing to pressures from above. *Below:* Terminus of Mica Creek Glacier, Caribou Range, British Columbia, showing its rounded, tongue-like edge (photo by R. T. Chamberlin). Débris is strewn about the edge, and streams of meltwater flow along the valley.

PLATE 13
GLACIAL SCOUR

Left: A glacial cirque marks the head of former glacier in Cottonwood Canyon, Utah. Cirques are among the most impressive features formed by glacial erosion.

Right: Glacial Lake McDermott, Glacier National Park, Montana. (Photo copyright by Hileman, Glacier Park Photo Shop.) Note the ice-scoured valley in the background, recognizable by its broad, rounded, U-shaped cross section. Grinnell Glacier is just visible in a notch through right center.

Left: Yosemite Valley, classic example of glacial scour. Half Dome rises prominently on the right, showing the broad rounded summit that we saw in Plate 3. (Photo by Ansel Adams; courtesy of National Park Service.)

Right: Glacial scratches and scours mark the former passage of glacial ice over this rock surface in Glacier Creek Valley, Montana. (Photo courtesy of Northern Pacific Railroad.) Note how the irregularities have been smoothed from the rock. Glacial action here has been so recent (geologically speaking), that the ice-eroded surface has not been appreciably weathered since.

Deposition from streams occurs when the velocity of the water is checked. Abrupt decrease of gradient is a cause of deposition at the base of mountain valleys. Thus are formed alluvial fans (*left*, aerial view of partially dissected alluvial fan; photo courtesy Spence Air Photos). Streams also deposit material on their flood plains; if later excavation lowers the stream level, remnants of earlier deposits may flank the valley walls as depositional terraces. *Right:* Terraces along Gardiner River (photo courtesy National Park Service). Note the level tops of the terraces, which are named from their steplike form. Terraces may also be formed by resistant beds of rock; the test applied is whether the feature is composed of sand and pebbles or of hard uneroded rock.

PLATE 14
STREAMS DEPOSIT

River bars (*center left*, Provo Canyon, Utah) often form at the inner bends of streams, where the velocity of the water is less.

Deltas form when streams carry débris into quiet waters. *Below:* Artificial delta (photo by Kaufman and Fabry) showing typical form and steep drop-off at edge. Levees, another stream deposit, aid in preventing floods; when breached, destruction may result. *Lower left:* Broken levee, showing torrent pouring over flood plain.

PLATE 15
EOLIAN AND GLACIAL DEPOSITS

Dunes result from deposition of sand by wind. With unlimited supplies, dunes form long ridges normal to direction of winds. Often, however, ridges are breached and *blowouts* (long tongues of sand that migrate with the wind) form. (*Left top*, Waverly Beach, Indiana Dune State Park, Ind.; photo courtesy American Air Lines. Large blowout in center foreground.) In contrast to dunes, wind-blown dust, carried in suspension, is deposited as a mantle over the landscape. *Left center:* Road cut section of loess trimmed to illustrate fineness.

Depositional land forms are associated with glacial ice *Above, right:* Mendenhall Glacier, Alaska, with outwash plain in center and moraine in foreground (photo courtesy Canadian National Railways). *Drumlins* are curiously shaped hills formed beneath the ice (*left bottom*, drumlin near Fou du Lac, Wis.). They are steep sided on stoss end, gentle on lee. Large boulders not uncommonly left by retreating glacial ice, as photo *lower right* shows (Yellowstone National Park; photo courtesy Northern Pacific Railroad).

PLATE 16 WAVES BEAT

Ocean waves approach shelving shores, increase amplitude, topple forward, surge against coast with battering-ram force. The turbulent zone within breakers sweeps sand, gravel to and fro, gnawing incessantly at edge of continents. *Upper left* and *left center:* Waves approach and strike coast along Gaspe Peninsula, Quebec. (Photos courtesy Canadian National Railways.)

The continual abrasion by moving sand and gravel cuts notches in the exposed headlands, forming sea cliffs. *Upper right:* Sea cliff at Perce, Gaspe Peninsula (photo courtesy Canadian National Railways). Note that the land surface slopes away from the cliff, showing that the greater part of a hill has already been eroded. As the attack goes forward, the sea cliff recedes, forming a wave-cut terrace just below sea level (*right center*, portion of terrace fringing sea cliff at Bay of Chaleur, Belledone, New Brunswick; photo courtesy Canadian National Railways). During the recession of the cliffs, more resistant or sheltered rocks may remain behind, forming stacks and chimney rocks (*lower left*, Perce Rock, Gaspe Peninsula; photo courtesy Canadian National Railways).

material picked up usually near the head of the stream is deposited farther along. Now, erosional land forms are cut *below* some upper land level, whereas the corresponding deposits are built up *above* some previous lower level. We may thus speak of erosional land forms as negative, and the deposited or constructed ones as positive, land forms. Whatever we call them, it is clear that they will differ rather markedly.

NOT only do streams develop erosional and depositional topographies, but the other geological agents also do so. We have given examples of each in our earlier chapters, and you undoubtedly can recall an example or two. We shall, for the sake of convenience, save you the trouble of compiling your own classification of land forms by building up a table which includes the commonest features of each agent. Here it is:

Agent	Erosional Land Forms	Depositional Land Forms
Wind	Wind-scoured basins, wind-eroded pinnacles	Sand dunes
Running water	Valleys, terraces, flood plains, peneplains	Alluvial fans, levees, flood plains, deltas
Ground water	Sinkholes, caves	Spring terraces, landslides
Glacial ice	Cirques, hanging valleys, glaciated valleys, rock basins	Moraines, till plains, outwash plains, kames
Shore agents	Sea cliffs, stacks, wave-cut terraces	Depositional terraces, bars, spits, beaches

When you merely glance through the table (as we suspect some of our readers may do), you fail to notice that *flood plains* are listed both as erosional and depositional features of streams. This is not a typographical error. As valleys pass from youth to maturity, a valley flat or flood plain is cut by sideward erosion. Later on, various stream deposits may be laid on this flood plain. For example, floods may cover the valley flat and deposit a veneer of mud and sand over it. In that case the flood-plain deposits may be so important that we classify the flood plain as depositional. Similar overlaps among land forms may occur among the other agents.

ONE of the more interesting things about erosional land forms is that in large measure they are independent of the particular types of rocks involved in their formation. Streams, for example, carve valleys in any kind of rock, ocean waves cut sea cliffs without regard for the kind of rock they sculpture, and valley glaciers excavate cirques with a fine disregard of the kind of material on which they work. To the experienced eye of a geologist, however, the detailed shapes and forms of the features tell much about the effect of rock hardness and resistance on the exact form assumed by the feature. We have already met the type of geologist who is specially trained to sweep such a knowing eye over the landscape. He is called a *physiographer*. When a stream valley, for example, becomes alternately wide with gentle valley walls, and then narrow with steep walls, as it is followed downstream, the expert will recognize that the stream is traversing first soft and then hard rocks.

The physiographer sweeps a knowing eye . . . .

IN SIMILAR fashion the depositional features are somewhat independent of the rock and mineral particles that form them. It makes little difference whether the stream has been carrying hard or soft rock fragments and pebbles in the case of alluvial fans, for example. The fan itself is a response to gradient changes, and whatever the stream carries at that point will tend to be dropped. There is, however, a marked influence shown by the sizes of the pebbles and fragments. Thus, some depositional features are made of coarse material and others of fine material. That, of course, is an attribute of the exact conditions at the site of deposition, as well as of the energy of the transporting agent. Wind, for example, could never form dunes made of boulders, because wind never has sufficient energy to sweep along a load of such large fragments.

NOT only is wind action dependent on certain sizes of débris, but to some extent the land forms that result depend on the *supply* of such débris. We have mentioned sand dunes as the typical land form made by wind. But there are dunes *and* dunes, and they may be long ridges or irregular hills or even quite symmetrical features called **barchans.** These barchans are interesting things. They are sand dunes shaped like a crescent, with the horns pointing with the prevailing wind. We do not care to exhaust the reader with technical details, but we may point out that barchans are apparently formed when the sand supply is rather limited, whereas ridges or lines of irregular hills tend to form when the sand supply is quite abundant. In either case there is a balance between the environment on the one hand and the land form on the other. The shape of a barchan remains fairly fixed, but the dune as a whole migrates slowly along the surface with the prevailing wind. Thus barchans are another of those cases of equilibrium in nature. Plate 18 includes photographs of barchan dunes.

Barchan dunes

THERE is another geological agent, in addition to wind, that is not independent of the material it has at hand. That is ground water, and for its work of making caves and sinkholes it requires special kinds of rock. Since ground water operates only by solution, the rock acted upon must be soluble. In nature there is only one soluble rock that is very abundant, and that is *limestone,* composed almost wholly of calcite. Hence ground-water land forms almost always are associated with limestone rocks.

THE curious lack of dependence of depositional features on some factors and their dependence on others forms an interesting chapter in geological processes. We said that, although in detail it made little difference whether a depositional feature was composed of hard or soft rock, it is true that the sizes of the pebbles and fragments have much to do with the matter. We may illustrate this in several ways: If a depositional feature is an elongate ridge, it could conceivably have been formed as a moraine by glacial ice, as a dune by wind, or even as a bar or beach by the sea. Which is it? That depends on the sizes of material in it, largely, and the shapes of the particles to some extent. Thus we already know that if it is a moraine it will be composed of glacial till, which ranges in size from

A ridge of débris—moraine, beach, or dune?

large boulders to very fine clay, all heterogeneously intermingled. A dune, on the other hand, is composed of sand exclusively; and a bar or beach may be largely of gravel. In other words, although all the agents may find similar materials to start with, certain ones select given parts of the weathered débris and, by virtue of their mode of operation, form land features with sufficiently different characteristics so that we may distinguish among them. In broad outline the land forms may be similar, but seldom will confusion regarding their origin linger if the feature is carefully studied.

EVERYONE is interested to some extent in the scenery that surrounds him, and this interest is heightened when there is some understanding of the processes responsible for the scenic features. With some understanding of the geological processes that wear down the land in one locality and build it up elsewhere comes a finer appreciation of even the more humble aspects of the landscape. Streams and their valleys take on a new and broader significance, and one is able in some measure to evaluate his own locality in terms of the entire region of which it is a part.

Naturally enough, an inexperienced onlooker finds much in nature that is confusing. That is because nature is complicated in detail, and becomes simple only when we look at the broader and larger features. Thus, just as depositional features may cover and modify previous erosional forms, so later erosion may attack

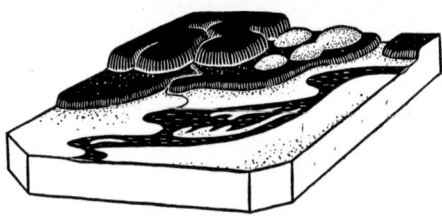

Composite topography along the Illinois valley. Valley walls, flood plains, terraces, and migrating dunes

the deposits themselves; and the result is a topography quite difficult to understand in all of its ramifications. A common example of such an intermingling of forms is seen on the flood plains of many large streams in the Northern United States. The Illinois River has an erosional flood plain on which glacial outwash gravels and sands were deposited. These were in turn eroded in part, and among the partially eroded terraces were laid normal stream flood deposits. The sand, both of the stream and from the glacial outwash, has been acted upon by winds, so that here and there dunes clothe the earlier features. Despite such complexities, a detailed examination of land forms, combined with a study of the actual deposits themselves, usually furnishes sufficient data to clear up the difficulties.

THE examination of the pebbles, sand grains, and mud particles that make up the sedimentary deposits forms a fascinating field in the application of laboratory methods to geology. Any laboratory procedure is technical, and we cannot take the reader through the entire gamut of details. Contrary to the usual policy of elementary books, however, we propose to let the reader look behind the scenes with us and watch the type of work that today is being carried on with sedimentary rocks. The subject, together with a description of some of the commoner rocks themselves, deserves its own chapter.

CHAPTER 14

EXCURSIONS AMONG THE SAND GRAINS

>Wisdom is ofttimes nearer when we stoop
>Than when we soar.
>—Wordsworth

IT MAY seem at first glance that of all the subjects we might choose to talk about, the least significant would be a grain of sand. Probably no one ever stopped to figure out how many individual grains he steps on in walking across a beach, but it must be a number well up in the millions. We shall not devote this chapter to a single sand grain, although each grain could tell a long story. But we may confine parts of it to a handful, which is almost as insignificant, in comparison with the billions of grains on even a small beach.

Millions of sand grains . . . .

It is part of a geologist's work to find out all he can about the rocks of the earth, and this includes the mode of their origin, the sources of the grains or crystals that compose them, and the nature of the environment in which they were formed. It is natural that we should first choose the simpler cases, and study rocks in process of formation, where the agent that forms them and the particular environment of formation are known. Now the sand on beaches, in dunes, and along rivers is rock in the process of formation, for, as the centuries and millenniums go by, these sands may be buried beneath later deposits and become hardened into sandstone.

Primarily, we are concerned with the problem of learning something about the ancient sands and other rocks that were laid down during the geological past. We have learned that if we study present-day sands we often are able to find some characteristics among the grains that furnish us with clues about the environment of deposition. These same criteria may then be applied to the more ancient sandstones.

How shall we go about finding the characteristics of sands? Obviously, we cannot study all the sand being laid down today, so that we must confine ourselves to samples, so chosen that they display the main features of an entire deposit. After collecting our sample, which may be a pound or so, we take it into our laboratories and look it over.

ONE of the first things that may be done to the sample is to determine the sizes of the grains and the distribution of the sizes. To do this, we may shake a weighed quantity through a set of sieves, each with a particular sized mesh. We may then collect the grains that rest on each sieve, and weigh or count them to determine the percentage distribution of the grains over the sieves involved. The data may next be plotted on graph paper to yield a curve like the adjacent figure, which represents the curve obtained when *beach sand* is analyzed in this manner. The figure shows us that the grains range in size from about 0.5 mm. diameter to about 0.05 mm., because at these points the curve touches the horizontal or 0 per cent line. Next, the height of the curve over

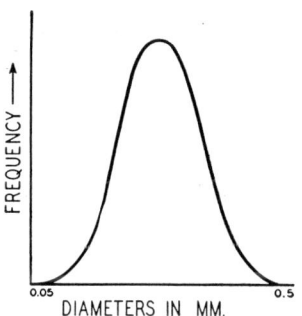

Size frequency curve of beach sand. (The horizontal scale is logarithmic)

any small interval is proportional to the abundance of grains in the interval, and we may see at a glance that the greatest abundance of grains is about equidistant from the ends of the curve. This maximum represents not only the most abundant grains in the sample; but in the curve shown it is also the *average size* of the sand grains, which usually is about 0.15 mm. diameter in these sands. Here, then, we have a fairly definite picture of the sand in terms of the grains composing it, and the relative abundance of the various sizes that are included in it. This is a **size frequency curve,** and it is of considerable value in the study and comparison of sedimentary rocks. Not only sands, but gravel and muds or clays, may be studied in this manner.

If we were to make a similar analysis of *glacial till*, we would find that the particles show a much wider variation in size and a much more irregular distribution over the sizes. The adjacent figure shows a size frequency curve of glacial till. See how strikingly it differs from that of sand. Instead of having only one maximum point, it has several. If we compare our two curves further, we see that the sand is pretty well confined to the immediate vicinity of the average size: it is, in short, quite **well sorted,** and consequently shows the selective effect of the agent that carried and deposited it. The till, on the other hand, is relatively **unsorted,** just as we should expect from the non-turbulent and non-selective movement of glacial ice. Thus our laboratory study of size distributions has confirmed what we should have expected from our knowledge of the agents themselves.

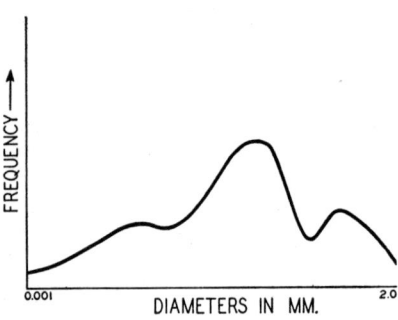

Size frequency curve of glacial till

WE MAY also study the individual grains under the microscope and find out how they vary in shape. The shape is defined in terms of the approach of the particles to true *sphericity* on the one hand and the degree of *roundness* of the grains on the other. These two measures involve different concepts, as the figures show. The measurements are rather complicated and we shall not bother the reader with them, but the net effect is that a numerical value of relative shape may be obtained for the grains. Such a study is called **shape analysis,** and it affords us much valuable information

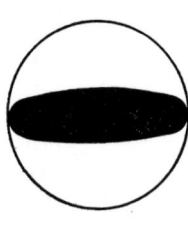

Low sphericity, high roundness

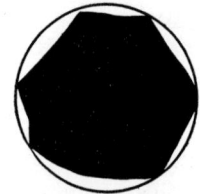

High sphericity, low roundness

about the past history of sedimentary rocks. For example, if most of the grains are well rounded and nearly spherical, we infer that they may have suffered considerable transportation; whereas if they are angular, with sharp corners, the presumption is that they have not traveled very far from the point where they were first formed.

NEXT we may identify the minerals that make up the sand grains. Such a study is called **mineralogical analysis.** Usually, each grain is composed of only one mineral, but there may be a number of mineral species in the sample. Many sands, for example, consist almost entirely of quartz. This mineral gives the predominant light color to sand. There is, however, commonly about 1 per cent of darker mineral grains in the sand, and these darker grains have a higher density than the quartz. We may easily separate out these heavier minerals for more detailed study. This separation is made by means of heavy organic liquids like *bromoform*, $CHBr_3$. Such a liquid, which may have a density of about 2.8, is chosen, and the sand is poured into it. The quartz floats on its surface, and the heavier mineral grains sink to the bottom. They are separately collected; and the heavy grains, usually called the *heavies*, are mounted on a microscope slide and studied.

The heavy minerals in sediments give clues to the nature of the source rocks, and so aid in reconstructing the whole series of vicissitudes through which the sand has passed. Not only may sand result from the weathering or erosion of rocks containing the primary minerals, like *igneous rocks*, but earlier generations of *sedimentary rocks* may have been eroded to yield the grains; and the heavy minerals of the specimen also shed light on that question.

WHEN we combine the results of our size analyses, shape analyses, and mineral studies, the lines of evidence afforded by these different approaches often converge in certain directions, so that the combined data make us reasonably certain about the conclusions we reach. The obvious application of these methods to the ancient rocks of the earth, supplemented by what we know of present-day processes, is often indispensable in our understanding of the past.

We do not wish to give the impression that such studies as these tell us in every case the detailed history of the rock. There are many situations in which the play of forces is so subtle that at present even our most refined methods of analysis are of little avail. The subject we are touching upon here is one at the very forefront of geological research today, but continual improvements in technique are rapidly advancing our ability to interpret Nature's complex ways.

A FULL understanding of sedimentary rocks is not to be obtained from the laboratory alone. Valuable as such studies are in furnishing us with quantitative information about the rocks, they must be supplemented by investigations carried on in the field, where the rocks occur. It is here that the forces acted on the rocks during their formation, and by observing the entire environmental situation we learn how the rocks fit into the whole regional picture. For example, we mentioned before that along ocean shores the sediments display a more or less regular gradation in size as they are followed out from the beach. This type of observation aids us in relating each type of rock to its environment and in understanding the significance of the data that our laboratory studies bring to light.

The field class gets down to earth

The field examination of the rocks includes studies of numerous features besides the physical and chemical properties of the rock particles. For instance, the nature and type of bedding in the rocks is important in evaluating their past history. Wind work usually results in types of **cross bedding** that may serve to distinguish it from the work of water. Currents of wind and water form certain kinds of **ripple marks** in their deposits, whereas wave action results

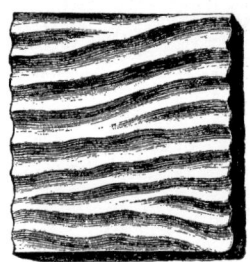

Ripplemarks, one of the criteria of sedimentary rocks. Old woodcut from Lyell's *Manual of Elementary Geology* (1855)

EXCURSIONS AMONG THE SAND GRAINS

in quite another type of sand ripple, familiar to bathers as hard ridges on the sandy bottom of lakes and seas. Some mudstones show **raindrop impressions,** made while the muds were exposed along tidal flats or valley bottoms. **Fossils,** which are plant and animal remains, commonly occur in rocks at the very place where they became entombed during the deposition of the sediment. From the nature of the forms thus preserved, much can be learned of the physical and chemical environment of deposition; and we shall show a bit later how all these data are used in reconstructing the past.

WE SEE here some of the lines of attack that distinguish geology from other physical sciences. Direct experimentation is often closed to us, because the processes of nature are so slow and involved. Rather, we must take the products of natural processes and from them reconstruct the events through which they passed. This results in an inferential type of analytical thinking which is best supported by a convergence of evidence. Hence the geologist strives to assemble a great deal of data all pointing to the same conclusion, since in such a case that conclusion is probably sound. Thus we must combine field studies of the larger aspects of the rocks with studies of the forms and arrangements of their beds, and supplement all these investigations with laboratory studies of the detailed characteristics of the rocks.

IN A purely descriptive sense we subdivide the sedimentary rocks on the basis of the sizes of the fragments that compose them. This size classification, for technical reasons, is based on a geometric scale. Thus **gravel** has pebbles larger than 2 mm. in diameter, while **sand** has grains ranging from 2 mm. down to 1/16 mm. **Silt,** the next finer material, ranges in particle size from 1/16 mm. to 1/256 mm. in diameter. Silt feels something like silk when rubbed between the fingers. To the eyes it looks like a very fine powder. **Clay** comprises material finer than silt, and the clay particles may range well down into the colloidal classes, in which each particle may be no larger in diameter than a wave-length of light is long. You notice that the word *clay* is used in two senses by geologists. In one sense we refer to the clays as minerals which result from rock-weathering, and in the other it has purely a size meaning. The double use introduces no confusion when we recall that clay minerals usually occur in crystals or fragments in the clay sizes. That is, the term as used for size really also covers the mineral composition, because most material that small in size is a clay mineral.

The range of particle sizes among sedimentary rocks is far greater than our previous paragraph indicates. **Pebbles** may range in size up to 64 mm. diameter, and **cobbles** and **boulders** are larger still. Boulders 10 m. (about 30 ft.) in diameter are not unknown. Such boulders have been found associated with glacial deposits.

AS WE already know, all the particles of a given rock need not be exactly of the same size. Usually, however, there is an *average size*, and most of the particles are near that average diameter. When gradations occur among the sizes, we call them, for example, *sandy gravels*, or *silty sands*. Among the poorly sorted deposits, such as *glacial till*, the particles may range from immense boulders down to the finest clays, and for that reason glacial till is often called *boulder clay*.

As long as the deposits remain in a loose or incoherent state, the names we have just applied to them hold. Commonly, however, as the sediments become buried beneath later deposits, their constituent particles are cemented together, so that the rocks become hard and *indurated*. This cementation is generally performed by ground water, which deposits various minerals in the interstices among the grains, and so binds them together. Among the commoner cements are calcite, iron oxide, and quartz or silica. When the rocks become indurated this way, the names applied to them change, to indicate the change in physical state. Thus, cemented gravel becomes **conglomerate;** cemented sand is called **sandstone;**

Conglomerate, a rock composed of cemented pebbles

EXCURSIONS AMONG THE SAND GRAINS

hardened mud or clay is called **shale;** and indurated glacial till is designated **tillite.** Among the mixed types of sediments, composite names are applied.

THE types of sediments thus far mentioned are called the *fragmental rocks,* because they result from the breaking-down, or fragmentation, of other rocks. In addition to these, however, there are other sedimentary rocks that result from the deposition of material carried in solution. We saw that calcite and some other minerals, such as common salt, are soluble. The calcium carbonate dissolved in the oceans is deposited in various ways. Perhaps the most important manner is through the agency of organisms. Animals withdraw the lime for use in their shells, corals build up their skeletal features, and certain bacteria are active in precipitating the carbonate from solution. In these various manners are formed lime muds; and when later they become hardened, **limestone** results. In a similar fashion, when enclosed basins of sea water evaporate, the dissolved salts are precipitated and may form extensive beds of common salt.

IN ADDITION to the descriptive aspects of the sedimentary rocks, there are *genetic* classifications, in which the particular agent that deposited them is involved. For instance, there are sandstones resulting from the action of wind, running water, or oceanic shore agencies. The distinctions made among these are included in the terms *dune sands, river sands,* and *beach sands.* Similarly, fine material deposited by wind constitutes the sediment called *loess,* and the word "loess" is confined to such wind deposits even though other agencies also deposit fine material. The only agent that deposits *till* is glacial ice, so that the term is at once both descriptive and genetic. It is the goal of geologists to be able to classify all rocks on such genetic bases, but this desirable result has not yet been reached because of the high degree of complexity among rocks and because of the many variations among rocks formed even by the same agent.

For our purpose it is most convenient to classify some of the more common sedimentary rocks on a purely descriptive basis. We may accordingly summarize our discussion in a manner like this:

THE SEDIMENTARY ROCKS
Fragmental Rocks
Well sorted
Coarse particles
Gravel, conglomerate
Medium-sized particles
Sand, sandstone
Fine particles
Clay (mud), shale
Unsorted
Glacial till, tillite
Non-fragmental Rocks
Lime-mud, limestone
Peat, coal
Salt
Gypsum

You will notice that although the table purports to be purely descriptive, we actually can read into it the influence of the geological agents. Thus the well-sorted sediments result from turbulent and selective agents, whereas the only unsorted sediment listed is due to glacial ice. (There are other unsorted sediments which are not included in our table.) Similarly, among the non-fragmental rocks, at least the peat and coal are due to organic agencies, and much of the limestone itself is formed with their aid. The salt (NaCl), and likewise *gypsum* ($CaSO_4 \cdot 2H_2O$), results largely from the evaporation of bodies of sea water.

A form of gypsum crystal called, interestingly enough, a "fishtail twin"

THE outstanding features of the sedimentary rocks, as we mentioned before, are their arrangement in layers or strata and their sorting according to size. Glacial till is, of course, an exception to this, as are a few other rocks. In general the sedimentary rocks are also transported rocks, and during the process of transportation minerals from several sources may become intermingled in the same sediment. All of them, however, bear traces of their origin, and it is seldom that sedimentary rocks are confused with rocks formed by other geological processes. Plates 19 and 20 illustrate the common sedimentary rocks.

Inasmuch as the oceans and seas are the receiving basins of the earth, it is there that the greatest amounts of the sedimentary rocks are deposited. Under marine conditions there may be formed widespread layers of sediment; and in the relatively quiet conditions that prevail away from the immediate vicinity of coasts, the strata of oceanic sediments are strikingly uniform over wide areas.

It may be well to reiterate that it is from ancient marine sediments now exposed on the continents that most of our data are gleaned concerning the past history of the earth.

CHAPTER 15

BRIEF INTERLUDE

"The time has come," the Walrus said,
"To talk of many things: . . . ."
—Lewis Carroll

WE HAVE now progressed in our story to the completion of one of the great processes of geology, namely *gradation*. We have traced through the changes by which entire land areas are worn down essentially to sea level by the relentless operation of the geological agents. During this process the original crystalline rocks of which the lands were composed have been weathered into soils, transported, and redeposited as sedimentary rocks beneath the seas. It behooves us now to pause for a while and raise a few questions regarding the present configuration of the earth in the light of what we have described.

Reflect on the past, and consider the future . . . .

If the efforts of the geological agents are continually directed toward wearing away land masses and sweeping them into the seas, why do we have any land left at all today? We have hinted here and there that the age of the earth is to be reckoned in billions of years. Surely, in all that time any original continents would long since have been submerged beneath a universal ocean. Indeed, if the work of degradation had gone on uninterruptedly, the surface of this universal sea would be standing some 600 ft. above the continental platforms—a watery grave, indeed.

Not only, however, do we have abundant land above the seas today, but these lands locally are in all possible stages of erosional development. Even allowing for differences in climate and in the nature of the rocks, it is obvious that all these land areas could not have begun their erosional histories at the same time. What is more,

BRIEF INTERLUDE 127

we have mentioned that we find ancient sedimentary rocks exposed on the continents, and these rocks bear unmistakable traces of their deposition beneath the seas. Unquestionably, then, these sedimentary rocks must have been uplifted from the sea bottom.

Q UITE clearly, there must be two opposing forces operating on the earth. Gradation, we know, tends toward a downcutting of the lands, and the opposing force must tend toward an uplift or renewal of the lands. Such renewals of the continents give them a new lease on life, as it were, and enable them to keep their heads above water for a while longer. As soon as they are renewed, however, the gradational agents resume their attack; and world-history, on this basis, appears to be a continual battle between the down cutting and uplifting agents.

Not only is this battle going on at present, but there is abundant evidence of its continued operation through all of geological time. Among the ancient rocks of the earth we find sedimentary cycles (repeated alternations of similar rocks) which are separated from one another by great time gaps in the record. These features attest to numerous occasions when the lands have been worn down nearly to sea level, only to be renewed again. Thus over long periods of time there appears to be an essential balance between the downcutting agents and the restorative process. Degradation may dominate for a while, but eventually it is interrupted by uplift, so that cycle after cycle of uplift and downcutting has occurred as far back as we can trace geological history.

The two series of sediments are separated by a time gap, as shown by their lack of conformity. The limestone below was partially eroded before the sandstone was deposited

W HAT is this process that renews and re-elevates the continents, and how does it operate? We shall answer these questions in detail in the several succeeding chapters. Before turning to them, however, suppose we relate our present discussion to the energy diagram that we presented way back at the end of chapter 4.

If we consider gradation in terms of the world's sources of energy, we see that it depends for its operations on the external energy of the earth, namely, that received from the sun. In short, it is the upper half of the energy diagram that has been receiving our attention thus far. The reader is urged to turn back to the diagram again and to convince himself that in their operation the geological agents are directly related to the flow of the sun's energy through the economy of the earth.

AT THIS point our alert readers will at once suspect that the opposite phase of the subject, the uplift and renewal of the lands, must be related to the earth's internal sources of energy. They are right. It is this internal energy that maintains the lands in their battle against the geological agents. It would be an interesting speculation to reflect on the possible end of the earth as a habitable globe were one or the other of these energy sources to fail. We shall not pursue the thread here, but some of our readers may find the notion entertaining.

CHAPTER 16

WHAT PRICE CONTINENTS?

> Where order in variety we see,
> And where, though all things differ, all agree.
> —POPE

WHEN the Romans built the Temple of Jupiter Serapis in Pozzuoli some hundred or so years before Christ, it overlooked the azure waters of the Mediterranean. It today overlooks the same sea, but its position has by no means been a stable one. For back in the late 1700's it was necessary to excavate the old columns, still upright, from a series of marine sediments, and on the ancient pillars at various levels were unmistakable boreholes made by marine mollusks. The old temple has undoubtedly been both above and beneath sea level at several times during its history. Only relatively recently has it been elevated to its present height.

Temple of Jupiter Serapis at Pozzuoli, Italy. Old engraving from Von Leonhard, 1844

Nor is this by any means our only example of changes in sea level during historic times. Along the coast of Sweden former submerged reefs now project above the sea. Along the west coast of Greenland the opposite situation is present: the land is slowly sinking beneath the sea, and old buildings and other works formerly above high tide are now lapped by the waves.

IN GENERAL, we need not even confine ourselves to historical times for proof that land and sea levels change. We have abundant evidence that during the past history of the earth such changes have been common phenomena. For example, we find sea cliffs with their associated terraces and chimney rocks high and dry, miles from the nearest coast. Our only possible inference is that this particular set of land forms, which can only be developed by shore agencies, marks the former position of the coast. Further, that the coast line must have been at the base of the cliff in comparatively recent geologic times, else wind and water would have obliterated the features long since. Even in the interior of North America are striking evidences of recent changes of level due to an actual warping of that part of the continent. During the Ice Age which closed some twenty-five thousand years ago, great ice-dammed lakes persisted long enough to develop distinct shore features, like beaches and cliffs. Everyone knows that water assumes its own level, and when we find that these old shore lines now rise to the north, we must conclude that within the last twenty-five thousand years the northern part of the continent has been rising relative to the southern part.

Drowned valleys are an evidence of changes in sea or land level. Part of an old map by J. Russell, 1795

If you wish one more striking example, examine a good map of Lake Superior. Along the north shore the lake is bounded by high cliffs, and many of the streams that flow into the lake do so over waterfalls. The south shore, on the other hand, is marked by drowned valleys or estuaries, as witness Keweenaw Bay and the harbors at Ashland, Marquette, and Munising. What is happening here is simply that the land is tilting up at the north and spilling Lake Superior over into Wisconsin and Michigan.

ALL of these phenomena are examples of **diastrophism,** or movements of the solid part of the earth. Such movements may be slow and imperceptible to the inhabitants, or they may be

WHAT PRICE CONTINENTS?

sudden and violent, as in earthquakes. Further, they may represent merely general uplifts or downwarps of broad areas, or they may involve upfolds of local areas into great mountain ranges, for in diastrophism we are dealing with one of the most impressive, inexorable, and overwhelming forces on the earth. Plate 21 shows some of the results of diastrophic forces.

WHEN we examine the face of the earth, we find here and there great wrinkles of folded and crumpled rocks, which we call mountains. These chains of distorted rocks follow, in general, several well-defined belts. Witness the "Backbone of the Americas": the Cordilleran chain that reaches from the Arctic to the Antarctic circles, and thus sweeps nearly halfway round the world. Consider also the great Eurasian chains that stretch across Southern Europe into and through Asia Minor and culminate in the Himalayas of Northern India. For the most part the present sets of wrinkles on the earth parallel coast lines, and so lie near the boundaries between continents and ocean basins.

From detailed studies of the mountains we find that they are composed in large part of ancient marine sediments which have been lifted out of the sea and wrinkled and crumpled into folds. Now sedimentary rocks, as we know, are laid down almost horizontally, so that the folding and crumpling must mean that they have been crowded together into less extent than they originally had. For, notice that if we have points A and B placed along a series of horizontal beds, and then fold up the beds into a series of crinkles, the points will be brought

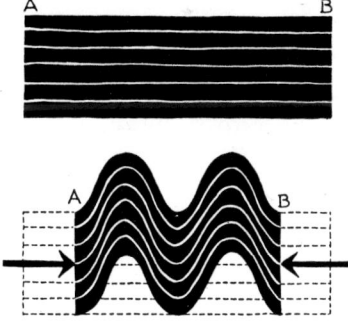

The distance between A and B decreases when the rocks are folded

closer together. On a world-scale, then, folding of rocks suggests a shortening of the earth's circumference. When the circumference decreases, the radius decreases; and with a sphere the volume also decreases. One may with some reason, then, conclude that the earth has shrunken at some time or other in the past. If shrinking actually has taken place, as this chain of reasoning suggests, certainly one of its results would be that the outer layers of the earth would wrinkle and fold, something like the skin of a drying apple buckles and wrinkles to adjust itself to a reduced volume.

WE MAY not be able to prove conclusively that the earth actually has shrunk, but the great mass of geological evidence seems to allow this as a reasonable conclusion. Furthermore, there is evidence that this shrinking has happened many times in the past, because we find remnants of folded mountains among the most ancient rocks of the earth, in some cases overlain by undisturbed younger sediments. In later sections of the book these points will be discussed in detail; for the present we merely record the fact. In addition to the occurrence of mountain-folding in the past, we have evidences of *recurrence* in the folding. That is, whole regions show evidences of having been peneplaned one or more times—evidences of a gradual wearing-away of the land by wind, water, ice. Now, in order to form a peneplain a long, long time is required—a time of comparative rest and quiet, during which the gradational agents may operate undisturbed. If then we find alternations of periods of quiescence separated by intervals of shortening of the earth's radius, the inference is pretty strong that the shrinkage of the earth, as manifested at the surface, takes place not continuously but rather as interruptions of otherwise quiescent times.

MANY have been the theories to account for the shrinkage of the earth. We shall not burden the reader with them, but rather proceed at once to current notions. Remember that no one has ever penetrated even 1 per cent into the earth's interior. Direct knowledge of the forces acting in the deeper parts of the earth may never be had. We may, however, reason from known physical and chemical laws, and we do have some important data on the interior furnished by earthquake waves. We are reasonably certain that the interior of the earth is not perfectly homogeneous, but that it is composed of stony and metallic materials, all under tremendous pressures. Near the center of the earth the pressures are estimated to be nearly 50,000,000 lb. to the square inch—a pressure more than three million times that of the atmosphere. We cannot produce anything approaching such great pressures in our laboratories, and so we cannot say definitely how matter behaves under such conditions. Our knowledge of atomic structure, however, shows us that the atoms of which all substances are made consist of a heavy central nucleus surrounded by electrons revolving about it. Small as these atomic systems are, there nevertheless is relatively much space between the nucleus and the electronic shells. Thus it may be possible that under the terrific pressures near the earth's center there may be some actual compression of the atoms into more compact systems. Compression generates heat, owing to forces that tend to resist the compression. As long as the heat remains as molecular agitation, actual shrinkage may be avoided; but if the heat is conducted away, the compressional forces may predominate and cause a reduction of volume.

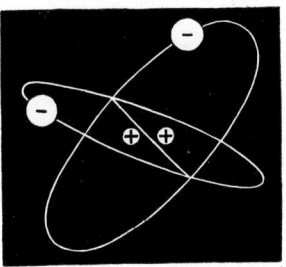

Small as these atomic systems are, there still is space between nucleus and electrons

DO WE have any direct evidence that heat is being conducted away from the interior of the earth? Remember the geothermal gradient, and recall that from volcanoes great masses of hot liquid rock are poured on the earth's surface. The earth is conducting heat to its surface and there radiating it into space. Furthermore, as far as we know, this is a continuous process. Hence our point has some support, and it would appear that shrinkage may well be taking place.

Such shrinkage would be a function of depth in part, however, since near the surface the compressive forces probably are not great enough to produce contraction. Thus we may picture an interior of this terrestrial sphere where a continuous reduction of volume takes place due to the continuous conduction of heat outward, upon which is superimposed a more rigid outer layer which tends to maintain its volume and areal extent.

IF OUR reasoning thus far is sound, how may we account for the recurrent aspects of diastrophism that we mentioned before? Well, consider an arch of a building. This arch is able to support itself as long as the strength of the building material is greater than the stresses operating on the arch. If the various forces are not balanced, the arch collapses. In a crude way this illustrates our point. The outer portion of the earth has a certain strength due to its rigidity and elasticity, and it is capable of withstanding a number of stresses. Eventually, however, the shrinking interior causes an accumulation of stresses of such magnitude that the outer part can no longer withstand them. Under such conditions this outer part yields, and there results a folding and crumpling of the outer layers of rocks.

Thus, the very fact of recurrence in mountain-making lends support to the theory we have built up. It would be possible to point out additional lines of evidence both for and against these current notions, but we shall not labor the point. We shall, however, pause to bring out an item or two that may not have been explicit in the discussion.

YOU will notice that we have a sequence of observations and inferences. Our observational data included mountains, the folding of rocks, recurrence of mountain-folding, heat conduction from the interior of the earth, and finally the strength of materials. Given these observations, how to account for them? We inferred from the first two observations the probability of earth shrinkage. Then we noted that our next observation required recurrent shrinkage on the outer part of the earth at least. The fourth observation demanded that shrinkage of the interior be continuous because heat conduction and radiation are continuous. To reconcile the apparent contradictions of the last two observations, we pointed out that the rigidity and elasticity of rocks in the outer part of the earth eliminated the apparent disagreement.

The lower sediments represent an ancient mountain range which was peneplaned before the upper sediments were deposited

WHAT PRICE CONTINENTS?

NOWHERE have we had recourse to direct experiment. Experiments on a world-scale are, of course, impossible; no one can take the earth into his laboratory and operate on it. What we do have, however, is a sequence of observations and inferences; and a given inference suggests other observations that may support it. This results in a seeking for further data; and these data are sought in the world of nature, among the rocks where evidences of past deformation are to be found. We have thus an illustration of the building-up of a scientific theory, and even without experimentation a convergence of the evidence at least does not exclude our conclusion as a reasonable one.

SO MUCH for observations and their probable explanation. What about the actual mechanics of earth adjustments at the surface, and what have they to do with our immediate story? The topic is even more involved than our previous analysis, and we shall postpone details until later. To pave the way, however, we shall mention the following points:

Gravity determinations, by which the value of g (the acceleration due to gravity) is found, show conclusively that the oceanic portions of the globe are denser than the continents; volume for volume, they are heavier. Near the surface at least, there may be a density discontinuity at or near the continental margins. We may expect, then, that stresses will tend to accumulate near the margins; and when failure occurs, it should usually be in the zone near the continental margins. This is actually where the most apparent deformation takes place. The tendency will be for the continents to be uplifted relative to the ocean bottoms, and during actual yielding the relative motion may result in the oceanic segments settling downward with respect to the continental blocks.

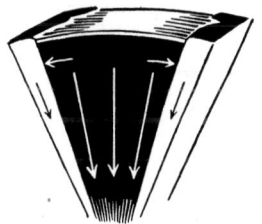

The heavier oceanic segments squeeze the continents

AS A result of these adjustments, two broad types of diastrophic effect are produced. There will be broad uplifts of the continents as a whole, and localized folding of some of the rocks into mountain chains. The intense folding of the ranges follows from wedge action on the continental blocks, which sets up forces that squeeze the continents from the sides. Now, while the continents are being uplifted and the mountains folded, what will happen to the oceans themselves? The waters will drain from the continents into the deepened basins, exposing and renewing the continents above the seas. Thus new gradients are given to streams, fresh rocks are exposed to the action of the gradational agents, and a new cycle of erosion is set into being.

The seas are drained from the continental platforms....

NOW we may see why the continents are given new leases on life, and why the world is not covered with a universal ocean, as we hinted in the last chapter. The land masses are renewed, relative to the oceans; but as a whole, the earth is reduced in size. In the light of modern theories, then, the price of continents is a shrinking earth.

CHAPTER 17

FOLDS AND FAULTS

>What can we reason but from what we know?
>—Pope

THERE is a curious kind of sandstone that can easily be bent by hand when it is in thin slabs. We say it is "curious" because everyday experience tells us that rocks are quite brittle. They may be chipped with a hammer, and one would be surprised to find a rock that merely bends under the blows. Our piece of sandstone, however, is a sort of trick rock. It is composed of quartz grains rather loosely held together with flakes of another mineral (mica); and as the rock is bent, the sand grains readily slide over the mica flakes.

Surprised to find a rock that bends....

Despite the brittleness of common rocks, we do find them in nature contorted into all kinds of intricate folds and plications. This is especially true in mountainous regions, where diastrophic forces have acted on the rocks. As a matter of fact, the conditions of pressure, temperature, composition, and the time element itself, are the factors that largely determine whether rocks will bend or break. Our direct experiences with rocks occur in the surface environment with its low pressures and relatively low temperatures. It is under such circumstances that rocks break rather than fold. Further, the time at our disposal is limited. Even so, a century or two is sufficient for some changes. For example, covers of old tombs occasionally are found to sag downward a bit. The rock (commonly limestone) adjusts itself to its unsupported condition by a slow bending, even under surface conditions.

The pressure and temperature effects may be brought out by experiments that have actually been performed in laboratories.

Certain substances, like salt, ice, and calcite—even some types of rock—can be made to flow when they are confined by great pressures. By *flow* we mean a fractureless change of form, for under great confining pressures the material cannot break or shatter, but must adjust itself to its environment by an actual plastic yielding. The flow of rocks within deep-seated environments is likewise a plastic yielding, and it takes place partly through a recrystallization of the constituent minerals. We shall learn more about such deformed rocks in chapter 23; for the present we shall examine their larger features, the folds and other **structures** shown by rocks that have been subjected to diastrophic stresses in the deep-seated environment.

SUPPOSE we consider a cubic foot of rock somewhere down beneath the earth's surface, say 10 or 15 mi., in what we have called the "deep-seated environment." Our cube is thus part of a column of rock 1 ft. square that extends down from the surface to the depths far below. Hence it will be under considerable pressure from the overlying rocks, and at a respectable temperature. The pressure under normal conditions will be static, and the cube will remain at rest because all the forces operating on it are equal. But now suppose that, owing to adjustments in the earth, the cube is acted upon by diastrophic forces from two opposite sides. What will

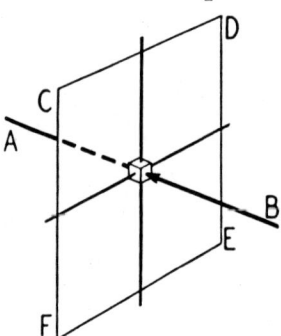

Deeply buried cube of rock acted on by diastrophic forces

happen? The rock cannot very well shrink into a smaller volume, because it is already quite compact. There may be a slight adjustment of volume due to a recrystallization of its minerals; but on the whole, the tendency will be somehow to ease the pressures that operate on it by escaping from them. The directions in which this relief will find expression lie in a plane at right angles to the diastrophic forces. In the adjacent figure the cube is acted on by diastrophic forces along the line *AB*, and the plane *CDEF* contains the directions of maximum relief. But will the relief be upward,

FOLDS AND FAULTS

downward, or sideward away from the compressive force? In actual mountain-making, of course, there is a *zone* subjected to these forces, rather than some fixed cube. Hence, whatever yielding takes place affects the entire zone. Our cube, however, serves to fix attention on a particular part of the zone affected.

The question of the direction of relief cannot be answered positively, because we cannot get down to examine the situation. All we can do is to examine the eroded belts of folded rocks along mountain chains and see what they tell us. This much we know: The belts of folded rocks are usually linear, and in many cases they are symmetrical on both sides of the belt. Further, they show an elevation relative to their surroundings, so that in the folding the whole series of rocks has been uplifted.

Portion of an elevated belt of folds, showing the direction of compression

The accompanying block diagram sketches the usual relation of a mountain belt to the diastrophic forces that formed it. We are reasonably certain that some relief has been upward and sideward along the plane of maximum relief. As to possible downfolding simultaneous with the other adjustments, we are not certain.

THE reader will have noticed long ago that our chapters on diastrophism are fraught with pitfalls for the unwary. We have to pick our way slowly and painfully, distinguishing between what we know and what we infer, for, in dealing with this subject we are approaching one of the outstanding problems of present-day geology. The nature and operations of diastrophic forces are not fully known, partly because the scale on which earth operations take place is too great for us to encompass in experiments; nor can we make direct observations at the localities where the deep-seated disturbances take place.

Confined, as we are, to surface environments, our observations include mainly the surficial aspects of these deep-seated diastrophic adjustments. The upward relief of rocks under differential stresses results in those great elevated belts of rock that we call mountains; and usually within them the rocks are found to be highly folded, or greatly mashed, and quite commonly altered from their original appearance. Fortunately, gradational agents attack these elevated tracts and incise deep valleys among the mountain folds, so that the very roots of the mountains may be laid bare. Then, by observing how many layers of rocks are involved in the folding, and their thicknesses, we are able to tell how deeply such rocks were buried before the folding occurred. **In fact, it is by examining eroded rock structures like this that we learn most of what we know about deep-seated environments.** Although we are able thus to reconstruct many of the environmental factors involved in deep-seated diastrophic adjustments, we are handicapped by the fact that the show is over. The adjustments have been made, and we are forced to reconstruct the entire picture from the remnants we find. Fortunately again, rock-weathering is relatively slow, so that many deep-seated rocks, exposed at the surface by recent erosion, still retain most of their deep-seated characteristics.

IN SOME cases we find that the rocks have been folded into tremendous arches, thousands of feet high, and equally broad. Such upfolds we call **anticlines,** and they may be rounded, dome-like affairs or elongate bulges of rock. Corresponding downfolds are called **synclines.** In some cases we find an alternation of upfolds and downfolds occupying a wide belt. The Appalachian Mountains are a case in point. See Plate 22 for examples of anticlines and synclines.

Anticlines and synclines in the Jura Mountains. Old woodcut from Lyell, 1855

Now these folds of rock need not be symmetrical or even upright. Nature is complicated in her ways, and we find folds with all degrees of complexity. That is, if we imagine a plane through the center and along the length of a fold (the *axial plane*) the plane will be vertical

FOLDS AND FAULTS

if the fold is upright; but it may be inclined at any angle, so that the folds may be inclined or even recumbent if they lie over on their side. Similarly the *axis* of the fold, which is the line along which the axial plane cuts the top or bottom of a folded stratum, may not be horizontal but may dip away at an angle. In that case the fold is *plung-*

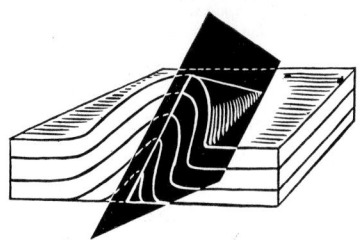

Complexities of rock folds. The axial plane and the axis of an inclined, plunging anticline

ing; and by the time several of these complexities are put into the picture, the reader may be left far behind.

WHILE an uninterested reader may dismiss the topic as abstract, the geologist dares not. On just such complexities may depend the location of valuable ore bodies or extensive pools of petroleum. For the geologist, in order to understand and interpret nature, must devise ways and means of untangling her complexities and making them available for economic exploitation or for scientific study. But these complexities of the folds are only part of the story. We said that erosion promptly attacks the uplifted and folded belts, so that the picture is further obscured by the absence of much of the structure and the burial of other parts of it beneath mantle rock and soil. Fortunately, the remnants of folds afford ample evidence of their former extent, and usually the bedrock is sufficiently exposed at the surface so that the essential features of the picture may be reconstructed.

WHEN a geologist studies an area, he hunts around for **outcrops,** or patches of the bedrock exposed at the surface. A common place to find them is along valleys where streams have cut away the superincumbent layer of mantle rock. Ditches, railroad cuts, and other man-made excavations may also expose the bedrock. Hence you may find geologists in the most out-of-the-way places.

Several observations are usually made on the rock outcrops. For our immediate purposes we are interested in the rock structures; and to work them out, two rather precise measurements are made on each outcrop. For simplicity let us assume that we are dealing with sedimentary rocks, because then we may speak about the *attitude* of the beds. Sedimentary rocks are best adapted to such studies because they are originally laid almost horizontally, and any subsequent tilting may easily be seen. The first measurement is made along the plane of the bed, and the angular inclination of this plane from the horizontal is called the **dip**. At right angles to the direction of dip, where the horizontal plane intersects the bedding, is the **strike** of the bed. The strike is expressed as the compass direction of this horizontal line. The adjacent diagram will give you the complete picture. Thus, dip and strike afford the necessary data to tell us what the direction and amount of inclination of the bed is at that point. In the meantime the exact location of the outcrop is also noted on a map.

Sedimentary rocks showing direction of dip and strike. The eroded beds are partially reconstructed

AFTER the dips and strikes of all the outcrops have been measured and plotted, some conspicuous rock layer is chosen as a standard. Then, by connecting the indicated strikes of this rock layer, the structure of the region is determined. It may be a single syncline or anticline, miles in extent. On the other hand, there may be a succession of small folds or any combination of large and small. Furthermore, the surface expression of the land may not reflect the structure at all, because erosional agents, cutting away in terms of the relative resistances, attitudes, and elevations of the rocks, may so disguise the original landscape that hills exist over the synclines, and valleys over the anticlines.

WHILE plastic deformation, and its attendant rock-folding, is taking place within the deep-seated environments, diastrophic adjustments may also be taking place near the surface. Here the confining pressures are less, and the relative nearness to

FOLDS AND FAULTS

the surface may permit sudden crushing effects to operate on the rocks. Hence they may yield by fracturing instead of by flow, so that joints and fractures may be formed in the rocks. If there is any slippage along the fractures, as a result of the rock on one side moving relative to that on the other, the result is a **fault. Fault-planes,** then, are surfaces along which relative movements take place.

When the compressive forces are unusually great, or when the rocks are unusually resistant to folding, nearly horizontal fractures may develop; and along these fractures whole mountain masses may be shoved over the underlying rocks. We call these features **thrust faults,** and in our own Northern Rockies such faults occurred, in one of which a great slice of rock, 10,000 ft. thick, was shoved eastward for at least 18 mi. over the rocks beneath. This thrust fault resulted in a shortening of the earth's crust by some 18 mi. Thrust faults, then, like mountain-folding, move hand in hand with the concept of a shrinking earth.

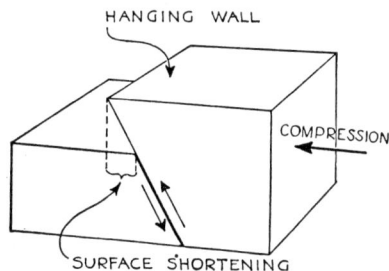

Block diagram of a thrust fault, showing surface shortening

IN ADDITION to thrust faults, in which the main result is a shortening of the surface, there are **normal faults,** which result in surface elongation. In order that you may clearly see why shortening or elongation of the surface accompanies faults, we have included the adjacent figures. Note that in both cases the fault plane dips down at an angle, so that the two surfaces which bound it may be called the **hanging wall** and the **foot wall.** Notice that in the thrust fault the hanging wall has gone *up* relative to the foot wall, while in the normal fault the hanging wall has gone *down*. The resultant shortening or elongation is indicated in the figures. In actual practice it is this relation between the walls that is observed by the geologist; and from it, when he finds the evidence that shows the relation clearly, he is able to tell which kind of fault is involved and in which direction to continue the search for a faulted ore body or coal seam. Examples of faults are included in Plate 23.

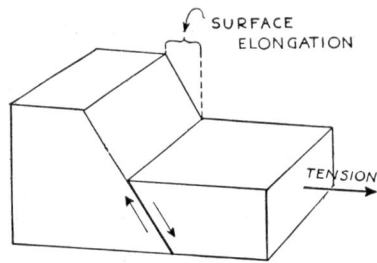

A normal fault, showing surface elongation

Faults may take place horizontally as well as vertically. They, in common with folds, may display any degree of complexity. There are some regions so intricately faulted that they defy full understanding. There also are regions in which folds and faults are so intimately intermingled that the task of untangling them is well-nigh hopeless. Futhermore, folded and faulted regions may later be subjected to more folding or faulting or both, and the final picture may be one that gives nightmares even to a specialist.

NOW while folds and faults are the two main effects of diastrophism in a broad sense, there are other effects in detail. Thus chemical rearrangements that we mentioned in passing have profound effects on the nature of the rocks. New minerals are often formed, and the rocks take on new appearances to such an extent that we consider them a separate group of rocks. They are called the **metamorphic rocks,** and we shall become acquainted with them in chapter 23. Either of the two broad classes of rocks, igneous and sedimentary, may be altered to metamorphic rocks.

WE WANT to point a few cautions for the reader before we leave this chapter. He must not gain the impression that mountains are folded up over night. No, it is an extremely slow process, as most geological phenomena are. We have pretty good evidence for believing that some mountain ranges are today in process of growing, but of course that growth is not revealed to a casual glance. While folding may be extremely slow in its development, that need not necessarily be true of faulting. In fact, we know it isn't true, because faults often take place with enough suddenness to set up distinct shocks in the earth. When these shocks occur, the region adjacent to the fault suffers an earthquake. So important are earthquakes to mankind, both because of their destructiveness and their scientific value, that we shall introduce them to the reader in some detail.

CHAPTER 18

EARTHQUAKES

> .... The labouring mount
> Is torn with agonizing throes.
> —MALLET

AMONG all the cataclysms of nature, none is more terrifying than an earthquake. It has been said that fear of earth tremors increases with increasing experience, whereas in almost any other peril familiarity breeds a certain degree of indifference. Whatever the reason for this, assuming it to be true, there doubtlessly is involved our unconscious acceptance of the earth as a firm and unyielding foundation. When even this moves beneath our feet, it is not surprising that one's consternation should exceed his judgment. Combine with this the jumble of shattered houses, of gaps and fissures that open in the earth, of springs and streams gone suddenly dry, the low rumble of earth noises, and the dust, confusion, and conflagration that are summed in the word *earthquake*, and the cumulative effect is hardly surprising.

.... none is more terrifying ..... Old engraving of an earthquake, from Von Leonhard, 1844

SO COMMON are earthquakes that the instruments used to record them show several thousand a year. Most of these are imperceptible tremors, of course; but it must be true that somewhere on the earth every day some people are subjected to greater or less disturbances. How is it, then, that so few of us have ever experienced an earthquake? Simply because earthquakes, despite their abundance, occur in certain restricted areas on the earth and because, beyond a relatively small radius, the effects of the shocks are not perceptible to our senses. Records covering many years show that most earthquakes are confined to certain belts about the earth. Many of these belts are close to continental margins. The

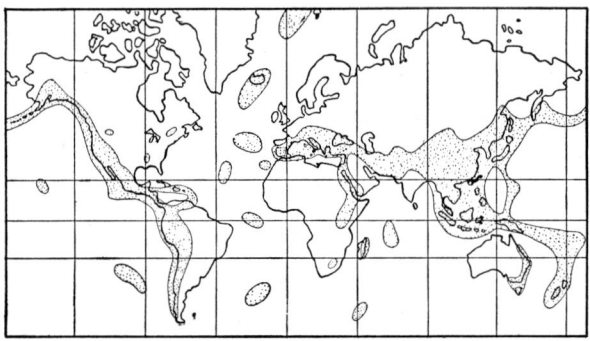

Map showing world-distribution of earthquake belts

accompanying map indicates the extent of these belts; and if you will examine it in detail, you will realize that practically every earthquake you have ever read about has occurred at some locality within these zones. We cannot say that no earthquake will ever occur outside of these zones; but beyond the limits of the heavy belts, earthquake insurance is pretty much of an idle gesture.

A GREAT deal of evidence collected from many parts of the world shows quite conclusively that most major earthquakes are due to faulting. Quite commonly fissures are formed in the ground during major earthquakes, and these fissures are found to be the surface expressions of faults. The movement along fault planes may be relatively sudden, and this movement sets up powerful vibrations, which give rise to waves that radiate outward from the center of the disturbance. It is the sudden shock of these waves that does some of the damage

Earthquake fissures often mark the position of faults. Old woodcut of Calabrian earthquake of 1783, from Lyell, 1842

during earthquakes. We shall examine the waves more closely in our next chapter; at present we are concerned with the causes of earthquakes and their surface effects.

In many cases the movements during earthquakes take place along already existing fault zones, which may bear evidences of numerous displacements. One of the most noteworthy earthquakes of recent times was that of San Francisco in April, 1906. This earthquake resulted from a displacement of the rocks along the nearly vertical *San Andreas fault*, a zone of weakness that can be traced for many miles through California. The maximum movement along the fault was 21 ft. in a horizontal direction, as shown by displaced fences, roads, and the like. Faults, we may recall, can take place in any direction, depending on local conditions, although it is true that vertical movements are more common than horizontal ones.

ROCKS, contrary to popular fancy, are quite elastic. When an external force is applied, the rocks suffer an elastic deformation, or strain. When the strain due to these forces becomes too great, the rock yields. Near the surface the yield is quite sudden and may result in fractures or actual movement along fault planes. The conditions leading up to an earthquake may then be pictured as the gradual accumulation of stresses within the rocks until they are so great that the rock must yield. Imagine the sides of a fault tightly pressed together, so that movement can only take place through the overcoming of tremendous frictional resistance. If forces tending to produce movement along the fault plane are set up, the rock resists the movement until its elasticity is exceeded. Then a sudden movement takes place, and the vibrations set up during these movements may be quite intense.

This interpretation of earthquakes has direct evidence to support it. Certain slow displacements of landmarks, demanded by the theory, have been reported along the San Andreas fault; and such slow displacements represent the accumulation of stresses within the rocks adjacent to the fault, so that, when the release occurs, the rocks "snap" into their new positions by an *elastic rebound*.

The accompanying three block diagrams represent the picture as a whole. In the first block a pre-existing fault plane is shown. As forces accumulate in the zone adjacent to the fault, we may visualize a tendency on the part of the rock to adjust itself by a slight deformation, as the second block fancifully suggests. No slip has yet occurred along the fault plane, however, but as the elastic limit is reached, the rocks yield by relatively sudden or jerky movements, and the earthquake damage is done. The third block pictures this final stage, after the faulting has occurred. Once relief is afforded, the stage is set for a repetition of the process. These examples refer to an earthquake in which the movement is horizontal, but obviously a similar set of conditions could apply to movement in a vertical direction.

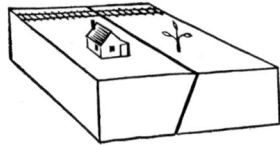

Elastic rebound theory of earthquakes. I. The inclined line marks a pre-existing fault plane

Elastic rebound. II. Exaggerated view of accumulating horizontal stresses. No movement along the fault plane yet

Elastic rebound. III. After the earthquake

The theory just presented is relatively recent; the whole subject of earthquake study is still rather new, but the mass of data now available fairly well establishes the broad outlines of the subject. In addition to direct field observations, scientists nowadays have instruments of such sensitivity and precision that even the feeblest earth tremors are faithfully recorded, and cases are not unusual where the center of the disturbance is determined from these records before even the news reports tell of the damage. It was not always thus, of course; until shortly before the turn of the present century, the study of earthquake phenomena was confined largely to qualitative observations. Some of the data obtained from these earlier investigations are interesting and important, and we shall examine them during the course of our chapter. The exact geological significance of earthquakes was but dimly realized, also, and even today there is much that remains to be explained. One thing is certain, however: the advent of modern **seismology,** or the scientific study of earthquakes, has opened up a field which has shed important light on the question of the nature of the earth's interior. More of this anon; at present suppose we journey backward in time about a century and pick the story up there.

IN 1846, Robert Mallet, an Englishman, published a memoir on the dynamics of earthquakes. Here, for the first time, were applied the laws of wave motion in solids to the phenomena of earthquakes. In England, where Mallet lived, however, earthquakes are rarities, and so his experience was more theoretical than actual. It is reported that when he was awakened one night by the tremor that stirred Britain in 1852, he did not recognize it as an earthquake. Despite this lack of actual experience, Mallet believed that he held the key to many earthquake secrets. His opportunity came late in 1857, when the great Neapolitan earthquake occurred.

So great was the destruction in this Italian earthquake that news of the catastrophe did not reach the outer world until about a week later. Mallet at once proceeded to the stricken area and began his researches. His principal aims were to determine the exact center of the earthquake disturbance and to find out how deep within the earth the disturbance occurred, for Mallet argued that both of these ends could be gained by a study of the surface effects of the quake.

YOU undoubtedly know from newspaper accounts that an earthquake may first present itself as a rather sudden shock, accompanied by billowy movements, as though waves were passing along the surface of the ground. The very suddenness of the shock may contribute in part to the damage, although the amplitude of the earthquake waves, the duration of the quake, and resonance in building structures all fit into the complete picture. The general result is that after an earthquake of destructive violence, there usually is a center of maximum damage, surrounded by areas in which the destruction diminishes. The sequence ranges from towns that witness the dramatic events sketched in our opening paragraph to more distant towns and villages in which the greatest damage is perhaps the stopping of clocks or the rattling of a few dishes. Naturally enough, any disturbance originating at some center dies down as it moves outward, much as the ripples in a pond diminish in size as they move away from the point where the stone is dropped in.

NOW, what Mallet did was to set up a scale of damage suffered, a comparative scale of the intensity of the quake at any point. A tour of inspection of the towns in the stricken area then afforded him data for drawing on a map a series of curves which enclosed areas of equal damage. These lines are called **isoseismals,** and they enclosed elliptical areas increasing in size as he moved away from the center of greatest damage. The innermost isoseismal included the zone in which the towns were prostrated, the next larger one those where major parts of the towns were destroyed, and so on with decreasing effect outward.

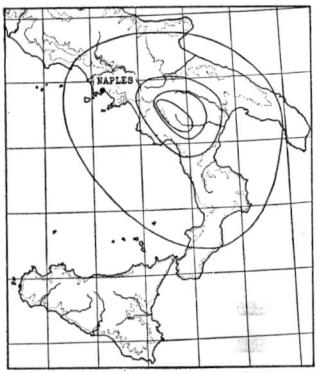

Isoseismal lines of the Neopolitan earthquake of 1857, after Mallet

For locating the center of the disturbance and its depth below the surface, Mallet devised an ingenious method. Suppose a disturbance occurs beneath the surface at O (called the **focus** of the earthquake), situated a distance OA beneath the surface. Waves are sent out in all directions from O, and we shall consider one of them, OB. This strikes the surface at B, at an angle b with the surface. This angle Mallet called the *angle of emergence.* Also, if B is connected with A, the line AB has some given compass direction; that is, it has an azimuth a with respect to the line AN, supposing N to be the north point.

Now Mallet argued that the direction of travel of the earth wave OB, its angle of emergence b, and the direction AB should be determinable from the effects of the shock at B. The cracks in damaged buildings or walls should be at right angles to the wave direction OB; and overthrown monuments and the like should be

EARTHQUAKES

found lying in the direction BC, with their bases toward the point A, which is called the **epicenter** of the earthquake.

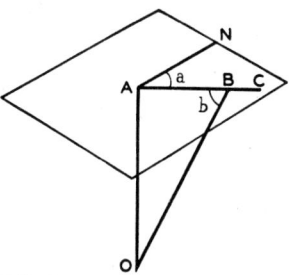

Mallet's method for locating the epicenter and focus of an earthquake

As a result of this study he found that the epicenter of the Neapolitan disturbance was not a single point, but a somewhat linear zone (as we would expect if movement along a fault plane caused the quake). Neither was the depth to the focus the same for all measurements, but the average depth was found to be about $6\frac{1}{2}$ mi.

A NUMBER of technical objections can be found to Mallet's methods; but the point remains that his was the first fairly successful quantitative study of earthquake phenomena, and Mallet's work laid the cornerstone for many investigations since his time. Nowadays, instruments are used much more extensively, and reliance is placed on mathematics to determine the depth of focus. It is an interesting verification of Mallet's work that most earthquake foci are rather shallow, so that his results were at least of the right order of magnitude.

Despite the shift of emphasis toward more instrumental methods, there is no doubt that Mallet gave a decided impetus to earthquake study, and demonstrated that earthquakes, in common with most other earthly phenomena, were amenable to investigation by observation and inference. As time went on, Mallet's methods were extended and improved upon; and with a growing realization of the nature of the disturbances, instruments were devised for recording many features of quakes that could not be determined from the surface effects of the quake itself in the stricken area. The great advantage of instrumental observation is that it obviates the necessity of being on the ground during or after the earthquake; the instrument may be located far from the disturbance, and yet from the record it traces the distance to the center of disturbance and the violence of the shock can be determined.

THE instruments used to detect earthquake waves are called **seismographs,** and in principle they are quite simple. A seismograph is essentially a pendulum with a long period. A much simplified diagram of one is shown herewith. A heavy weight, P, is suspended by a spring from a rigid frame, E, firmly fastened to the earth, often by means of a pillar of concrete resting directly on bedrock. On a platform of this frame is a revolving drum, D. A stylus, S, attached to the weight, rests against a sheet of paper wound around the drum. Now imagine an earthquake suddenly vibrating the frame E. The frame, including the drum, will then oscillate a bit, while the mass P, because of its great inertia, will momentarily remain at rest. Thus a wavy line will be traced on the paper, recording the passage of the earth wave. This is called a *vertical-component seismograph;* and in practice two instruments are used, one to record the vertical components of the wave and the other to detect the horizontal components.

Simplified diagram of a vertical component seismograph

Actually, of course, modern seismographs are much more refined and complicated than this. Timing devices are put on, so that the exact moment of the wave's passing is recorded, and usually the record is made by a beam of light on photographic paper instead of with a stylus. Further, the natural vibrations of the pendulum must be severely damped, so that it does not unduly distort the earthquake waves with its own periodic motion. We shall not bother the reader with these technical details, but specialists must take careful account of them in actual practice. Plate 24 illustrates modern instruments and their records.

IN ORDINARY operation, as long as no disturbance is passing through the instrument, the record consists of a straight line, interrupted once a minute for timing purposes. When an earthquake occurs, however, the waves sent out from the center of the disturbance pass through the machine, and the line becomes wavy, with varying amplitude. The record thus obtained is called a **seismogram,** and from it is read much more than one might suspect at first glance. Here is shown a sample of such a record from an earthquake a thousand or so miles from the instrument. Notice the letters p, s, and l that are indicated below the curved line. They mark the appearance of the three wave trains that make up the

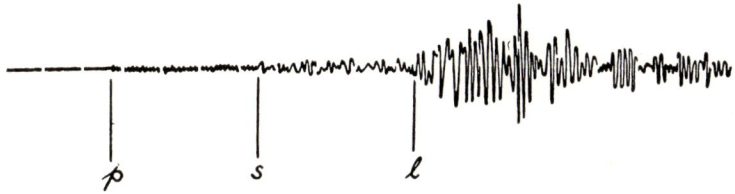

A seismogram, or earthquake record. The letters p, s, and l mark the arrival-time of the three sets of waves

usual record. The first is called the **primary wave,** because it travels fastest and arrives first; the second is called the **secondary wave,** and the third is called the **long wave.** The exact nature of these waves will be developed in our sequel because they afford such an interesting example of agreement between theory and practice. Aside from any theory, however, it was noted that these waves are separated from each other on the seismogram by time intervals which increase as the distance from the disturbance to the observatory increases. Thus, at an observatory a few hundred miles from the earthquake the times of arrival of the waves may differ by only a few minutes, whereas from more distant earthquakes they are separated by greater and greater intervals.

AN IMPORTANT practical application is made of this time-lag between the earthquake waves. It was found by plotting the data of many seismograms that there is a definite relation between the time of arrival of the p waves and s waves, for example, and the distance from the observatory to the quake. This relation holds regardless of the geographic location of the earthquake or of the observatory; in short, the time-lag is a function of distance only. Hence, to determine the distance between the observatory and the disturbance, it is only necessary to observe the time in minutes, say, that elapses between the arrival of the first two waves.

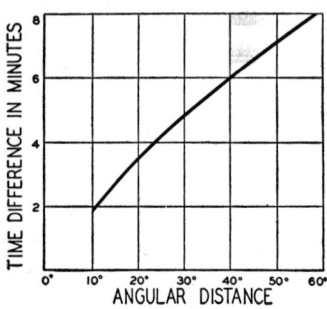

Curve showing the relation between distance, and the difference in time of arrival of the *p* and *s* waves

Graphs have been prepared to show this relation between timelag and distance, and the adjacent diagram is one somewhat simplified. The time difference in minutes between the arrival of the *p* and *s* waves is shown along the vertical axis, and the angular distance* to the epicenter is given along the horizontal axis. Suppose, for example, that the *p* wave is recorded on the seismogram at 7:02 A.M. and the *s* wave at 7:07 A.M. Here is a difference of 5 min.; and by finding this value on the vertical axis, the corresponding distance is seen to be 30°, or about 2,100 mi.

NOW, for any one station to determine the *direction* to the epicenter is somewhat tedious. It requires quite a bit of calculation, so that several observatories usually co-operate in locating the stricken area, for, notice that if each station can determine the distance, then three stations working together may locate the center of the shock. This is how it is done. A circle is drawn about each observatory as a center, with a radius equal to the computed distance to the quake. The intersection of the three circles is the desired location. The adjoining figure illustrates how observatories at Pasadena, Chicago, and Washington may locate an epicenter in Mexico.

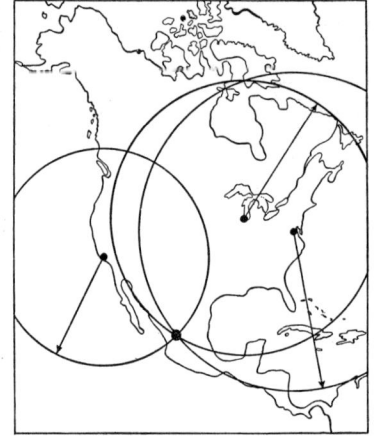

Three observatories co-operate in locating a quake in Mexico

The example just given serves as a striking contrast to the much less refined methods that were available in Mallet's time. During

* The angular distance is measured in degrees around the circumference of the earth. Each degree represents about 70 mi.

the earthquake of 1857, the area was so stricken that news of the quake was delayed nearly a week; nowadays the recording and location of quakes are a matter of hours only. Earthquake waves travel at the rate of several miles per second, so that almost immediately after the shock seismographs for many miles around record the disturbance.

NOT only have methods of studying earthquakes improved tremendously in the last third- or half-century, but much more is now known about the immediate causes of the destructive effects. It has been amply demonstrated, for example, that structures on mantle rock and on loose sedimentary deposits suffer much more than those on solid bedrock. In the great San Francisco quake of 1906 it was the buildings erected on filled-in land that suffered most; the same was true of the Messina earthquake of 1908 and the Tokio quake of 1923. From an engineering viewpoint, then, earthquake research has done much to avoid the extreme destruction of earlier shocks, both by new types of shock-resisting construction and by recognizing the probable centers of greatest damage in regions subject to earthquakes.

The reason why earthquake damage is greatest on loose or unconsolidated rock is not hard to understand. As long as the earthquake waves pass through dense rock, they merely vibrate the particles; but when they enter loose materials, their energy may be transformed to an actual shifting of the particles among themselves. We may illustrate this by placing a marble on the floor. If the floor is tapped smartly with a hammer several feet away, no movement of the floor is perceptible, but the marble is made to jump several inches into the air. The energy here has been partly applied to movement of the loose marble itself.

WE HAVE hinted at another aspect of earthquake research that has not been touched upon thus far in our chapter. Inasmuch as earthquake waves are partly transmitted beneath the earth's surface during their propagation from the focus, we may expect that a theoretical study of the waves themselves could shed light on the nature of the earth's interior. This is exactly the case: earthquake waves are our most powerful weapon for studying the deep interior of the earth.

CHAPTER 19

JOURNEY TO THE CENTER OF THE EARTH

It is a world of startling possibilities.
—Charles Fletcher Dole

JULES VERNE and other romancers have indulged their fancy with the interior of the earth and have even sent hardy explorers to face the real or imagined dangers that lurk there. In the interests of a purely scientific investigation, however, this path of direct exploration is closed to us. Direct observation of the earth is limited to an exceedingly shallow zone near the surface, and it is doubtful whether man will ever be able to penetrate directly even 1 per cent of the way into the vast interior.

The descent. Woodcut from Jules Verne's *A Journey to the Center of the Earth* (1874)

Despite these limitations, there are ways of exploring the innermost portions of the earth; and, curiously enough, earthquakes afford us the surest means of doing so. Thus, from a cataclysm that formerly inspired only terror, man now seeks to wrest the secrets of his terrestrial sphere.

IT WAS not much before the turn of the present century that scientists definitely recognized the three sets of waves that characterize earthquakes. Then in 1906 occurred that great world-shaking quake in San Francisco; and the data from that event, recorded by seismographs all over the earth, did much to establish understanding on a firmer foundation. The Carnegie Institution in

JOURNEY TO THE CENTER OF THE EARTH

Washington undertook a detailed study of all the records of that earthquake, and we shall use their data in developing the story of earthquake waves.

A study of the seismograms collected from all over the earth showed several interesting things: first, that the record of the quake increased in its time of arrival with the distance between the shock and the observatory. The precise moment of the major shock at San Francisco was pretty well established by pendulum clocks at the naval observatories in California, which were thrown from their supports by the violent movement. Hence the starting time of the disturbance was known down to about a second.

IN GENERAL, three separate items could be read from the seismograms. These were the times of arrival of the primary, secondary, and long waves. The arrival time of each wave was plotted on a graph which showed *time* along the vertical axis and the *distance* from the stations to the earthquake center along the horizontal axis. Curves were fitted to the three sets of points, and the resulting graph is sketched here. The uppermost curve is a straight line, and the other two are curved lines, convex upward.

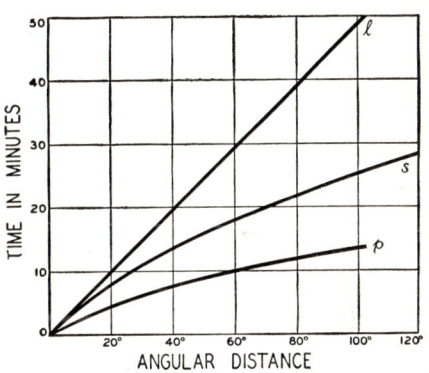

Time-distance curves of the p, s, and l waves, based on records of the 1906 San Francisco earthquake, after the Carnegie Institution's Report

Here, then, are the facts that were collected. What do they mean?

PHYSICISTS have shown, both by theory and experiment, that when any solid elastic body (and even rock is elastic) is deformed by a force, two types of waves may be set up. They are called *elastic waves*, and the first is a compressional or *longitudinal wave* caused by contractions and expansions of the material body. The other set is *transverse*, caused by a shear, or sideward, deformation, which generates waves whose movement is at right angles to the direction of oscillation of the particles. Now it has been amply confirmed (although the proof is too abstruse for presentation here) that, of these two waves, the longitudinal wave travels faster. Indeed, if the density and elastic constants of the material are known, the exact velocities can be computed.*

It so happens that both of these waves vary their velocity in accordance with changes in the physical properties of the materials through which they are passing. The exact relation is that they are slowed down by an increase in density and speeded up by an increase in elasticity of the rock. In fact, if such waves show variations in their velocities, it is certain that the properties of the material transmitting them has changed.

IN OUR earlier discussions we saw that beneath the earth's surface the pressures increase, owing to the weight of the overlying rocks. Furthermore, physicists have shown that both the densities and elastic properties of the rocks must increase with increasing depth; in short, the characteristics that control wave velocities do change within the earth. Let us then consider earthquake waves in this light and see whether they reflect these changes.

Recall that two of the curves in our earlier diagram are convex upward. Thus these two waves show a change in their apparent velocities as the distance to the seismograph increases. If we assume for a moment that these are the elastic waves passing beneath the earth's surface, we see that they behave as they should. In other words, since the greater distances involve a deeper penetration of the wave into the earth, the apparent velocities should change, due in part to the very curvature of the earth itself. Furthermore, these two waves travel with different velocities, as is demanded by theory. It seems, then, that we may safely say that the first two waves to arrive are the elastic waves that follow more or less direct paths from the earthquake to the seismograph, via the earth's interior. Further, the first wave to arrive must be the longitudinal wave, and the second the transverse wave.

* R. J. Stephenson, *Exploring in Physics*, pp. 146–47.

SO FAR so good. But what about the third wave, which travels with a uniform velocity? The theory said nothing about that. No, nor need it. Since this wave travels with a uniform velocity, it must be moving in some zone where the pressures and densities are fairly constant. Where is such a place on the earth? Only along the surface. These waves are surface waves that travel around the earth from the center of disturbance. Such surface waves are a principal cause of the destruction during quakes, and now we shall consider how they are set up.

We saw that an earthquake is a sudden movement in the rocks beneath the surface and that during these displacements there occur bumps and jars, as well as friction and slipping along the fault plane. These movements momentarily compress and expand the adjacent rock, setting up longitudinal waves which radiate in all directions from the disturbance; and in addition they set up vibrations which are transmitted in all directions as transverse waves. Of these two waves, the longitudinal wave travels faster, and hence *is the primary wave;* the transverse wave is the *secondary wave*. These waves are called the **body waves** because they spread through the body of the earth. As these waves strike the surface they set up vibrations there which give rise to surface waves. These surface waves constitute the *long waves* of the seismograms, and in motion they are quite complex; but for the most part they appear to be transverse waves.

The surface waves, which cause the damage, are formed when the body waves strike the surface. Note the effect of the angle of emergence

THUS far, our reasoning has brought us several important results. It has enabled us to identify and explain the several waves that are registered by seismographs. Once we have these points clinched, we may take the known behavior of such waves in general, and then, by studying the changes they undergo as they pass through successively deeper portions of the earth, we may be able to say something about the nature of that interior. This is exactly what geophysicists and seismologists have done, and their researches have shed much light on the dark interior of the earth.

WE CANNOT pursue in detail all the researches that have been carried out with earthquake waves, because they bristle with mathematical symbols; but we may at least point out some of the current hypotheses that seem to explain the behavior of earthquake waves in the earth's interior. Essentially, the idea is this: to set up a world of such characteristics that if elastic waves are sent through it, the behavior of the waves will be the same as that observed on the earth itself. The setting-up of such a world is mainly a mathematical proposition, involving the postulation of certain physical constants at various depths, which, when put into the equations, give results which check well with observed earthquake records.

THIS much appears to be certain: the behavior of the waves beneath the surface is very much the same regardless of the geographic portion of the globe that they happen to be passing. This points to a single conclusion: namely, that whatever changes in physical properties take place are functions of depth mainly. Hence the material inside the earth must be homogeneous in concentric shells.

The detailed studies that have been made also demonstrate that the densities of the earth materials increase in the deeper shells. This certainly checks with our earlier observations, because we know that the average density of the globe is 5.5 times that of water, while that of the surface rocks is only 2.7. Hence the inside must have a density higher than 5.5 to counterbalance the lightness of the exterior. Furthermore, having approximately determined the elasticity and density of the materials in the successive shells, we may then seek for known rocks or materials that have these same properties, and thus attempt to identify the materials composing the earth's interior. But it is about here where doctors begin to disagree; how do we know that the characteristics of surface materials may not completely change when subjected to the tremendous pressures of the interior? We do not know, because we cannot produce such pressures in our laboratories to check them. At best we may say that there is some possibility that material of such and such composition constitutes the successive shells.

BEFORE we talk about particular materials, let us see what the waves say when they are treated mathematically. The primary wave appears to be most amenable to this treatment, and the result of the computations is that this wave changes its behavior several times in going through the earth. Graphs have been prepared of this change of velocity with depth, and we are showing one here. Notice that during the first 750 mi. or so the velocity steadily increases. At the 750-mi. point there is a change in the slope of the curve. The meaning of this is that there is a change in the nature of the earth material at a depth of 750 mi. In short, the outermost 750 mi. of the earth constitute one of the main shells of which it is composed.

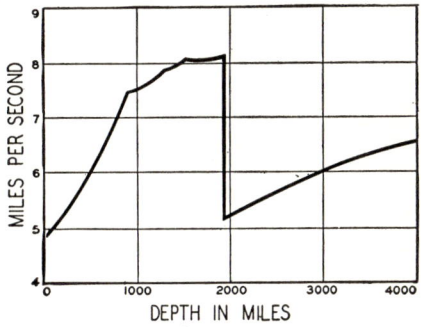

Graph of the p-wave, showing changes in velocity with depth below the earth's surface. (After Gutenberg)

Next we follow the curve to about the 1,800-mi. point. Here the velocity takes a toboggan ride downward, and thereafter increases slowly again toward the earth's center. At about 1,800 mi., then, there is a pronounced change in the earth's constitution. Hence there is another shell of material from the 750- to the 1,800-mi. levels. Beyond the 1,800-mi. level there is no other abrupt change in velocity, so far as is known at present, so that the inside 2,200-mi. of the earth may constitute an inner or central core.

LET US next consider the transverse waves. What do they tell us about the earth's interior? In seeking an answer to this question, seismologists have faced several difficulties. In the first place, until rather recently no transverse waves could be detected with certainty in the records of very distant earthquakes. From the study of numerous records it appeared that the earth's inner core of 2,200 mi. radius stopped the transverse waves by absorbing them. On this basis it was concluded by many that the central core of the earth was liquid, because transverse waves cannot be transmitted by matter in the liquid state. Very recent researches, however, are pointing more and more definitely to the conclusion that transverse waves actually are transmitted through the entire earth; but they appear to be quite feeble, so that certainly the waves are severely damped during their passage through the central core. Except for this difficulty of the innermost core of the earth, the transverse waves indicate the same concentric arrangement of earth material that was shown by the longitudinal waves.

WE ARE here on the very forefront of present-day research; but one thing is certain: if the passage of the transverse waves through the central core is a fact, then there can be little doubt that the earth is solid from center to circumference. It is a matter that perhaps must rest until more data are collected, but it may be said that today most seismologists incline to the opinion that we inhabit a completely solid earth.

NOW what about the exact composition of these several earth shells? We do not certainly know, but various lines of evidence afford some suggestions. We cannot detail all the evidences here, but we shall at least point out some of the present conclusions of experts whose entire time is devoted to the solving of the fascinating riddle of the earth's deep interior.

FROM the depths of the earth we may emerge for a moment to the surface and contemplate the heavens. Everybody has seen those curious things called "shooting stars," and most people know they are not stars at all, but only small fragments of cosmic matter heated and consumed by friction as they pass through the earth's atmosphere. What have they to do with our story of the earth's interior? Just this: Sometimes such *meteorites* actually strike the surface; and when they do, we may analyze them and see what they are made of. It is found that for the most part they consist of iron and nickel in metallic form, with some stony material in varying amount. Now after all, this terrestrial sphere of ours is an astronomical body, presumably made of the same stuff as other astronomical bodies. If so, the composition of these meteorites may shed light on the nature of the interior of the earth. Our table of chemical compositions of the lithosphere way back in chapter 3 showed only 5 per cent of iron. If the earth has a composition something like an average meteorite, where is the rest of the iron, and nearly all of the nickel? Nobody knows, but many scientists suspect that through long processes of earth adjustments these heavier parts of the primary earth-substance have been concentrated at the center and now make up the inner core. Certainly, the high density of the inner core agrees with this notion; and on the whole the idea is reasonable and shrewd. Remember that it is outside the realm of demonstrable fact, but nevertheless is quite within the bounds of reason.

Shooting stars . . . .

AS TO the shell of earth-material between the inner core and the outer 750 mi. of the earth, various lines of reasoning suggest that it is composed of sulphides of iron and nickel. Sulphides are combinations of metals and sulphur, and such features as density and other physical constants suggest them as possibilities for the position. The outer 750 mi. of the earth are undoubtedly made of rocks and stony materials, and this itself is differentiable into more than one zone. We shall not pursue the matter in closer detail; we have already shown the lines of attack used by earth scientists, and have demonstrated how, step by step and fact by fact, the structure of our ideas is built.

On the basis of the picture we have sketched, we may indicate present ideas about the earth's interior by the adjoining diagram.

Simplified section of the earth's interior, showing the central core and the main earth-shells

It is somewhat simplified, and shows only the three main shells included in our discussion. This picture, as the reader knows, may be tentative in part, because, in taking this journey to the center of the earth, we have marched along the frontiers of present-day science, and it may be that our views will have to change as new evidence here and there modifies the details. See Plate 25 for an interesting point in this connection.

Our chapter thus far has shown us how an intelligent and directed study of earthquake waves opens up entire new fields of knowledge concerning the earth's interior. But no sooner had scientists learned the value of natural earthquake waves in these studies than they saw possibilities in the creation of little earthquakes of their own. Clearly, if earthquakes set up shocks that can literally be recorded around the earth, why shouldn't smaller shocks, such as are set up by dynamite explosions, yield information about the surface layers of the rocks?

New methods of prospecting for oil and ores have in fact been developed, based on a study of such artificially created elastic waves. Charges of dynamite are set off in the region studied, and the waves sent out from the explosion are reflected back to the surface from rock layers at various depths. These reflected waves are detected with seismographs of a special type; and from the data obtained, the structures of the underlying rocks can be worked out. Similarly, much information is gained from the

Geophysical prospecting

velocities of the waves themselves, as they pass through different types of rock. *Geophysical prospecting* is the illuminating name applied to this new technique: it is a perfect blending of physics and geology.

WE HAVE now fairly well covered the subject of diastrophism, except for two topics: a discussion of the metamorphic rocks that are formed by the diastrophic process, and a summary of the process itself. We shall, however, change the order a bit and save these topics for a later section, where they may serve as a general summary of all the geological processes. Hence we propose to turn our attention for a while to that third and last great process of geology, which concerns volcanoes and the phenomena associated with them.

CHAPTER 20

VULCAN'S CHIMNEYS

> Forth from whose nitrous caverns rise
> Pure liquid fountains of tempestuous fire,
> And veil in noonday mist the ruddy skies; ....
> —John Auldjo, in *Sketches of Vesuvius* (1833)

IT WAS about one o'clock in the morning of September 29, 1538, that the inhabitants of Pozzuoli first became aware of a disturbing event. Flames were seen shooting from the earth between the town and the hot baths farther along the bay. The startled natives snatched up household oddments and fled toward Naples. The flames increased to such a degree that within a short time the earth burst open and spewed forth great quantities of ash and pumice. The ejecta fell back into and around the fissure and by morning had accumulated into a mound some hundreds of feet high.

Eruption of Monte Nuovo, from an old woodcut by Falconi, 1538

For two days and two nights the eruption continued. On the third day the activity ceased, and never since has it annoyed the somnolent town of Pozzuoli. The inhabitants, appropriately enough, christened this new addition to their landscape *Monte Nuovo*, and to this day the name remains.

THIS startling event marked the birth of a volcanic cone. Naturally, local conditions were suitable for the sudden apparition, for Pozzuoli is situated in the Phlegrean Fields, that great volcanic region dominated by Vesuvius. Scattered about Pozzuoli are numerous old volcanic cones, and we are fortunate that ample evidence exists for the verity of this unusual occurrence.

VULCAN'S CHIMNEYS

MOST volcanoes with which we are acquainted have been with us since the dawn of history. Certain famous volcanoes, like Vesuvius, Stromboli, and Mount Etna, have been known since ancient times, although Vesuvius first impressed itself on the Romans as an active volcano in the eventful year 79 A.D. Fanciful legends sprang up about such fiery mountains, and the smoke was commonly attributed to Vulcan's forge, where the thunderbolts of Jove were fashioned.

PERHAPS no other volcano is so well known as old Vesuvius. Visitors to Naples return with a vivid recollection of the dominating mountain across the bay, set off by its plume of steam. More intrepid visitors take the tram railway to the top of the cone, and shudder on the crater's lip while flames, steam, and blobs of lava are emitted from the smaller cone in the center. It was not ever thus, however, because during Republican Rome the mountain was a quiet and inoffensive feature, clothed with a fertile soil and supporting a luxuriant forest. The

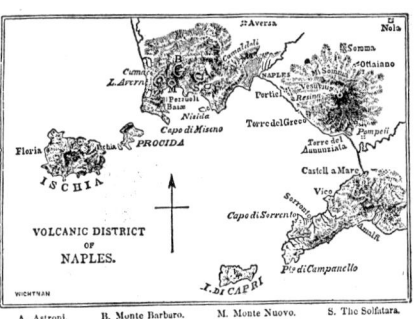

Map of the Vesuvius region, from Lyell, 1855

Romans recognized it as an extinct volcano from its general appearance, but within the memory of man it had lain dormant. Nestled at its base were the populous towns of Herculaneum and Pompeii.

The first rude awakening occurred in 63 A.D., when a violent earthquake rent the region. Comparative quiet reigned again until 79 A.D., when in the month of August shocks again occurred, finally culminating in a burst of activity from Vesuvius. During the eruption a cloud of ashes was scattered far and wide, and both Herculaneum and Pompeii were buried from the sight of man. At the same time, a dense cloud of vapor rose vertically from Vesuvius, spreading out at the top into a bulbous mass. The cloud was pierced by appalling flashes of lightning, succeeded by darkness most profound.

The elder Pliny, that great natural historian, was hard by, and could not resist the temptation of studying this phenomenon in detail. He paid for his pains with his life, however, for he was suffocated by sulphurous vapors. In a century or so the Romans forgot about the buried cities, and it was not until comparatively recent times that the cities were excavated.

AFTER its first recorded catastrophic eruption, Vesuvius alternated between long intervals of quiet and shorter periods of activity. Since the year 1036, fairly authentic records of its activities are available; and at present it appears to be in a state of quiescent ebullition, broken now and then by more or less destructive eruptions. The most recent eruption of destructive significance occurred in 1906. This eruption lasted some 18 days, during the first part of which great floods of lava were poured from the crater and from fissures on the mountainside. Later came a stupendous emission of gas and steam under tremendous pressures: the clouds of vapor rose to a height of 8 mi. and spread out at the top into a vast cauliflower-shaped mass. In the final stages of the eruption quantities of ashes were thrown out of the crater and disseminated over the countryside.

Vesuvius in eruption, 1784. Woodcut from Gray and Adams, 1853

MOST of us, perhaps, think of volcanic eruptions as involving mainly the emission of lava, which flows down the sides of the cone as a fiery liquid. As a matter of fact, volcanic eruptions are of several kinds, and range from violent explosions on the one hand to quiet outwellings of lava on the other. Occasionally no lava at all escapes, but clouds of ash and of volcanic fragments are scattered about. The eruption of Monte Nuovo was of this latter class.

Perhaps the most violent eruption of the purely explosive type occurred at Krakatoa, a volcano in the Indian Ocean near the western end of Java. In the year 1883, with little warning, the en-

VULCAN'S CHIMNEYS

tire cone exploded, blowing the island to pieces, and generating an immense water wave that spread out across the wide expanse of the ocean and among the islands of the East Indies. The detonation of the explosion was heard for some 3,000 mi., and the force of the explosion projected dust upward about 20 mi. There it was disseminated by the winds over the entire earth, causing numerous spectacular sunsets before it finally settled back to the surface. The greatest damage was done by the waves, which moved out from the center in a wall 50 ft. high and swept away nearly three hundred coastal villages.

In contrast to this explosive type of volcano is that of Mauna Loa in the Hawaiian Islands. This great volcano is nearly 14,000 ft. high. During its numerous eruptions great flows of lava quietly well out from fissures along its flanks and flow downward toward the sea. No explosive outbursts accompany these eruptions, although occasionally fountains of lava develop within the crater, sending up columns of the fiery liquid to heights of several hundred feet.

Mauna Loa, after a woodcut from Boccardo, 1869

THE examples we have cited illustrate the geologic process of **volcanism,** the third of the great geological processes. As in the case of diastrophism, volcanism is an awe-inspiring phenomenon; and when nature breaks into activity in one case or the other, humankind can do little except get out of the way.

As inhabitants of the earth's surface, we are naturally more familiar with the spectacular surficial aspects of volcanism. In its broadest sense, however, the process refers to any movements of lava, whether they reach the surface of the earth or not. There is ample evidence that great movements of lava have taken place within the earth, with no external activity. The solidified masses of such lavas are later disclosed when erosion exposes them to view.

These rocks are such that they have clearly solidified from a liquid state, and the size of the crystals indicates that cooling was very slow in contrast to the rapid cooling of lavas at the surface. Hence we assume that these coarsely crystalline rocks were formed under a cover of other rocks, which served as a thermal blanket and permitted only a slow radiation of heat.

Our evidence that some movements of lava are confined below the surface impels us to divide volcanic activity into two types. We speak of **extrusions** when the lava or other ejecta appear on the surface and of **intrusions** when the activity is confined to subterranean regions. We shall devote the present chapter to phenomena of the first type, reserving for our next the intrusive aspects of the subject.

CONTRARY to our first impressions, perhaps, is the fact that among the most abundant products of volcanism are gases and vapors. Every volcanic eruption involves the liberation of large volumes of gas, and it is from the associated gases that the explosive and non-explosive types of eruption arise. If the confined gases can readily escape from the lava, the eruption is quiet; whereas, if the gas is retained under considerable pressure within the volcano, slight shifts of the confining masses may enable the gases to escape with destructive violence. It is to some occurrence of this sort that we may attribute such explosions as wracked the East Indies when Krakatoa blew up.

In part, the escape of the gases is related to the fluidity of the lava. When the lava is thick and viscous, the gases cannot escape so readily and there is more opportunity for pent-up pressures. In thinly fluid lavas, on the other hand, the gases escape quietly, something like the bubbles from soda-water, and explosions are uncommon. Mauna Loa has lava of this type.

AMONG the gases that accompany volcanism, two are of outstanding importance. They are water vapor and carbon dioxide. Accompanying these is a host of others, such as sulphur dioxide, hydrogen chloride, and hydrogen itself. During eruptions like that of Vesuvius in 1906, the liberation of the hot gases develops the columns of vapor that rise above the cone. The condensation of the water vapor into water sometimes causes violent rains, and the friction of the rising vapors and dust causes the electrical phenomena that result in lightning flashes playing about in the column.

Submarine volcanic outburst near the Azores, 1811. Woodcut from De la Beche's *Geological Observer* (1851)

Some of the gases are very active chemically and enter into chemical reactions with the oxygen of the atmosphere. Such reactions usually involve the liberation of heat, and thus serve in part to keep the lava fluid. In general, the temperature of lava ranges from about 750° C. to something over 1,000° C. Naturally enough, as soon as the lava reaches the surface its heat is rapidly dissipated and the fluid tends to solidify. In some cases the solidification occurs only at the surface, and then a stream of the lava flows along beneath the hardened crust. In this manner lava streams have often reached for considerable distances from the volcanic vent.

IT WAS formerly supposed that the explosive liberation of steam and other gases resulted from the influx of sea water into the volcanic duct. The presence of chlorides supported this view, but we know that, while such occurrences are possible, they are not the general rule. The Hawaiian volcanoes, situated directly on oceanic islands, contain no chlorides in their gases, although they are common enough in the Italian volcanoes.

RAPID cooling of lava at the surface tends to produce rather glassy rocks that look much like the slag from blast furnaces. In the thicker lava sheets the cooling may be slow enough to permit the development of small crystals, but quite generally these extrusive rocks are very fine-grained. If the lava hardens before all the enclosed gases are liberated, a porous rock is formed, like *pumice*. During the somewhat explosive stages of eruptions, like those displayed by Vesuvius, blobs of liquid lava may be thrown into the air. Here the whirling masses solidify, forming spindle-shaped masses called *volcanic bombs* or *lapilli*, depending on their size. Clouds of *volcanic dust* and *volcanic ash* are usually intermingled with these larger fragments. We must not, incidentally, interpret the word *ash* here to mean that the rocks were burned. There is no actual combustion of rocks during eruptions, but the small gas chambers that commonly remain in the fragments give them a cindery appearance. There may, however, be some combustion of inflammable gases from the lava.

NOTICE that, just as the gradational agents have associated with them the sedimentary rocks, so volcanism tends to produce typical kinds of rocks. These are the **igneous rocks,** as we already know; and they will be discussed a bit farther along.

THE essential part of a volcano is the opening which leads downward into the earth and from which the lava and gases emerge. As time goes on, and the lava streams radiate out from the center, a low cone is built up around the mouth, as a result of the solidification of the fluid rock. If explosive action accompanies the eruptions, many of the fragments fall back in the vicinity of the vent and pile up into a mound. Thus arise **volcanic cones,** which we may visualize in general as conical mounds with a depression, or **crater,** at the top. This simple form is much varied in nature, due to smaller vents opening up along the slopes, giving rise to smaller cones. Occasionally, also, the cone may be wrecked by an explosion, with the result that a new cone is built up within the walls of the old. Vesuvius is an example of this; and Monte Somma, an irregular semicircular wall partially surrounding the present crater, is the remnant of an older cone. Plates 26 and 27 include several illustrations of volcanic cones.

Volcanic cones

VULCAN'S CHIMNEYS

JUST as gradational land features pass through a series of stages, so does volcanic activity display a cycle of events. In the early stages the volcano is active, a cone is built up, and eruptions are rather common. Gradually the activity dies down and gases, rather than lava, issue from the vent. In this stage the volcano becomes a *solfatara*, named from a famous volcano in this stage of activity. Likewise, *hot springs* and *geysers* tend to develop in the vicinity. During these dying stages the cone itself is subjected to erosion by geological agents, so that in some cases only the hard and resistant interior of the cone, made of dense and massive lavas, stands up as a lone sentinel at the former site of impressive phenomena.

WE MENTIONED geysers back in chapter 9 but deferred a discussion of them to here. Hot springs and geysers occur in the later stages of volcanic activity, usually after eruptions have ceased but while there still remains hot rock not far beneath the surface. If the rocks are near enough to the surface for water to reach them, the water naturally becomes heated. Density differences between the hot and cold water may then set up circulations, and some of the water may return to the surface and give rise to hot springs. Geysers, while dependent for their heat on the same set-up, require special conditions for their operation.

For obvious reasons, no one has ever descended down into a geyser tube; still a fairly sound and reasonable theory has been advanced for geyser action. Suppose there is a fissure extending down into the rock, with perhaps a few constrictions here and there along it. The bottom, we shall say, is in or near a mass of hot rock. Rain water or ground water fills up the fissure, and the bottom of the water column becomes heated. Convection currents are set up, which, however, are impeded by the constriction of the tube. Thus the water down below, under some hydrostatic pressure, becomes heated to a temperature higher than the usual boiling-point of water. Under the influence of pressure, boiling-points rise; but with the pressures we may postulate in geyser tubes, the new boiling-point is not excessive. Thus the time comes when the water down in the tube begins to boil. A few steam bubbles are formed, and the surfaces of these offer an opportunity for more water to vaporize. Since steam occupies much more volume than the original water, some of the water in the upper part of the tube is displaced.

The displacing of water from the upper part of the tube is often seen as a spilling of some water over the lip of the geyser basin. This removal of some water from the column reduces the pressure on the water below, so that it finds itself hotter than the boiling-point under the now reduced pressure. Therefore the whole mass of water vaporizes rather suddenly and forces up the overlying liquid, which rises as a jet from the geyser tube. When the water has been forced out, the stage is set for a repetition of the process.

AT PRESENT there are about four hundred active volcanoes on the earth. They are not scattered uniformly over the surface, but rather are confined to certain belts and regions. The belts

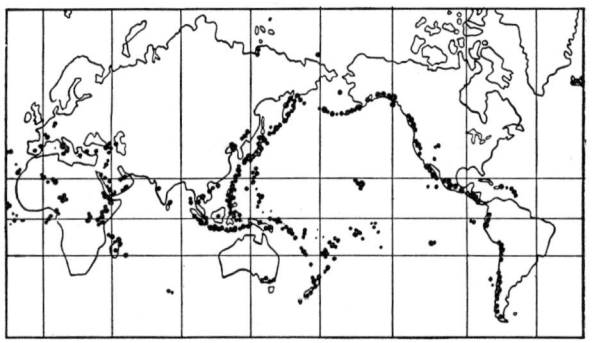

Map showing world-distribution of present or recently active volcanoes

are, in fact, usually associated with mountain chains; and they thus occur, on the whole, along the continental margins. The famous "belt of fire" encircling the Pacific Ocean is shown on the accompanying map. In addition, certain chains of islands are made up of volcanoes, notably in the Pacific Ocean. Further, volcanic regions are usually those in which earthquakes are common, as witness the Italian Peninsula. This close relation between earthquakes and volcanoes was formerly taken as evidence that the quakes are due to volcanic activity. We now know there is no necessary relation between them, but it is significant that volcanism and diastrophism are confined essentially to the same regions of the globe.

A study of past and present volcanic activity discloses that in many cases the volcanoes are arranged along fault lines; and the suggestion is strong that these weakened zones, in which the rocks have suffered dislocations, afford an easy means of access of the lavas to the surface. We have yet to consider, however, the ultimate causes of volcanism, and to explain how it is possible for bodies of rock to become liquefied within the earth.

IT WAS formerly supposed that the entire interior of the earth was liquid and that the surface was a thin floating crust. Earthquake waves have taught us that this is definitely not the case. The essential solidity of the earth precludes the possibility of a common liquid source for all lavas. Even so, we still commonly refer to the outer part of the earth as the *crust*, although we no longer mean a solid exterior about a molten globe.

OUR present knowledge of volcanic activity leads us to the conclusion that local reservoirs of liquid rock are developed within the earth, not very far beneath the surface. We saw in an earlier chapter that during the compressive stresses of diastrophism heat is generated. This heat may be concentrated locally through radioactive processes, with the result that bodies of earth-material may fuse, despite the pressure to which they are subjected. Such fusions are, in part, mutual solutions. Once the material is in a liquid state, its density is decreased and it tends to rise toward the surface. We may well imagine such threads of liquid being gradually squeezed and kneaded outward from the deeper parts of the earth. As they rise, they continually enter regions of less pressure, and consequently the heat they carry with them may be sufficient to liquefy some of the rock they pass through, thus adding to their volume. Furthermore, coalescence of the liquid threads may lead to fair-sized reservoirs of the liquid rock.

Threads of liquid lava rise toward the surface . . . .

All of these liquid masses may not reach the surface; indeed, perhaps the greater part of volcanism is a subterranean phenomenon. If the liquid bodies should reach near enough to the surface to enter the fractured zones, then they may well take advantage of zones of weakness for their escape to the surface. Thus would arise the extrusive aspects of volcanism.

THERE is another possible cause of liquefaction, which is associated directly with mountain-folding. Mountains arise during the operation of great diastrophic forces; and once the mountain belt has been elevated, the pressures tend to be relieved. Now, during the high-pressure stages, before the rocks yield by folding, the heat content of the rocks rises; but the melting-point also rises, owing to the increased pressure. Thus one factor tends to oppose the other. However, when the pressure is relieved by uplift, the superheated masses of rock may pass over into the liquid state. We actually find in nature that the central cores of many mountain ranges are composed of rocks which hardened again from such a liquid state, so that the local relief of pressure may well be among the causes of volcanism.

IN ALL these cases of migrating lavas within the earth we see evidence of the transfer of heat from the interior of the earth to points at or near the surface. The process is thus directly related to the gradual shrinkage of the earth, because volcanism serves to bring heat from the interior of the earth to the surface. The transfer of heat by the liquid lavas aids the compressive forces in overcoming resistance to compaction in the deep interior of the earth. Volcanism, viewed in this light, is seen to be diastrophism's henchman in the struggle between *status quo* and a smaller globe.

CHAPTER 21

IGNEOUS INTRUSIONS

Search Nature's depths and view her boundless store.
—YALDEN

THE Henry Mountains of Utah seldom are mentioned in travel literature, partly because of their inaccessibility; but they have had a most curious and interesting origin. For there, in seeking access to the surface, a reservoir of liquid lava became entangled among the horizontal sediments that composed the region. The fluid rock exerted its hydrostatic pressure against the confining walls, the overlying beds were pushed upward into a series of "blisters," and the Henry Mountains were born. We

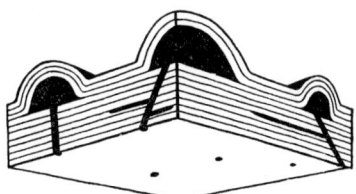

The "blisters" of the Henry Mountains, in odd perspective

may visualize these blisters as enormous rounded domes of sedimentary rocks confining within themselves a lenticular volume of liquid rock. In the course of time the lava hardened into igneous rock, and a long period of erosion enabled running water to cut through the overlying sediments. The result is that along the flanks of the mountains may be seen the edges of the upturned beds, and in the valleys within them are the masses of igneous rock with the sediments arched up over them.

THUS are the Henry Mountains today; and by reconstructing the great blisters, the origin just described has been inferred. Since the monumental work of Gilbert on these mountains, other examples in various parts of the earth have been recognized. To them is applied the name **laccolith,** and they are a type of intrusive volcanic action in which the lava did not succeed in penetrating to the surface while in a liquid state. The fluid rock was injected into the sediments under tremendous pressures, and thus succeeded in bowing up the surface of the land. In general, the igneous rock is found to be conformable with the surrounding sediments; that is, the lava did not cut through the beds to any great extent except where the conduit led the lava up from greater depths below the surface.

TRUE, laccoliths are not the most common type of intrusion; but they illustrate an interesting adjustment of fluid lava and solid sediments to one another in regions below the surface of the earth. Laccoliths are often large igneous intrusions, and the enormous blisters of the Henry Mountains were perhaps 5 mi. across and a goodly portion of a mile high.

Much more common types of intrusion are certain smaller features that generally accompany volcanic activity. In volcanoes or in the underlying sediments the lava that rises through the crater may find fractures and fissures along its path and, in penetrating them, form intrusive features. In a broad way we recognize two types of such smaller intrusions. If the sheets of lava penetrate between the beds and lie parallel with them, they are called **sills;** while, if they cut through the beds at an angle, we call them **dikes.** Where the sediments or other layered rocks are horizontal, the sills will also be horizontal sheets of lava, while the dikes will be more or less vertical sheets of lava that flowed up along fractures and cracks.

Block diagram of dikes and sills

ALTHOUGH we refer to dikes and sills as *minor* intrusions, the term is a relative one. In contrast with some intrusions that form the cores of entire mountain ranges, it is true that these two features are small. Yet dikes are known to extend for miles along the surface of the earth, where they often project above the surface as walls, because of their greater resistance to erosion. Similarly, sills may constitute sheets of rock with areas of many square miles. In thickness a dike or sill may run from a fraction of an inch to hundreds of feet. They are thus quite thin in contrast to their depth and length. Plate 28 illustrates a number of intrusive igneous features.

IGNEOUS INTRUSIONS 179

BY FAR the most important intrusive feature is associated with mountain-folding. We touched on this toward the end of our last chapter, when we considered the causes of volcanism. We may now examine these features themselves in more detail. Since mountain chains are generally much longer than they are wide, we may think of the fold in a broad way as an elongate arch elevated at right angles to the direction of pressure. If the relief of pressure that follows uplift permits some of the rock beneath to liquefy, this liquid lava will penetrate into any joints or fractures present in its vicinity. The result may be that the intruded rock will bear rather complex relations to the folded and fractured rocks, with dikes and sills leading off in all directions. Further, the zone of liquefaction may extend under most of the length of the folds, thus forming a sort of core to the entire mountain chain.

These intrusions that constitute the cores of mountains are called **batholiths.** When the mountains are later eroded to the point where the igneous rock is exposed at the surface, the structure of the mountain core, and its relation to the folded rocks, may be studied in detail. Even when the mountains have been eroded away to peneplains, the central core is discernible among the folded roots of

Batholiths form the cores of mountain ranges. Block diagram showing a partially eroded batholith

the range; and by mapping its trend, ancient zones of diastrophic adjustment may be followed in some detail. The central core is consequently of considerable importance in tracing former mountain ranges that diversified the globe in the distant past.

As to size, batholiths may stretch for a thousand miles along the mountainous belt. In every direction dikes and sills cut and interfinger the folded sediments, bearing witness to the intimate relations between the intrusive features and the diastrophically affected sedimentary rocks.

A CONTRAST between batholiths and other igneous intrusions may prove instructive. Dikes, sills, and laccoliths are *injected* into cracks or fissures among the rocks, or they penetrate between the beds and so force them apart. Batholiths, on the other hand, apparently *transgress on* and take over the space formerly occupied by other rocks. One explanation of this is the actual liquefaction of masses of rocks in the cores of folded mountain tracts, as we suggested. Many other factors are probably also involved in their formation. As a matter of fact, the detailed explanation of these tremendous igneous features, involving hundreds of cubic miles of rock, with their evidences of transgression and actual replacement of the surrounding rocks, constitutes another of the current problems of geology.

WE MAY expect that when masses of hot liquid lava are intruded among sediments, the heat conducted outward, as well as the new chemical compounds that are present, may profoundly affect the surrounding rocks. As a matter of fact, they do. The rocks are often "baked" and altered by the heat, and valuable ore deposits are formed by the fluids that escape from the lava. This is mainly true of the larger igneous intrusions; near the margins of batholiths there often are large ore-bodies, and this is one reason why so much metal-mining goes on in mountainous areas.

The alterations due to heat and chemical change around intrusive bodies often enable us to infer the presence of batholiths or other igneous masses even if the intrusions themselves are not exposed by later erosion. In general, however, our direct knowledge of intrusions is confined to cases where the now hardened rocks have been exposed at the surface by subsequent erosion. It is from such exposures, and from the nature of the igneous rocks themselves, that the very process of igneous intrusion has been inferred. For we must understand that no one has ever seen a batholith being formed; yet the evidence of intrusion is so clear that no doubt of its existence is entertained. We shall examine other lines of this evidence in succeeding chapters.

WHEN we compare extrusive and intrusive rocks in the field, we are sometimes involved in complexities. Consider, for a moment, a lava flow on the surface of the land or on the sea bottom.

IGNEOUS INTRUSIONS

In either case a suitable environment may result in the flow being buried by sediments later. When erosion again attacks the rocks, we may find them exposed along some valley wall. With sedimentary rocks above and below the igneous layer, how can we tell whether it was an extrusion or an intrusion, a flow or a sill?

The question is usually not difficult to answer, because each situation carries with it some clues to its origin. In this case the heat effects of the lava furnish us with the key. Obviously, when

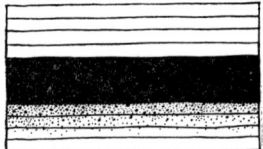 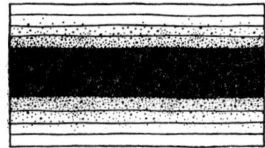

The lava flow (*left*, black) baked the sediments below it, whereas the sill (*right*) affected the rocks above and below

the lava is extrusive, only the rock beneath is affected by the heat, and sediments deposited later show no such alterations. But when a sill penetrates between two layers, both the lower and upper beds are baked, attesting to the fact that the lava was intruded between two beds already present. In similar fashion we know that, when a dike cuts a series of beds at an angle, all the beds so cut must have been present before the dike was formed. These inferences are so obvious that it may seem to be an unnecessary laboring of the point, but it is from such known relations of one rock to another that the history of the earth is largely read.

THERE is another aspect to our contrast of extrusive and intrusive rocks that deserves mention. Extrusions, as we saw, are exposed directly to the atmosphere and therefore cool rapidly, yielding glassy or very fine-grained rocks, sometimes with a porous, cellular texture due to gas bubbles imprisoned in them. In similar fashion dikes and sills, because of their sheetlike forms, cool rather rapidly by conduction, and commonly are somewhat fine-grained but seldom glassy or cellular. The larger intrusive masses, like laccoliths and batholiths, being deeply buried, cool quite slowly and thus develop rather large crystals that may easily be observed with the unaided eye. In general, then, the resulting igneous rocks are of two kinds: coarse-grained and fine-grained. Chemically a given set of coarse and fine-grained rocks may be essentially identical; but, because of the sizes of the crystals, they may present entirely different characteristics to the eye. Thus the environment of cooling plays an important rôle in the final product.

SO IMPORTANT are the igneous rocks in the geological story that they deserve a chapter for themselves. Part of their importance, from our point of view, arises from the fact that they so well illustrate the types of inductive reasoning used by geologists.

CHAPTER 22

FROZEN SEAS OF LAVA

> And where the vineyard spread its purple store,
> Maturing into nectar; now despoiled
> Of herb, leaf, fruit, and flower, from end to end
> Lies buried under fire, a glowing sea!
> —Mallet

THE little town of Niedermendig might conceivably have a prettier geographic setting, but its paving blocks certainly cannot be improved upon. For there the fortunate visitor may occasionally catch the bright gemlike glisten of blue crystals in the old paving stones. Ages ago that part of the Rhineland was the site of volcanic outbursts; and among the extrusions were lavas which hardened into dark fine-grained masses containing cubical crystals of *hauynite*, a comparatively rare mineral. The rock is extensively quarried round about,

Jeweled pavements . . . .

and thus arises the curious paradox of peasants' carts rumbling over "jeweled" pavements.

We could draw other illustrations from the interesting topic of rocks and their uses, but this example will serve to introduce us once again to the general subject. After all, rocks are the stock in trade of geologists, and a natural bent drives them ever and anon to discuss the subject.

ROCKS formed by the hardening of lava are by no means rare, nor are they confined to regions in which active volcanoes now exist. Niedermendig has no active cones anywhere about. In other parts of the world even more extensive areas are underlain by ancient lavas. Large parts of the state of Idaho, for example, are covered with hardened lavas that poured from numerous vents in the dim past. No active volcanoes now exist in Idaho; but at the time of the activity, vast seas of liquid lava inundated the area, only to freeze into rigid rocks as their heat content was dissipated into the atmosphere. They are literally frozen seas of lava; and even today the rough, tumbled-about surface is much the same as it was shortly after Vulcan held his court among those barren hills.

Although these lavas hardened literally millions of years ago, there is no doubt that they once were liquid. All rocks carry with them some traces of their origin, just as sediments bear ripple marks and have bedding to attest to their origin. A common attribute of hardened lavas is a certain denseness of texture occasioned by the finely crystalline or glassy body of the rock. Fluid lava rapidly cools by radiation when it reaches the surface, and the rapid solidification that ensues results in very fine crystals or even a glassy solid. Sometimes the lava hardens while bubbles of gas are still rising to the surface, so that the bubbles become frozen into the rock. The presence of such bubbles may be an important criterion in proving that the rocks are lavas.

NOT only do volcanoes pour out molten rock, but usually during explosive phases great clouds of dust and volcanic fragments are thrown into the air. These settle back to the ground, where they may become incorporated into the lavas; or they may form deposits of their own. All in all, the whole picture of volcanic rocks is quite complex because of the many kinds of phenomena that occur simultaneously during eruptions. The net effect, however, is almost always a composite picture that leaves no doubt about the general set of events that transpired. Thus in Idaho the tortuous surface of dense volcanic rocks, with fragments of ejecta scattered about in the most promiscuous manner, leaves no doubt that volcanism formed the rocks.

Old volcanic cone and lava flows, after Scrope

FROZEN SEAS OF LAVA

WE SHOULD pause here to clear up one or two terms that we shall need. Strictly, the word *lava* refers to liquid rock or its hardened forms on the surface of the earth. It has also been applied to liquid rock that never sees the light of day—those bodies of lava that constitute the igneous intrusions. To avoid confusion, however, we shall use the term *magma* for liquid rock beneath the surface of the earth, confining *lava* to that part which actually escapes to the surface and there hardens into the volcanic rocks.

NOT only, then, are there rocks that form from the hardening of lava at the earth's surface, but also a whole series of rocks that form from the crystallization of magmas beneath the surface. Just as the lavas bear evidence of their former liquid condition, so do the intrusive igneous rocks. **It is, indeed, from the evidence that such rocks once were liquid that we infer the whole process of igneous intrusion.** It is readily possible to watch volcanic rocks form at the earth's surface, because a lava poured out on the surface may be studied in full detail a few days later. Samples of the liquid lava may even be collected, and the gases that accompany the eruption may be studied. In this manner a large body of direct evidence can be built up. No one, however, has ever seen a deep-seated magma solidify, so that here we must rely on indirect evidence.

THERE is one main attribute of the intrusive igneous rocks that differs from those of the hardened lavas, and that is the size of the crystals. Lavas generally do not have crystals large enough to be seen clearly with the unaided eye, while the intrusive rocks seldom have crystals too small to be seen as distinct entities. The hardened surface lavas range from volcanic glass (which has no crystals at all) to dense rocks that show innumerable glints to the eye from myriads of tiny crystals. When the latter are studied under the microscope, it is found that the minerals present may be identical with

Two main kinds of igneous rock....

those in the coarser-grained intrusive rocks. Furthermore, the crystals in intrusive rocks show a continuous gradation in size from the tiny crystals of dikes and sills to the large crystals of batholithic intrusions. In a similar fashion, the chemical compositions of the rocks furnish striking evidence of the essential similarity of the two kinds of rocks. It is possible to find hardened lavas with chemical compositions essentially identical with those of certain deep-seated rocks. Thus in many series of rocks from lavas to coarse intrusives, the only major difference seems to be one of crystal size. What may we conclude from this? Simply, that both the surface lavas and the intrusives of the series had a common origin but differed in the environment of their solidification.

Lava flowing from Vesuvius. Old woodcut from De la Beche, 1851

When a lava is poured on the surface, the loss of heat is extremely rapid because of the much lower temperature of the air and the large surface area of the lava sheet, so that it tends to solidify rapidly. With rapid solidification comes a tendency toward minute crystals, because the growth of crystals requires a gradual change of conditions, rather than an abrupt one. Allow the body of magma to remain deeply buried, however, and the loss of heat is much slower. This is due to the heated condition of the surrounding rocks themselves, as well as their poor thermal conductivity. Among the larger intrusions there is often relatively less surface area, also. That there is some loss of heat is attested by the geothermal gradient, which shows that heat is slowly being radiated out into space by the earth as a whole. Hence the magma will gradually cool, but so gradually that in terms of human comparisons it may appear to be infinitely slow. A million years? Perhaps; but at any rate, long enough for all the magma to crystallize completely into crystals, often of considerable size.

IT IS from reasoning like this that we infer the origin of coarsely crystalline rocks from bodies of magma deeply buried and slowly cooling. Here again is an example of adjustment to environment, and the tendency toward states of equilibrium. If we can clearly understand that these two broad classes of igneous rocks, the extrusives and the intrusives, are products of environments, the details of classification become quite simple. If we were inclined to the use of tables, it would be possible to contrast the two environments and their results something like this:

Surface Conditions	Deep Burial
Rapid radiation of heat content	Relative insulation from loss of heat
Low pressures	High pressures
Small crystals or none	Large crystals usual

We have not touched upon the effect of the difference of pressure in the two environments, nor shall we do so in any detail. Of primary importance, however, is the fact that under the reduced pressures at the surface, the gases and vapors contained in the lava may readily escape, because the solubility of such gases is decreased under lessened pressure. In deeply buried magmas, on the other hand, the gases are held within the magma during much of the crystallization, and escape finally as watery fluids which give rise to valuable deposits of ores and gems. While they are held within the magma, they are a very important factor in maintaining it in a liquid state until crystallization is essentially complete.

ONE of the most fertile lines of inquiry among the coarse-grained rocks is a study of the crystals themselves. It may seem surprising, but it is nevertheless true, that even the order in which the crystals appeared can be determined, even though the rocks may have crystallized millions of years ago.

Suppose we stop along the way here and see just how the crystals of igneous rocks are studied and what they tell us about the process of solidification of the magma. Consider a liquid magma which will form several minerals as it cools. Each of these minerals will have its own properties, and consequently will behave differently from the others. We may accordingly expect that, as the temperature drops, the minerals will not crystallize simultaneously, but

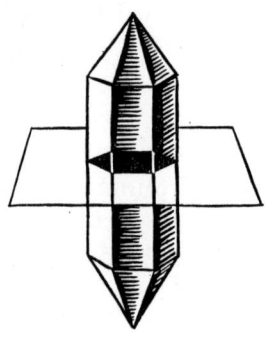
Crystal sections

rather a combination of solubility and concentration will largely determine which crystals appear first. These first crystals to appear will be bounded by plane crystal faces because nothing interferes with their complete growth. The same may be true of the second mineral; but as the liquid becomes nearly filled with crystals, later arrivals must adapt themselves to somewhat constricted surroundings. The later crystals may therefore have only some of their faces developed; or conceivably they may have none, but must form in the interstices among the earlier crystals. Furthermore, the earlier crystals may be imbedded in later crystals which simply grow around them, so that the final result is a close interlocking of all the minerals present.

Now, in the igneous rocks we use these principles in determining the sequence in which the several minerals formed. A slice of the rock, ground down to extreme thinness, is studied

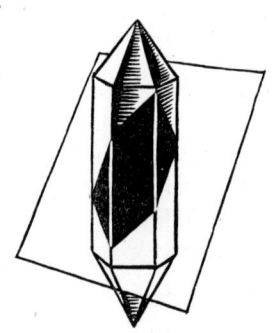

under the microscope. These thin sections are essentially in two dimensions; they are so thin that they are quite transparent. Hence we usually do not see the entire crystals but only sections through them. However, when we cut through a solid bounded by plane faces, the resulting section is an area bounded by straight lines, as the adjoining diagrams show.

WE SAID long ago (chap. 3) that the deep-seated igneous rocks are composed of primary minerals; and these, you will remember, include the ferromagnesians, the feldspars, and quartz. Suppose we choose a rock that contains all three minerals. **Granite** is such a rock; and when we study a thin section of it under the microscope, we find that the order of crystallization shows that the

FROZEN SEAS OF LAVA

first minerals to appear (except for various minor constituents that do not concern us) were the ferromagnesians. These, for the most part, are bounded by crystal faces. The next minerals that form are the feldspars, and finally the quartz fills in the spaces between the earlier crystals.

WE HAVE inserted a somewhat simplified diagram of a granite here to illustrate the sequence of crystals. The ferromagnesians (one is marked FM) can be recognized because they are bounded on all six of their sides by crystal faces. The feldspars (large crystals with alternate black and white bands, marked FEL), have most of their faces developed; but locally they grew around the ferromagnesian crystals, showing that the ferromagnesians preceded them. Finally, the quartz (dotted areas identified by QU), have essentially no crystal faces developed, but grew in the spaces among the earlier crystals.

It is difficult to find the entire sequence of crystallization in a thin section of granite small enough to photograph in one setting; but if you will refer to Plate 30, you will see two photomicrographs that illustrate the sequence.

Diagram of a thin section of granite (as seen under the microscope), showing the order of crystallization

The photograph (upper right), shows several dark crystals, bounded by fairly straight lines, imbedded in a matrix of a light feldspar. The dark crystals are ferromagnesians (plus a few rarer *accessory* minerals), and they preceded the feldspar because they are imbedded in it. The photograph below this shows some of the later stages of crystallization. Notice the light triangular patch near the upper center. This is quartz; and it is easy to see that it is completely bounded by crystal faces of other minerals, showing that it appeared last.

INTERESTING as this may be, it is only the beginning of the story. This *order of crystallization*, as it is called, tells us a lot more. It tells us, in fact, something about the chemical reactions that probably took place in the magma before or while it cooled. To develop this point we shall have to recall some of the things we said a long time ago about the chemical composition of the primary minerals. Remember that the quartz is crystallized silicon oxide, SiO_2, and that the ferromagnesians and feldspars are silicates, and so have silicon oxide, or silica in their chemical makeup. Now if the ferromagnesians crystallize first, that must mean that some of the silica in the original melt first unites with the iron oxides and the magnesian oxide (plus a few others) to form the ferromagnesians. Next, more of the silica must combine with the oxides of aluminum, calcium, sodium, and potassium, to form the various feldspars. After all of these oxides have combined with some of the silica, then the **excess silica crystallizes out as quartz.**

We have almost leaned over backward to make this story simple. Actually, the complications are far beyond the scope of this book; but the sequence given includes the salient features of the story. Granite, we see, is a rock with a high percentage of silica in its ultimate chemical composition—so much so, that after all the other oxides have combined with it, there still is some silica left to crystallize out as quartz.

NOW, although granite has a fairly high silica content and a relatively low content of iron and magnesium oxides, it is only one of a large number of coarsely crystalline igneous rocks. If we were to pursue the subject over the entire field, we would never finish this book. Hence we shall restrict ourselves to essentially the *normal series* of igneous rocks, which includes three main types. To bring this point out, we may say that from a chemical standpoint it is possible to contrast the percentage of silica in the rock with the combined percentages of iron and magnesium oxides. That is, we may find rocks low both in silica and in iron-magnesium oxides, or we may find them rich in one and low in the other. Look at the adjacent diagram and notice that the vertical axis shows the total amount of silica in the rock and that the horizontal axis shows the total amount of iron and magnesium oxides. We have arbitrarily divided the area into nine squares, and have numbered them. At a glance we may see that square No. 1, for example, has much silica and little iron-magnesium oxide; No. 9 has little silica and much iron-magnesium oxide; and so on. Thus we find in nature that all sorts of rocks are possible; but fortunately, most igneous rocks fall into three of the squares, which we have outlined more heavily. As you may notice, they form a diagonal of our chart, which means that from rocks high in silica and low in iron-magnesium oxides they descend to rocks high in iron-magnesium oxides and low in silica. It is this series that we shall examine. As a starter, we may say that granite is the typical rock of square No. 1.

Since granite has quite a high percentage of silica, we may call it an **acidic rock,** because silicon oxide is

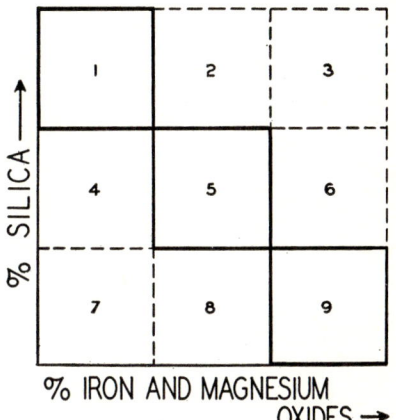

Diagram of possible igneous rock types. The total silica is contrasted against total iron and magnesium oxides. The other oxides are not included

an acidic oxide. Likewise the square on the lower right, being high in iron and magnesium oxides and low in silica, is called **basic,** and the middle square is called **intermediate.*** The table below shows more explicitly how the rocks within these three squares vary in their percentages of the oxides. This variation in the silica and iron-mag-

Rock Type	Percentage of Silica	Percentage of Iron-Magnesium Oxides	Percentage of All Other Oxides
Acidic...............	70	3	27
Intermediate...........	57	12	31
Basic...;............	48	17	35

nesium oxides is shown in the accompanying graph. Now, from the relation between the percentages of silica and of iron-magnesium

* Specialists in the field of igneous rocks seldom use the terms *acidic* and *basic,* largely because, when the entire range of igneous rocks is discussed, the terms may lead to confusion. We have retained them here, however, because they fit in well with the acidic nature of silica and the basic nature of the iron and magnesium oxides, on which our discussion is largely based.

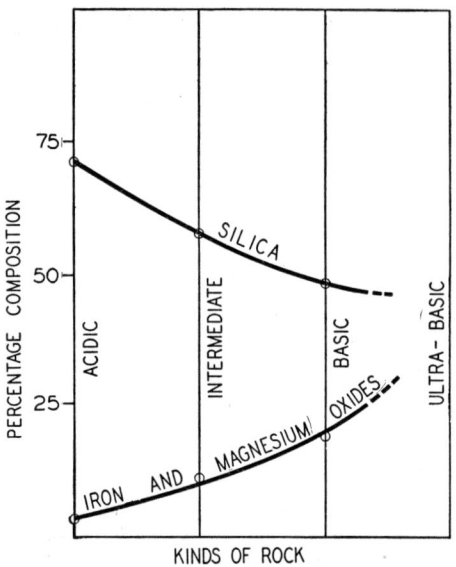

Curve of oxide content in the normal series. The vertical lines represent the main rock types

oxides in this normal series, what may we anticipate about the presence of quartz in the basic rocks? Well, if the amount of silica in their chemical make-up is relatively so low, maybe there will be no free silica left after the silicates are formed, so that no quartz can crystallize out. This is exactly the case. The basic rocks of the normal series do not contain quartz and, in extreme cases of our series, may not even form the feldspars. In that case they are practically all ferromagnesian minerals. On the same basis the intermediate rocks, with their intermediate amount of silica, may not have any quartz present; but they do have enough silica to form the feldspars, and thus they consist largely of ferromagnesians and feldspar. If we summarize all these data, we find that a table emerges from the welter of material presented:

ORDER OF CRYSTALLIZATION AND MINERALS
PRESENT IN THE DEEP-SEATED IGNEOUS
ROCKS OF THE NORMAL SERIES

Acidic Rocks	*Intermediate Rocks*	*Basic Rocks*
Ferromagnesians	Ferromagnesians	Ferromagnesians
Feldspars	Feldspars	(May have feldspars)
Quartz	(May have some quartz)	No quartz

As we pass from the acidic to the basic rocks, the diminished amount of silica in the original magma reflects itself by the absence of the minerals with high silica contents. Since the minerals of the rock reflect the chemical composition, a curve of mineral composition would be like that just below. If you study this for a few moments, you will see that it summarizes just about everything we have been saying.

Now that we are getting acquainted with these rocks, it may be interesting to know their names. The acidic rocks are **granites,** the intermediate rocks are **diorites,** and the basic rocks are **gabbros.**

But these, remember, are the deep-seated, coarsely crystalline rocks of the normal series. We said that each of the chemical types occurred in two kinds: the

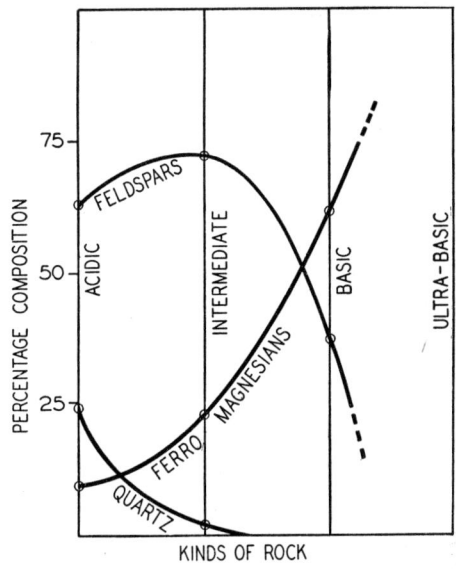

Curve of mineral percentages in the normal series

hardened surface lavas and the deep-seated crystalline forms. Hence, just as there are three kinds of deep-seated igneous rocks in our series, so there are three parallel groups of surface lavas. Each pair belongs to the same chemical category but differs in the environment of cooling. We may accordingly put all of our rocks in a single table, which will then afford us a classification of these igneous rocks.

Environment	Acidic	Intermediate	Basic
Surface (small crystals)	Rhyolite	Andesite	Basalt
Deep-seated (large crystals)	Granite	Diorite	Gabbro

The extrusive equivalent of granite, **rhyolite,** is light in color, whereas **basalt,** the extrusive equivalent of gabbro, is quite dark. As we may anticipate, **andesite** is intermediate in appearance. All three of these surface igneous rocks are fine-grained.

We may see the significance of the whole story now at a glance. The table tells us that if a volume of magma rich in silica rises from within the earth, and actually reaches the surface as lava, the rapid cooling will result in a rhyolite being formed; while if the magma does not escape, but hardens underground, a coarsely crystalline granite will result. And so on for the other types. Naturally enough, there are many individual species of rocks that we have not mentioned because of their relative rarity. Our simple classification tells the fundamentals of the story, and further details merely embellish and complicate it. Plates 29 and 30 illustrate these common igneous rocks.

THERE is, however, one peculiar type of igneous rock that deserves mention, because it bears evidence of having cooled in two environments. Sometimes a body of magma is intruded beneath the surface and there begins its cooling until a few of the earlier crystals have appeared in the still largely liquid magma. Later on, renewed activity may force this magma upward toward the surface. Here it cools more rapidly. The kind of rock that forms is easy to predict. It will be one in which some large crystals are imbedded in a fine-grained or glassy groundmass. Such a rock is called **porphyry**. Any of our three chemical types of rock may have a porphyritic phase, since that is controlled entirely by environment and is independent of composition.*

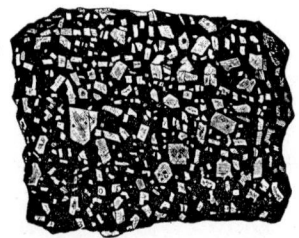

Porphyry, showing its coarse crystals imbedded in a fine groundmass. About $\frac{1}{3}$ size. Woodcut from Ludwig, 1861

IT MAY be interesting now to consider why there are these several chemical types of igneous rocks, but at most we can only touch upon the subject. Apparently, a whole series of complicated adjustments to environment on the part of the deeply buried magma results in a separation of parts of the magma during cooling. Among the many factors that may be operative is one that stands out prominently. Imagine a slowly cooling magma of about the composition of an intermediate or a basic igneous rock. The earliest crystals to form are the ferromagnesians; and these crystals, having

---

* Even the rarer rocks that we are not discussing here may occur as porphyry. In that case the first large crystals that form may not be ferromagnesians or even feldspar.

a rather high density, tend to settle downward in the still liquid magma. In the course of time they gradually concentrate toward the bottom, and this leaves the top of the magma relatively more concentrated in the other oxides, which tend to form feldspars and quartz. If now the liquid portion is squeezed out by diastrophic processes, it may lead ultimately to igneous rocks of a more acidic nature. Thus the normal series of rocks tends to be formed. We shall have to leave fuller details to more advanced treatises, to our sorrow and, perhaps, the reader's relief.

OUR story of the igneous rocks is nearly finished, although actually we have hardly glanced at the subject. Further detail involves greater complexity, because nature is seldom as simple as the pictures we draw. In fact, we have already oversimplified what is essentially a difficult subject bristling with *if's, and's,* and *but's.*

By rights we ought to afford the reader a breathing spell after this session and discuss a light subject for a change. Unfortunately, we cannot do so. We must still take up one other great group of rocks, which in complexity easily parallels the igneous rocks. There is no help for it; we shall have to take a deep breath and dive below the surface once again.

CHAPTER 23

HEREDITY VERSUS ENVIRONMENT

> Purged from their dross, the nobler parts refine,
> Receive new forms, and with fresh beauties shine: . . . .
> —Yalden

A PALIMPSEST is a document that had its writing erased so that the parchment could be used again. Traces of the earlier words may possibly be discerned, however; and the historian, poring over the musty parchment, may bring to light startling and important data. Just so the geologist, bending over his microscope, studies rocks for evidences of their earlier histories. The fruitfulness of this search was demonstrated by the study of igneous rocks. In like manner, the débris from the weathering of crystalline rocks may contain traces of the original minerals, from which the nature of the parent rock can be determined.

. . . . bends over the palimpsest . . . .

An even wider field for this search is among a large class of rocks so altered by environmental conditions that the original rock may never be identified. Heredity may thus yield almost completely to the irresistible impact of environment. What these environments are, and how they effect rocks, will be developed in the present chapter.

PROCESSES of sedimentation are taking place practically continuously, and sediments now beneath the seas will later be buried beneath other sediments, and so *ad infinitum*. As the load of overlying sediments increases, the earlier materials are squeezed

HEREDITY VERSUS ENVIRONMENT

and compressed by the increasing overburden. They are, in other words, entering a new environment. Just as deep-seated igneous rocks undergo alterations to weathered débris when erosion exposes them at the surface, so the débris itself, laid down as sediments, undergoes a new cycle of changes due to removal from surface conditions—changes that return to the material some of the original deep-seated characteristics.

Among the first changes that take place are a greater compaction of the particles, a cementation of the loose grains into a coherent rock, and the chemical alteration of some of the clayey minerals to new chemical compounds. Thus muds become shales, sands change to sandstones, and gravels to conglomerates. In short, they have entered the domain of the *indurated* or hardened sedimentary rocks. To all these changes we may apply the term **metamorphism,** or *a change in form;* but it is metamorphism in its gentlest and mildest state. The pressures are predominately static, that is, they depend merely on the overburden of the later rocks; and what rise of temperature there may be is not excessive.

THERE are other and more dynamic aspects of metamorphism, and to them we shall devote most of our discussion. Here are included deep-seated environmental changes of such magnitude that the rock may be wholly altered: its physical appearance may be completely changed, and in some cases its chemical composition as well. Such highly altered rocks are called **metamorphic rocks,** and they constitute the third and last great group. Metamorphic rocks arise from either sedimentary or igneous rocks, in environments where stability or equilibrium demands a complete alteration of the parent rock, in the direction of greater density, both physical and chemical.

AN EXAMPLE will help to crystallize our notions. Suppose a body of magma is intruded into a series of sediments, deeply buried beneath the surface. The great heat content of the magma will be conducted outward through the confining rocks and will accordingly raise their temperatures. Furthermore, the gases and vapors contained within the magma will penetrate the pores of the surrounding sediments and may there enter into chemical reactions with the minerals composing them. If the surrounding rock happens to be limestone, profound alteration may take place. The heat of the intrusion will facilitate chemical reactions, and the silica contained in the hot magmatic liquids will promptly react with some of the calcite to form calcium silicates. The other constituents of the fluids will simultaneously develop a whole series of special minerals adjacent to the contact between magma and surrounding rock. Thus may arise valuable ore deposits. In addition, the physical appearance of the remaining limestone will itself change, because under the influence partly of heat it recrystallizes into a sugary rock familiar to all as **marble.**

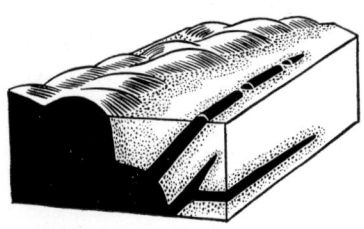

The zone of contact-metamorphosed rocks about an igneous intrusion

AGAIN we must point out that these events transpire without human witnesses, but detailed studies of the results of intrusions clearly indicate the sequence of events. We would expect the temperature to decrease away from the magma, and the magmatic liquids to diminish in chemical activity as they penetrate farther and farther into the sediments. These expectations are borne out by observation. The greatest alteration is near the contact between the limestone and the magma (when studied, of course, the magma has become a coarse-grained igneous rock); and as the limestone is followed away from the contact, the alterations gradually decrease. Indeed, from the characteristics of the new minerals formed during the metamorphism, it is possible, in a qualitative way at least, to sketch the decrease in temperature as one moves away from the contact. This is because certain minerals are known to occur only in restricted temperature ranges, and they thus furnish us with a *geological thermometer*, as it were, that enables us to evaluate the conditions extant during the metamorphism. Other lines of evidence also converge to the same conclusion, namely, that intruded bodies of magma induce profound effects in the surrounding rocks. In general the most striking changes are confined to a local area about the magmatic body, and seldom do the effects penetrate very many miles into the adjacent rock. Such metamorphism as this is called **contact metamorphism,** in distinction to more widespread metamorphic changes that partake of a regional nature. We shall come to them later.

WHEN we analyze contact metamorphism in environmental terms, we see that the magma brings along with it a new set of conditions: a higher temperature and a whole swarm of new chemical compounds. Under the impact of these new conditions, the adjacent sediments cannot remain indifferent. Rises in temperature may in some cases cause a reorganization of the sedimentary particles into new crystals; and in some cases the minerals of the sediments may themselves not be stable in the new chemical environment. When that is so, new minerals are formed which are stable. As we may expect, however, some minerals remain relatively unaffected under almost any conditions, and among them the outstanding example is quartz. A sandstone composed of pure quartz grains may show little change due to the increased temperature; whereas limestone, we saw, is markedly affected; and shale is also changed to a considerable extent.

EVEN dikes and sills cause some contact metamorphism, but to a much lesser degree than large intrusions. This is due to their smaller volume, and to their relatively larger surface area, which enable the heat to be dissipated more rapidly. Thus the conditions for a prolonged series of physical and chemical changes are not present. For *time* itself is an important element in metamorphism, and the time element may be easily as important as heat or pressure in the final result.

The time element may be illustrated by considering the effect of pressure on blocks of rock. If a block of rock is struck a smart blow with a sledge hammer, it fractures and shatters. Here a great force has been applied suddenly. If, on the other hand, the rock is deeply buried, and thus subjected to great pressures on all sides, it may yield by a process other than fracture if it is subsequently acted on by diastrophic forces. The yielding in such cases is similar to a plastic flow, but it involves physical and chemical changes as well. The exact changes that take place may most conveniently be studied among the rocks in nature, where the time element may have been thousands of years and the pressures thousands of pounds to the square inch. We saw in chapter 17 that diastrophic forces operate during mountain-making, and consequently we may look for our evidence among rocks that were involved in these major upheavals of the earth's surface.

A S ONE approaches a mountain range, he may first ride over a wide plain built up of sediments carried down by mountain streams, and finally enter the foothills of the range. These foothills are composed of stratified rocks turned up on edge: they are the marginal rocks involved in the upbulge of the range. As our traveler enters the mountain valleys proper, he may see further traces of these same rocks, but here they have been highly contorted and folded, or mashed and shattered, by the terrific forces of nature. Even the appearance of the rock changes, and a particular layer that in the foothills is a shale, for example, may be traced continuously into a shiny, glistening rock with a wavy banded appearance. Here, certainly, is a high degree of alteration, and we are interested in learning how it came about. In attacking the problem of the changes that took place in the shale, we may reason this way: if the shale originally had a fairly uniform composition, then the changes that progressively take place from the foothills into the mountains ought to be due primarily to changes in environment. Thus, from the mountain structures as a whole we may be able to learn something

Approaching the mountains . . . .

HEREDITY VERSUS ENVIRONMENT

about the environment of folding; and by combining this knowledge with a study of the alterations that took place in the rock, we should be able to evaluate the one in terms of the other.

Thousands of studies have been made on such metamorphosed rocks, and certain general conclusions regarding the problems have been developed. In this type of metamorphism we are dealing with much greater forces than in the case of contact phenomena, and the greater areas affected give to this aspect of the subject the name **regional metamorphism.** The changes that may take place in rocks under the stresses of regional metamorphism are quite complex, and we cannot stop to detail them all. Rather, we shall content ourselves with the transformations that take place in a shaly type of rock, since it illustrates all the principles we shall need.

SEDIMENTS, like all other rocks, have a multitude of gradations; and it is not unusual to find them intermediate between shales and sandstones, or shales and limestones. Suppose we choose a sandy shale for an example, so that we have more than one mineral to deal with in our story. Sandy shales are usually composed primarily of quartz grains and clay particles. Now imagine this rock subjected to the pressure of overlying sediments, to a reasonably high temperature, and finally to the directed stresses of mountain-building forces. These mountain-building forces operate mainly from certain directions, so that there will be a direction of maximum stress and one of least stress. We shall assume the simplest

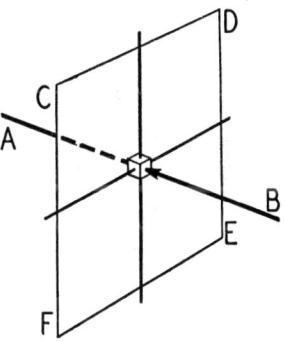

Block of rock under directed diastrophic forces

case, in which there is no tendency toward rotation. Then the direction of least stress will lie in a plane at right angles to the direction of greatest pressure. The same diagram we used in chapter 17 will be useful here. The line *AB* indicates the directed diastrophic stress; and the plane of maximum relief, *CDEF*, is perpendicular to it and contains the lines of least stress. Sometimes some particular direction within this plane affords the maximum relief from the forces.

That may depend on the local conditions. In any event, the great pressures tend to compress the rock into a smaller volume, but it already has been compacted by the overlying rocks. What will happen then? Well, by recombining its elements into denser chemical compounds, the rock may have its volume reduced somewhat; and by developing new crystals oriented along the plane of maximum relief, the situation may be eased still further. Curiously enough, the adjustment of the rock to its new environment involves these very things.

One of the first effects of regional metamorphism is a recrystallization of the original minerals into denser varieties, and among the new minerals is usually an abundance of light-colored mica. Here we may pause for a moment to introduce one more group of minerals, the **light micas,** which we have not met before. They form a group of their own for our purposes, and they are of considerable importance in the metamorphic rocks. The micas are silicates of quite complex structure; one of them, *muscovite*, or common insulating-mica, has the formula $(HK)_2Al_2(SiO_4)_3$, or $K_2O.Al_2O_3.3SiO_2.H_2O$, a complex hydrated silicate of potassium and aluminum. It is a silicate mineral, just as the feldspars and ferromagnesians are; but it belongs to a different group which, up to now, we have not had to use. Sometimes light mica occurs in igneous or sedimentary rocks, but by far its greatest abundance is in the metamorphic rocks. The light micas are formed largely from the clay minerals by a dehydration process during metamorphism. That is, a certain amount of chemically combined water is driven off from the clays, and the more compact and crystallized micas result.

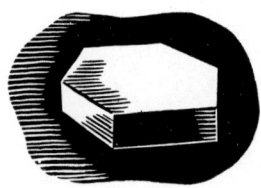

Crystal of mica

THERE is an interesting thing about the micas that you may not know. That is that they may be split into very thin sheets or flakes. This property is a significant one among the metamorphic rocks. The crystals and flakes of mica that grow in the rock tend to orient their broadest dimensions in the plane of maximum relief, something like the sketch alongside. Now as the pressures continue, or even increase (remember that mountain-making is a dynamic process), these mica flakes and crystals tend to separate into even thinner sheets along the plane of minimum stress. In this way the rock is able to adjust itself to the stresses of its environment and to come into balance with it by a complete internal rearrangement of its constituents. In the meantime the quartz grains are shattered into bits and strung along the same directions as the mica flakes.

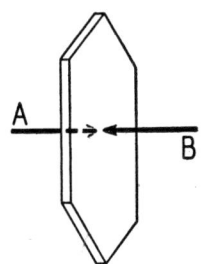

The broad dimensions of the mica are oriented normal to the stress

The net result is a rock which may easily be split along planes parallel to the direction of growth of the mica flakes. We say, accordingly, that the rock is *cleavable*. From our analysis it follows **that the cleavage of the rock tends to be at right angles to the compressive forces.** This is important, because in reading the history of folded areas the directions of cleavage in the metamorphosed rocks help us to determine the directions from which the operating stresses came. Obviously, this cleavage may develop at any angle to the bedding planes of the sedimentary rocks.

WE MAY now pause for breath while we tell you that, during the metamorphism of a shale, it first becomes **slate,** which is familiar to all as roofing material. Slate is in the stage of development where cleavage is prominent. That is why slate may so readily be split into thin slabs. In a later stage of metamorphism, when the micas become predominant, the rock becomes a **mica-schist,** which is a light-colored rock with a glistening appearance due to the myriads of tiny mica flakes distributed through it. Schist also tends to split somewhat readily along the surfaces in which the mica flakes lie.

We have purposely kept our story simple. In nature it is infinitely complicated. For accompanying the metamorphism of the rocks themselves there may be a folding of the entire series of beds into a mountain chain. These events may require thousands of years, and the reader should remember that what we find among the eroded mountains are only the end-products of these long and complicated processes. By a careful comparison of chemical compositions, by a study of the gradations among the metamorphic rocks, and by inferential reasoning, the story that we have presented has been built up. Moreover, new techniques of analysis are continually being developed to solve the infinitely complex problems of the metamorphic rocks. Modern methods involve the very grains and crystals of the rock, and their exact orientations in space. Optical methods and X-rays may be used in this work; and the net effect is that the very *fabric* of the rock is determined, from which much is learned concerning the nature of metamorphic adjustments. These newer techniques are somewhat beyond the scope of our book, but they illustrate the wide diversity of methods used by geologists in reaching their conclusions. The emphasis throughout is ultimately upon *environment*, **because the conditions to which the rocks are subjected determine the kind of metamorphic changes that occur.**

THE energy involved in these metamorphic changes probably has something to do with the story. Certainly in the formation of metamorphic rocks, potential chemical energy is stored in the new minerals that develop. Our chemically minded readers will understand by this that metamorphic reactions are *endothermic*. In fact, many deep-seated silicate minerals contain potential chemical energy, and thus some of the earth's internal energy is locked up in the deep-seated rocks. When, later, the rocks are subjected to exposure and decay, this energy is again released. It will be well, at this point, to refer once again to the lower half of the energy diagram at the end of chapter 4.

The terms "heat" and "pressure" also loom large in any discussion of metamorphism. These two factors of the environment are perhaps the most important, although the composition of the original rock, the new chemical compounds introduced, and time itself play their parts in the complete story. We obviously cannot pursue each of these variables in any detail if we ever expect to finish our chapter.

HEREDITY VERSUS ENVIRONMENT

Even as it is, we have not mentioned the metamorphism of the igneous rocks at all. As a matter of fact, they too become metamorphosed under suitable conditions; and we may mention, as a single example, that when granite is subjected to a set of conditions similar to those we postulated for the shale, it also becomes metamorphosed, and yields a banded rock called **gneiss**.

Gneiss, showing its banded appearance. About ⅓ size. Old woodcut from Lyell, 1855

WE MAY now summarize these metamorphic rocks in a sort of table which indicates for you the original rocks and their metamorphosed equivalents:

Sandstone ➤ Quartzite
Shale ➤ Slate ➤ Mica schist
Limestone ➤ Marble
Granite ➤ Gneiss

These metamorphic rocks are mainly those that result from regional metamorphism. We shall not introduce the specialized types that follow in the wake of contact metamorphism, because they are not as common as the others. Marble, you will notice, is included in the list; it happens to be one of several kinds of rocks that tend to be formed under both conditions. **Quartzite** is the only one of these rocks that we have not mentioned in detail. It is an example of the effect of original composition on the final metamorphic product. The quartz grains in sandstone are so indifferent to chemical reactions, and the number of elements present (silicon and oxygen) is so limited, that the sandstone may not greatly change during metamorphism. Usually it becomes harder and more dense, owing to cementation of the grains by silicious solutions. The net effect is a rock even more quartz-like in appearance. A number of common metamorphic rocks are included in Plate 31.

In general, there is a tendency for the metamorphic rocks to be visibly crystalline in appearance, even when the parent rock is quite fine-grained. Thus marble is usually coarsely crystalline, whereas limestone may be rather dense in appearance. Hence the metamorphic rocks, as well as the igneous rocks, are included among the *crystalline rocks* that we mentioned on several occasions. The *fragmental rocks*, on the other hand, are composed of sediments only.

NEAR the beginning of our chapter we mentioned quite an opposite aspect to the story of metamorphism. Instead of passing into environments of greater pressure and heat, we said, rocks may enter regions of lessened pressure and lower temperatures, as when they are exposed to weathering and erosion. Under these circumstances the rocks tend to break down. From one point of view, then, we may think of rock changes as being either *constructive* or *destructive*—constructive when the rock is further compacted and recrystallized, destructive when it decomposes and decays. Thus it is possible to look at all rock processes as cyclical phenomena in which the products are the results of the particular environments in which the rocks find themselves. Such a concept at once unifies and co-ordinates much of our knowledge, and to it we shall devote a closing chapter on the subject.

PLATE 17
SHORE DEPOSITS

The pulsational beating of waves on shores develops a zone of turbulence. Here material is swept to and fro, often eroding the shore, but not uncommonly building up shore features. Obviously material eroded at any point must be deposited elsewhere. Beaches are among commonest shore deposits. *Upper left:* Waves and beach, Pictured Rocks, Lake Superior. *Left center:* Pebble beach at Porte des Mortes, Door Peninsula, Wis., showing convex profile of the deposit. The pebbles here are almost entirely limestone, broken from exposed ledges along the shore. The flat, disklike shape of the pebbles may be discerned.

During transport of current-borne material, deepening waters at mouths of bays cause deposition, so that long tongues of sand and gravel extend from one headland toward the other. Such features are spits, shown in part at *right below* (Cape Madeline, Gaspe, Quebec; photo courtesy Canadian National Railways).

Sand spits show a characteristic form, easily recognized. *Below, left:* Small spit at end of island, Sturgeon Bay, Wisconsin (photo by J. S. Griffith). Here shore currents, sweeping along the island with sand in transport, build up the spit as deeper water is reached beyond the island's edge.

Shore deposits include sand, gravel. Fine material (mud) is carried to quiet waters. *Right:* Ancient marine beds exposed along valley wall.

PLATE 18 ENVIRONMENTAL CONTROL

Above: Delaware Water Gap, a notch cut through resistant rocks, affords passage for the Delaware River (photograph by courtesy American Airlines). Although streams carve valleys in any kind of rock, relative resistances and attitudes of the beds determine the configuration of the valley walls. The photo shows the valley constricted due to hard outcropping layers.

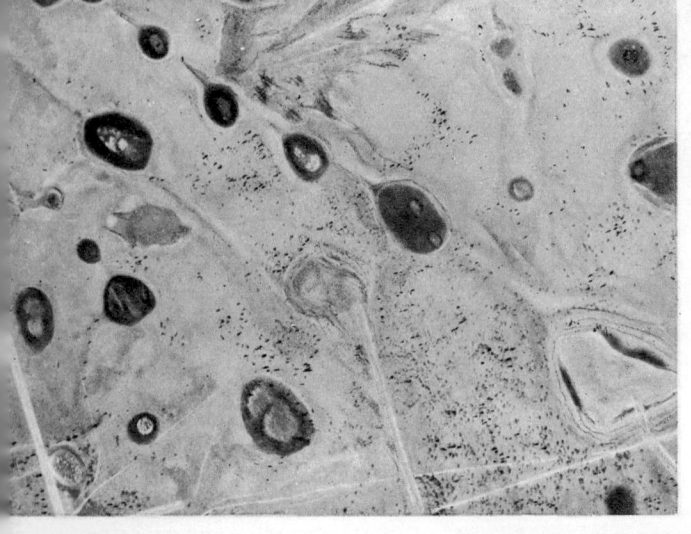

Above: Vertical view of sinkholes near Sarasota, Florida (photo Fairchild Aerial Survey; courtesy Geo-Aero-Photo). The sinkholes are occupied by lakes, some arranged along straight lines, indicating control by joints in underlying limestone. *Right:* Barchan dunes in southern Peru (photo courtesy of Geo-Aero-Photo). Barchans develop when the sand supply is limited; their form represents an equilibrium with environmental conditions. The hills move with the wind (wind direction from the right), but as they move they retain their crescentic form.

PLATE 19
SEDIMENTARY ROCKS

The most prosaic roadside ditch or rock quarry discloses sections of sedimentary rock. Streams, too, commonly cut through soil, expose the underlying rock. *Top, left:* Slablike black shale along creek near Havana, Illinois. (Photo courtesy Illinois State Geological Survey.) *Top, right:* Sandy calcareous shale, projecting from soil-cover, affords small but scenic waterfall. (Photo courtesy Illinois State Geological Survey.) *Upper left:* Ripple marks formed when this ancient sandstone lay beneath the sea. Geological revolutions have since turned them on end. *Above, right:* Sandstone along roadside ditch near Ableman, Wisconsin. *Left, lower center:* Limestone in quarry near Bloomington, Indiana. The sinuous line near the surface is a curious feature called a stylolite. *Left, bottom:* Sandstone near Russellville, Arkansas, shows "cross bedding" due to shifts in direction and strength of geological agent that transported the sediment.

PLATE 20 SEDIMENTARY PETROLOGY

Laboratory study of rocks includes size distribution of grains. Loose grains are sieved; when the rock is indurated (*top left*, outcrop of conglomerate), it is studied in thin section or by removing pebbles from matrix. Several photomicrographs illustrate common sediments. *Left:* Sandstone, showing close interlocking of grains. *Center:* Thin section of glacial till, showing heterogeneity (*above, right*, exposure of till along highway; photo courtesy Illinois State Geological Survey). Circle on right shows grains of highly polished sand; *lower right*, thin section of Clinton iron ore, with organic remains.

PLATE 21 — EVIDENCES OF DIASTROPHISM

Sedimentary rocks, known to have been deposited beneath ancient seas, are now found exposed in lofty mountains, attesting to great changes in land and sea level. *Above:* Majestic Mount Robson, British Columbia (photo courtesy Canadian National Railways), showing the sedimentary layers exposed along its wall. These beds contain evidences of ancient marine life, and must thus have been deposited beneath the seas. The mountain now towers more than 2 mi. above sea level. Other evidences, less spectacular, prove more recent changes. *Right:* Famed Temple of Jupiter Serapis, Pozzuoli, Italy, showing columns bearing bore holes of mollusks (text, chap. 16).

Another evidence of diastrophism is found in raised sea terraces; *left,* La Jolla, California; courtesy Spence Air Photos. Here older sea cliffs may be seen in background; a new sea cliff fringes the old terrace.

PLATE 22   ROCK FOLDS

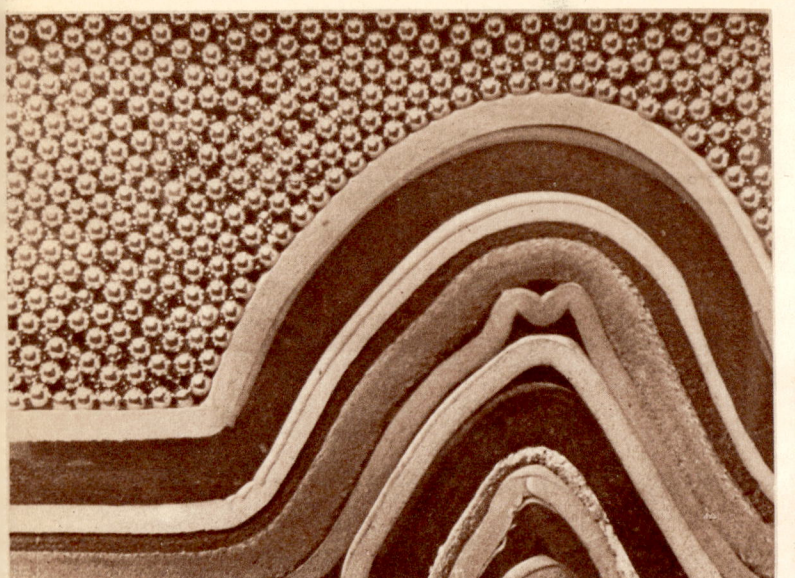

Beneath the earth's surface, in the deep-seated environment, rocks yield to diastrophic forces by folding. *Above left:* A pressure-box experiment shows how rocks fold. The steel bearings represent the weight of overlying rocks, and the sponge-rubber layers simulate an upfold or anticline. *Below:* Steeply dipping shales in western Virginia (photo by F. J. Pettijohn) show one limb of a large rock fold.

Below: Highly contorted and mashed metamorphic rocks result from the diastrophic adjustments that accompany rock folding. Note how intricately some of the layers have been folded (photo courtesy of National Park Service). During such metamorphism the original characteristics of the rocks and minerals may be greatly changed.

From a detailed study of rock folds it is possible to determine the structure of the bedrock beneath the earth's surface. Readings are made of the dip and strike of the beds, and these data are recorded on a map. The folds are reconstructed from the trends thus found. *Right:* Reading the strike of vertical quartzite beds near Ableman, Wisconsin.

PLATE 23   FAULTS

Left: The abrupt rise of the mountains above the plain (Grand Teton National Park; photo courtesy of National Park Service) is an evidence of the faulting that formed them. The truncated spurs of the range mark the remnants of the original fault plane. Such faults are the result of many small separate slips through a long period of time. The spurs may be seen in the photo.

Right: Diagram illustrating terms used in describing faults. The block at the left has moved downward; the distance along the fault plane is the *displacement*. The vertical distance is the *throw*, and the horizontal distance the *heave*. Often it is not possible to read entire story of faulting from two-dimensional view, such as a valley wall, owing to difficulty of determining the relations of the plane of observation to the plane of the fault.

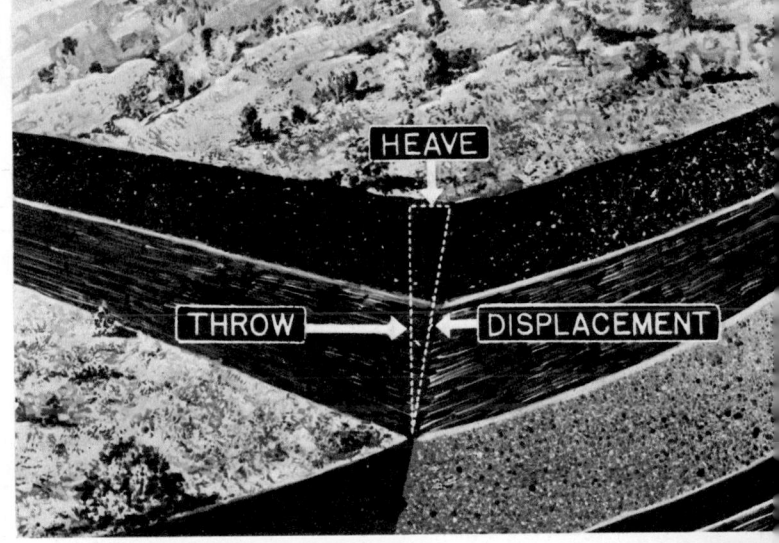

Left: Aerial view of the San Andreas Fault, California (photo courtesy of Spence Air Photos). It is perhaps the most widely known fault line in the world. Horizontal movement along it caused the San Francisco quake of 1906. The photograph admirably illustrates the abrupt topographic differences along the fault line, owing to the relative displacements on either side. This fault may be traced for many miles through California.

PLATE 24
EARTHQUAKES

Earthquakes are among the most dreaded of natural calamities, because of their destructive effects on houses and other works of man. *Upper right:* Shattered building at Helena, Montana, caused by the earthquakes of 1935 (photo by courtesy of J. Hough).

Seismographs are instruments that record earthquake waves. *Above:* Milne-Shaw seismograph at University of Chicago, showing upright arm and heavy mass (text, p. 152). The record, traced by a wavering light beam, indicates intensity and distance to source. *Upper left:* portion of such a record. Popular fancy may place cart before the horse, as appended clipping shows (*Chicago Tribune*, Nov. 3, 1935). *Lower left:* Adjusting relay that marks time on seismogram by extinguishing light once a minute.

Woman Blames 'Machine' for Quakes; Threatens Suit

Buffalo, N. Y., Nov. 2.—(P)—There's an indignant woman in Buffalo who thinks Canisius college's seismograph was the cause of yesterday's quake and she threatens suit for any damage in the next temblor. The woman phoned Joseph V. Schneider, Canisius treasurer. "The quake shook the footboard off my bed," she said. "Canisius has a lot of nerve starting earthquakes with that machine. You'll pay for the next damage you cause."

PLATE 25
SHOOTING STARS

Visitors from space are meteorites, stony or metallic masses that sear the atmosphere as they fall to the earth with high velocities. The upper insert illustrates a polished section of an iron meteorite (about natural size), showing the interlocking crystals that compose it. This meteorite is fancifully pointed toward Diabolo Canyon, Arizona (photo courtesy of Atchison, Topeka, and Santa Fe Railroad), among world's largest meteor scars (compare size of crater with large buildings on its floor). Ages ago a first-magnitude cataclysm must have accompanied the terrific impact of this meteorite on the desert plain.

Alert geologists and engineers, scenting vast iron and nickel deposit from this meteorite, have drilled holes in center, seeking to tap it. Thus far all efforts have failed, but magnetic surveys support the validity of the search. Angular impact would render successful venture a matter partly of chance.

The composition of these meteorites sheds some light on the probable composition of the earth's deep interior. Their high percentage of iron and nickel suggests the possibility that the central core of the earth may have a similar composition; certainly the high density of the earth's interior supports this view. (See text, chap. 19.)

Left: Small spatter cones of an ancient volcanic field (Craters of the Moon National Monument, Idaho; photo courtesy National Park Service) offer irrefutable evidence of former volcanic activity. The entire landscape is strewn with volcanic fragments.

PLATE 26
VOLCANOES AND LAVA

Above: Aerial view of active volcano El Veeyo, Central America (photo courtesy Pan American Air Lines), showing steam and gases emitted by the crater. The flanks of the cone show signs of erosion. The great Sierras, which stretch from Arctic to Antarctic, are marked by many such cones, especially in Central and South America.

Right: Impressive Mount St. Helens, Washington (photo courtesy of Northern Pacific Railroad), showing its symmetrical form, now carved by glaciers and streams. Well-developed volcanic cones commonly display such symmetry of form. The slope of the cone depends on whether it is composed of lava or of cinders or both.

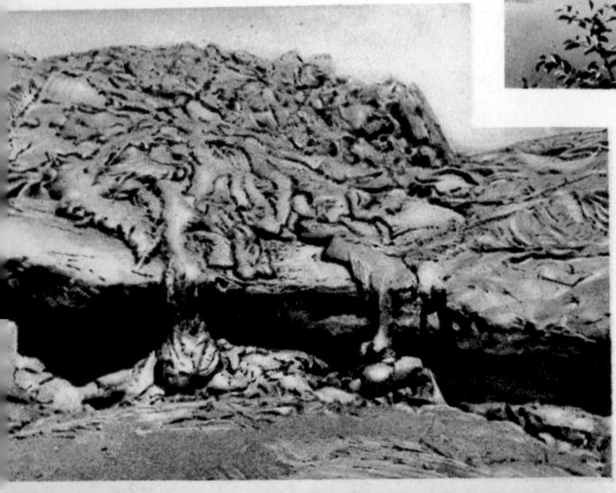

Lavas may be classed as two principal types, blocky, or clinkery, and smooth, or ropy. The former, highly gas-charged, presents a rough and tumbled appearance (*center left,* Craters of the Moon area; photo courtesy of National Park Service), in strong contrast to the smooth surfaces of ropy lava. The latter contains little gas, and solidifies as shown at *lower left* (lava from Kileauea, Hawaii; photo by R. T. Chamberlin).

PLATE 27 FIERY PITS

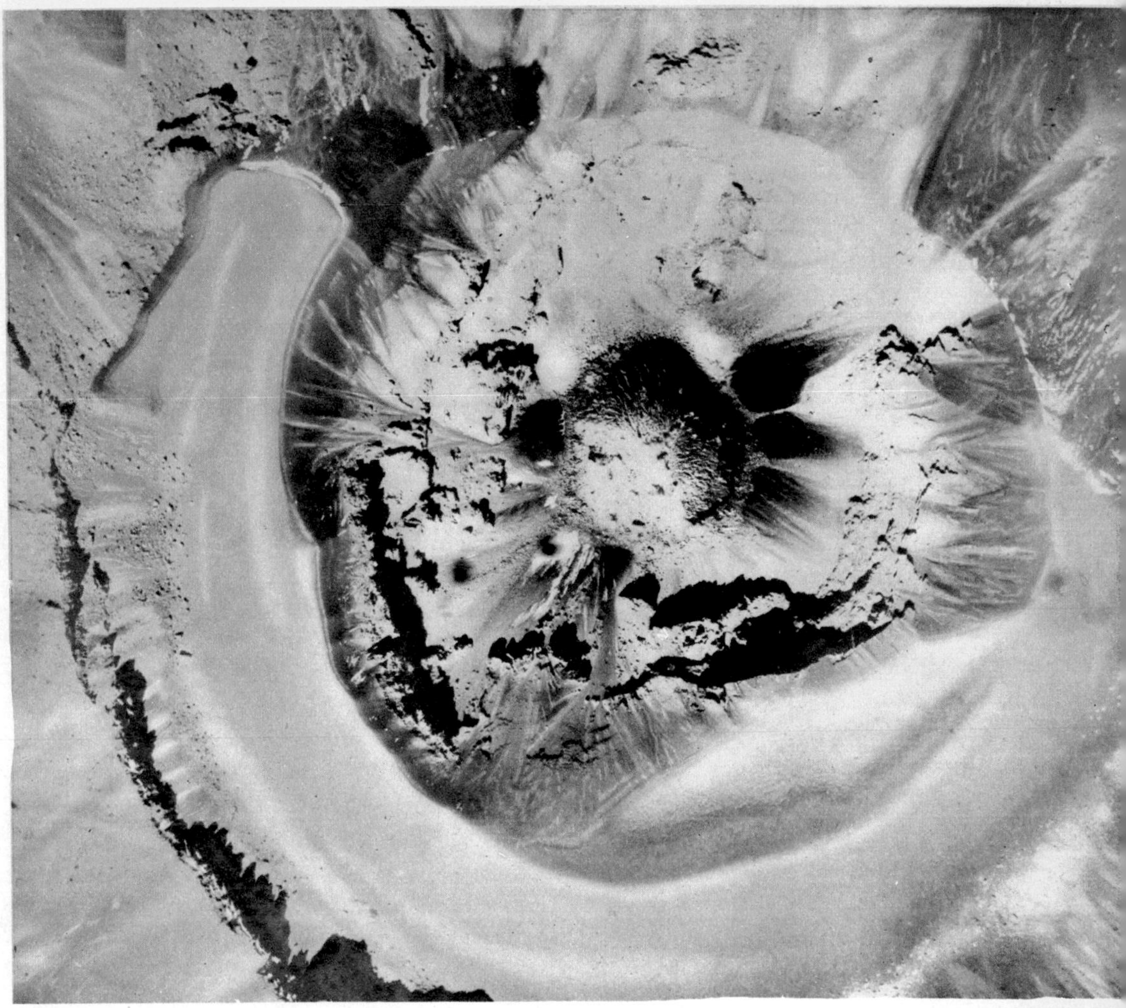

The photograph above, a vertical view into the crater of El Misti, Peru, was taken by Lt. G. R. Johnson (Shippee-Johnson Peruvian Expedition) from an elevation of 21,000 ft. This photograph was displayed in the Photographic Salon at the Chicago World's Fair and is reproduced here with the permission of the Museum of Science and Industry, Chicago.

Left: A purchased photograph of Mount Vesuvius (eruption of June 30, 1924) illustrates the inner cone of that famous volcano. The walls of the principal cone may be seen in the background; the present cone rises from a floor of hardened lava.

PLATE 28
INTRUSIONS

Sills *Right:* Yellowstone) lie parallel to surrounding structures. Dikes cut rocks, remain in relief as softer adjoining rocks are eroded. *Left center:* Vertical dike, Bulkley's Gate, B.C. (photo courtesy Canadian National Railways). *Right center:* Dike cuts granite, St. Cloud, Minn.

Major intrusions are giant batholiths, exposed to view by erosion. *Right:* The Sierra batholith, showing part of its roof as darker layers. (Center foreground shows rock débris moving downslope. Photo copyrighted by Spence Air Photos.)

PLATE 29
IGNEOUS ROCKS

Left: A "devil's post pile" of six-sided prisms. Dark lava developed this structure while cooling. Columns crack, fall, yield angular blocks in foreground. (Devils Post Pile National Monument, photo courtesy of National Park Service.)

Trimmed hand specimens (about two-thirds natural size) illustrate some attributes of the three main classes of igneous rocks. *Upper right:* A dark, fine-grained basalt, example of a rapidly cooled extrusive igneous rock, formed in the surface environment. Rock such as this often develops columnar structures while cooling.

The adjacent photo on the right is a typical granite, in which may be seen the crystals (light and dark patches) that resulted from slow cooling in the deep-seated environment, beneath protective coverings of other rocks. The basalt above it, while crystalline, is too fine-grained to show many details.

Left: Porphyry, an igneous rock that shows evidence of cooling in two environments. Under deep-seated conditions the larger crystals formed (light, feldspar; dark, ferromagnesians) in the still liquid magma; changed conditions drove this partly crystallized fluid to the surface environment, where it solidified rapidly into a fine-grained matrix around the larger crystals. Such porphyries may be formed from magmas of any chemical composition; the environmental conditions rather than the chemical composition determine the texture.

PLATE 30
CLUES FROM THE PAST

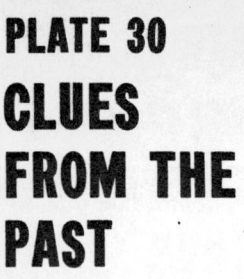
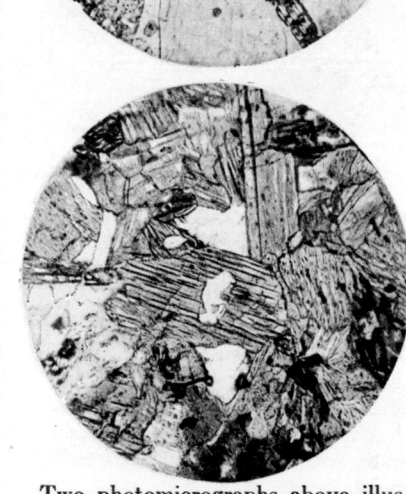
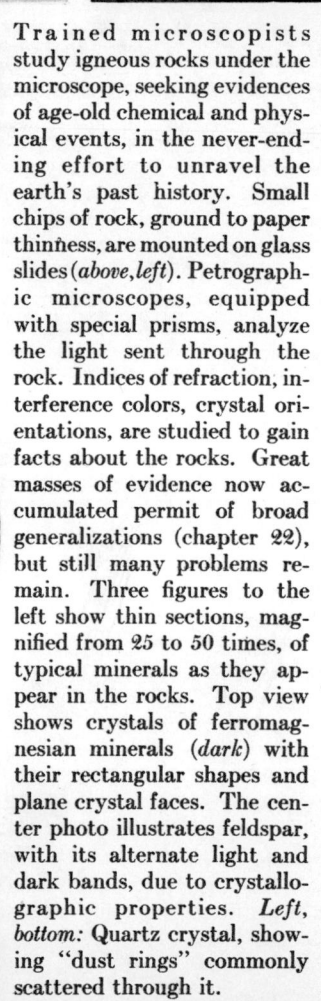

Trained microscopists study igneous rocks under the microscope, seeking evidences of age-old chemical and physical events, in the never-ending effort to unravel the earth's past history. Small chips of rock, ground to paper thinness, are mounted on glass slides (*above, left*). Petrographic microscopes, equipped with special prisms, analyze the light sent through the rock. Indices of refraction, interference colors, crystal orientations, are studied to gain facts about the rocks. Great masses of evidence now accumulated permit of broad generalizations (chapter 22), but still many problems remain. Three figures to the left show thin sections, magnified from 25 to 50 times, of typical minerals as they appear in the rocks. Top view shows crystals of ferromagnesian minerals (*dark*) with their rectangular shapes and plane crystal faces. The center photo illustrates feldspar, with its alternate light and dark bands, due to crystallographic properties. *Left, bottom:* Quartz crystal, showing "dust rings" commonly scattered through it.

Two photomicrographs above illustrate order of crystallization in igneous rocks. *Top*, outlines of earliest crystals (accessory minerals and dark ferromagnesians), imbedded in light feldspar, which crystallized around the earlier forms. *Bottom*, quartz (triangular patch) filling interstices among earlier minerals, proving its late crystallization. (These photos by M. E. Marsh, courtesy A. Johannsen.)

PLATE 31
LIGHT MICA AND METAMORPHIC ROCKS

Above: Thin sheet of light mica, showing its transparency. At right a darker crystal of the same mineral. Micas are complex hydrated silicates of potassium, aluminum (text, p. 202). The light micas find use in insulation and are an important constituent of the metamorphic rocks.

The hand holds a specimen of gneiss (metamorphosed granite), showing its parallel bands due to mineral alignment under diastrophic forces. The photomicrographs illustrate magnified views of thin sections of slate (*left*), mica schist (*center*), marble (*upper right*), and quartzite (*right center*). Marble is composed of recrystallized calcite; the quartzite shows rounded grains that composed the metamorphosed sandstone. The mica schist (*center*) illustrates the extremes to which metamorphism may proceed. The original rock minerals have been almost entirely altered to light mica and other metamorphic minerals. The crystals are predominantly oriented in a given direction (*toward upper left*), approximately normal to the direction of the diastrophic forces.

PLATE 32
RECEDING WATERS

This excellent photograph by W. O. Yates was displayed in the Photographic Salon at the Chicago World's Fair and is reproduced here with the permission of the Museum of Science and Industry, Chicago.

In many ways the photo epitomizes aspects of physical geology. We see the ever restless ocean in contact with relatively inert land, illustrating the process of gradation as the work of the earth's fluid films on the solid lithosphere. The ripples too, illustrate an equilibrium surface between water and sand, in response to conditions requiring a minimum of friction.

CHAPTER 24

THE EVOLUTION OF ROCKS

> And found no end, in wand'ring mazes lost.
> —MILTON

THE concept of cyclical changes in rocks has been implied throughout our previous discussion, but it deserves to be made more explicit. For the inorganic world, much like the organic world, is dynamic, and moves through cycle after cycle of activity. It would have been quite as logical for us to have started our geological story with the details of the igneous rocks, say, and eventually to have returned to them. Rock-weathering formed a more suitable topic for our present organization, but the point is not important. The important thing is that all rock processes are aspects of a great **rock cycle**—a scheme of evolution of the rocks which helps to clarify our ideas of unity in the inorganic world.

Cycles and epicycles....

WE HAVE made numerous references to the approaches toward equilibrium on the part of natural phenomena. It is this tendency toward balance in nature that largely determines the course of events in a particular situation. The tendency is itself an aspect of the ebb and flow of energy on and through the earth, however, so that in the final analysis it is the physical setup of the earth that forms the background on which the details of the picture are etched.

The tendency toward equilibrium seems to be inherent in all nature. It represents an adjustment to environmental factors—a contest between energy and matter, in which the energy usually wins, but is itself often degraded to useless forms in the struggle.

THE concept of equilibrium deserves further discussion. In chemistry or physics we may experiment with *closed systems*, in which certain factors only are allowed to enter the situation. Thus we may enclose liquid in a vessel free from air, allowing some space to remain above the liquid. If the temperature is held constant, a definite point of equilibrium is reached between the liquid and its vapor. At the start of the experiment the more rapidly moving molecules escape from the liquid and as they increase in number, some of them will return to the liquid. The stage will finally be reached, accordingly, in which as many molecules enter the liquid as leave it; and at this point equilibrium is established. As long as conditions remain precisely the same, this state of equilibrium persists.

A liquid and its vapor. *Above*, the start of the experiment; *below*, at equilibrium

For any given temperature there is a definite equilibrium point in our experiment. If, however, we allow the temperature to change slowly but constantly during the experiment, the equilibrium point will be continually shifted. The liquid and its vapor will then be always approaching, but never quite reaching, equilibrium. We cannot, in the latter case, speak of any single equilibrium point; but the conditions may be so controlled that the liquid and vapor are always nearly at equilibrium, even though the exact point itself continually shifts in position.

IN NATURE we find fixed equilibrium points only rarely; much more often they shift because of changes in environmental conditions. The reason for this continual change is that the earth is a dynamic body, across and through which there is a constant flow of energy. This energy may be present as blowing winds, as the heat in volcanism, and so on. Never is the ebb and flow about a given point precisely the same for any length of time, and hence whatever materials may there be striving toward equilibrium never quite reach equilibrium. That is one of the reasons why direct experimentation is of limited application in geology. However, true equilibrium may frequently be approximated, and such situations can profitably be studied by direct experiment. We refer here to experiments on molten minerals, which have contributed much to our better understanding of the igneous rocks: a topic, unfortunately, that is somewhat beyond the scope of our present work.

WHAT approximate equilibrium points may exist in nature are determined largely by the environment. We have used the term *environment* to mean the sum total of external conditions. In general these may be divided into physical and chemical factors. The physical factors are mainly temperature and pressure; and the chemical factors are the other minerals, the gases, and the liquids with which the rock is in contact. In some cases pressure may be the main factor; in others it may be the temperature or the chemical compounds round about. We have already had examples of many of these. Most commonly, all the factors are involved; but their exact effects are difficult to determine. For example, pressure may be the predominating factor at first; and as the rock, say, strives to adjust itself to it, the temperature rises owing to compression. Here the equilibrium point shifts because of the increasing effect of heat; but as the temperature rises, the tendency toward chemical reaction may increase, so that still another factor may enter that was not active at the start. When one reflects on the complexities that are thus possible, it is no wonder that simple mathematical expressions have not been found to explain the situations that arise in geology.

SUCH is the nature of man, however, that he is challenged by these very complications; and despite the tangled skein that he holds in his hands, he is able here and there to unravel an end, and eventually some of the worse snarls are reduced to smaller knots and twists. These may then be turned over to other specialists, each to occupy himself with his own particular tangle. There is much that remains without answer, but we have reached the point where some generalization is possible. One of these generalizations is the rock cycle.

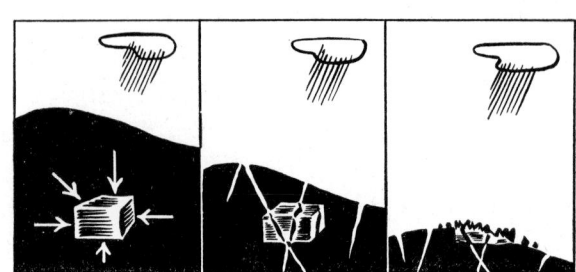

A hypothetical cube of rock in the deep seated environment finally becomes exposed to weathering agents, as the land surface is worn down by erosion

The outstanding thing about the rock cycle is that it at once ties together many apparently unrelated phenomena. By its means we may obtain a single comprehensive view of the three geological processes, and at the same time we may summarize and review the entire picture. Suppose we start the discussion with a deeply buried rock. The environment is one of relatively high temperature and pressure, and active chemical agents may be present. Whatever the final product of the environment is—be it a metamorphic rock, an indurated sediment, or a coarsely crystalline igneous rock—it will be stable as long as these conditions hold. But in geology the conditions may not hold for (geologically speaking) long. Suppose that erosion wears down the surface above, so that the overlying burden of rocks is gradually reduced. This affects the pressure and the temperature of the immediate environment, and induces in the rock a tendency toward change. There may be reasonably wide limits in the conditions before definite changes occur; but as the surface is further reduced by erosion, new factors enter the environment. Lowered pressures allow joints and fractures to open in the rock, and ground water begins to circulate through them. Dissolved atmospheric gases or other chemicals then begin a series of physical and chemical changes that mark the incipient stages of rock-weathering.

As erosion goes on, the rock itself reaches the surface. It may become completely weathered, and the débris be swept away by

THE EVOLUTION OF ROCKS

wind, water, and ice. During the removal of the débris, a whole series of erosional land forms is developed, depending on the agents of transportation. Here is a tie-in with the physiographic aspects of the subject. While the material is being shifted from its original site to the seas, there may be built up depositional land forms also; and during transportation the material is largely sorted out on the basis of size and other attributes. These sorted sediments finally come to rest as the stratified sedimentary rocks. As time goes on, these sediments become buried beneath later deposits and again approach environments of high pressures and high temperatures. The

Our weathered cube of rock is transported from its original site, deposited, and buried by later sediments. Thus it returns finally to the deep-seated environment

buried rocks are compacted, the pores are closed, mineralogic changes are produced, and the rock is once again in adjustment with the deep-seated environment.

THIS essentially simple scheme may be subject to any number of variations. Diastrophic forces may operate during a given cycle, so that the rocks may also pass through a metamorphic stage. Igneous intrusions may introduce other situations, and it is even possible that magma itself may originate from the fusion of some of the rocks. The ultimate sources of magmas are not certain yet, but there is a strong likelihood that much of it comes from relatively deep within the earth. However, it is also likely that some magma actually arises from the melting of other rocks not so very far beneath the surface. We may recall a possible clue to these cases in the great batholiths that form within the cores of mountain ranges. If the rocks fuse during the later stages of folding, under partially eased pressures, they enter the liquid state and thus become magmas, which may then penetrate among the surrounding solid rocks as intrusive bodies. Obviously, any kind of rock may be involved in such fusions, and hence it becomes a part of the general rock cycle.

It is often more convenient to represent our ideas graphically, and the appended figure illustrates the generalized rock cycle. Here

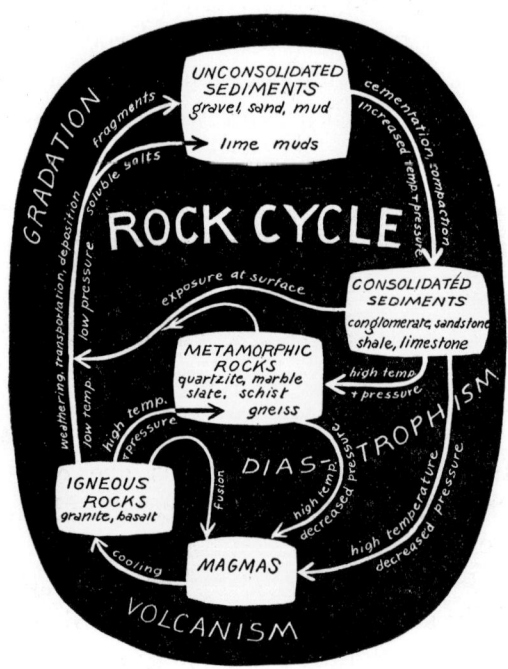

The generalized rock cycle

are the evolutionary paths that may be followed by any rocks; since the cycles are closed curves, it follows that no rock is ever a completely final product, unless the globe itself becomes lifeless and inert. As long as the earth is a dynamical body, these changes in the rocks will continue. The earth will be dynamic as long as present sources of energy remain. It is thus these sources of energy, and the ebb and flow of the energy over and through the earth, that set up the conditions for the rock cycle and all its attendant phenomena. Furthermore, the closed curves followed by the rocks during their cycles is in direct accordance with the concept of the indestructibility of matter. Forms and properties may change, but matter always remains.

ASSOCIATED with this rock cycle, and intimately bound up with the ebb and flow of energy on the earth, are numerous other cycles, some of which we have touched upon during the course of our story. The **physiographic cycle,** for example, deals with the land forms that accompany geologic changes, and these land forms are associated not only with gradation but also with diastrophism and volcanism. We need here only mention mountains and volcanoes as land forms to conjure up details in the reader's mind.

Other cycles were less formally mentioned by us, but many could have been considered. There is a **water cycle,** for instance, that starts with the oceans, passes through the water-vapor stage, and returns to earth as rain and snow. During its movement back to the oceans it produces most of the gradational effects that we have considered. Then, too, there is a **carbon cycle** on this earth, which co-ordinates the biological world with our geologic processes. Plants take carbon dioxide from the air and store the carbon in their tissues. Some of this is converted to coal and thus enters the inorganic world. Rock-weathering withdraws carbon dioxide from the air and locks it within the carbonates. Some of these are carried to the seas

The water cycle

as soluble salts and are there used by marine animals to build their shells. Eventually limestone is formed when the animals die and leave their shells behind. A fruitful and fascinating field for study, this subject of cycles in nature.

WE HINTED, at the beginning of the chapter, that in some cases the energy involved in all these cyclical changes is degraded to useless forms. By "useless forms" we mean energy radiations of long wave-length which escape from the earth and thus play no further rôle in its processes. The reader will do well to refer once again to the end of chapter 4 and reconsider the energy diagram there drawn. It will be seen that the energy passes through the earth's economy and is gone, for the paths of energy are unidirectional—only matter displays the cycles.

CHAPTER 25

ANCIENT HISTORY

> Anything but history, for history must be false.
> —Sir Robert Walpole

THERE are numerous quotations to the effect that history is, after all, but fiction agreed upon. Certain it is that the prosecution and the defense seldom get together on their stories, nor have victors and vanquished ever told quite the same tale as to how it all happened. These are natural differences of opinion. But when we recall that no two supposedly disinterested newshawks ever entirely agree on their stories of famous trials, and remember that two neutral war correspondents can give such divergent accounts of the same battle that their readers feel sure they are talking about different engagements, we see that Walpole's dictum is no mere cynicism.

Moreover, the farther back one goes in human history the more inaccurate it becomes, until at last even the ablest historical sleuth becomes lost in a maze of contradictory accounts, bewildering legends, and mythical stories. Although the historian ordinarily uses libraries in his research work, he also realizes the importance of cultural data in recasting some of the earlier chapters in his history. For it must be apparent to all that a fairly well-preserved statue or monument may throw as much

Historians all

light on the period in which it was made as contemporary fragmentary writings.

Finally, the historian, in delving backward into the past, comes to a point in his researches beyond which even the crudest type of written material can no longer be discovered. But there are still older archeological data which can be used in unraveling the past; and when these at last become rare, or difficult to interpret, there always remain the geological evidences which make it possible to read the most ancient history of them all.

SINCE Mother Earth is the only unemotional historical scribe, it follows that the earth historian should be able to construct the one history which would in no sense be false, could he but read the earth account accurately. But here again the more ancient the story the more difficult it is to interpret the record. And, although there is now little controversy among the earth historians regarding the facts of earth history, there is still room for a great deal of difference of opinion regarding the meaning of the data.

Mother Earth. The unemotional scribe

The modern historian is interested in the past because, in a large measure, it enables him to interpret the present and prognosticate the future. The earth historian, on the other hand, studies the present in order that he may be able to decipher the past. Consequently, with this in view we have studied, in the first part of this volume, the present-day geological processes. The knowledge thus gained should give us the proper background to understand how the geological record has been read. In a sense, then, each topic we have studied heretofore has been not only interesting and important in its own right but also indispensable in the attempt we shall now make to read the riddle of the ancient past.

THE beginning of earth history, of course, goes back to this planet's prenatal days, when it was a minor part of its parent, the sun. The birth and early stages of the earth are therefore sub-

ANCIENT HISTORY

jects with a rich astronomic background. Nevertheless, the early events in earth history fall within the domain of earth science proper, and constitute a division of geology (and geological time as well) known as *Azoic*. Conversely, at the other end of the time scale, recent geological history overlaps the fields of archeology and anthropology, and involves the recent, or *Psychozoic*, era of mental dominance.

THE amount of time involved in this history is so vast (as we shall see in chapter 32) that it is not surprising that there are many important gaps in the record. In fact, in view of the millions of years involved, merely in some chapters of the book of earth, it is remarkable that certain parts of the story are so completely recorded. The entire earth history has been reconstructed by employing the following general procedures:

1. **By studying the igneous rock masses, both intrusive and extrusive.**—Such investigations usually make it possible to determine the *relative* ages of associated igneous structures, that is, to ascertain, for instance, whether an igneous mass, *a*, is younger than both masses *b* and *c*, or merely younger than *b* but older than *c*. The relative age of the  associated sedimentary rocks also is commonly determined. These are points we have touched upon in chapter 21, and we will have occasion to refer to them again in chapter 27.

2. **By the study of metamorphic rocks.**—These studies often make it possible to determine the original character of the unaltered rock, that is, they may indicate whether the metamorphic rock was once igneous or whether it was originally sedimentary; and they may make it possible to determine the relative age of the metamorphism in relation to the associated, but unaltered, sediments or igneous rocks.

3. **By examining the faults and folds which deform the rocks of the earth's surface.**—Such structures commonly can be dated, so that the relative age of the faulting and folding can be ascertained.

4. **By studying the evidences of erosion, by whatever agency, on the ancient rocks, regardless of type.**—Such studies give certain evidence of the ancient oscillations of land and sea. The "breaks" recording the ancient periods of erosion we call **unconformities.** In those cases in which there is little or no evidence of folding in the beds below the break, this erosional phenomenon is called a *disconformity.* If the lower beds are folded, the resultant structure is designated an *angular unconformity.*

5. **By examining the sedimentary beds of all different types and by studying their varying thicknesses and distributions.**— Such studies usually yield information concerning the ancient configuration of continents, since the detailed characteristics of the rocks, such as ripple marks, cross bedding, and the like, commonly show whether the sediments were deposited on land, near shore, or in the deeper waters of the sea.

6. **By studying the organic remains entombed in the sediments.**—These fossils tell the story of the ever changing life of the past, and enable the expert to date ancient geological events with considerable precision, as we shall see in chapter 27.

SUPPOSE we point out briefly the significance of these methods of studying the rocks. The accompanying diagram of a canyon wall portrays a number of rocks marked by a variety of geological structures. The black bands are igneous dikes and sills, and the other rocks are sediments. Now, if we examine the sediments, we see that they occur in two series—the lower folded set and the upper horizontal set. Between the two is a horizontal line along which the sediments above and below do not "jibe," that is, there is a lack of conformity between the two sets of beds. This, then, is obviously one of the angular types of unconformity which we have just mentioned briefly. It marks a very important break between our two series of rocks, for it is obvious that an entire sequence of events is recorded by it. The lower set of sediments must have been deposited, then folded, and finally eroded to a peneplain. Once more must submergence have occurred, with the subsequent deposition of the overlying sediments. Finally, the entire area was up-

ANCIENT HISTORY

lifted, so that recent erosion could cut the present canyon, which exposes these rocks to our sight.

That single line marking the angular unconformity between the two sedimentary series is extremely important, for it tells us that a long interval of time elapsed between the folding of the lower set of beds and the deposition of the upper series. Peneplains, as we recall from chapter 8, are not developed overnight, so that here is certain evidence of an interval of time great enough to wear down the folded beds essentially to sea level before the upper beds were deposited.

Nor is this all the canyon wall tells us. Notice that all the igneous bodies penetrate right through the lower set of beds, and that some of them even continue across the younger sedimentary strata to the present surface of the land. This can only mean that the igneous intrusions took place after all these sediments were formed, and thus igneous activity was one of the latest geological events 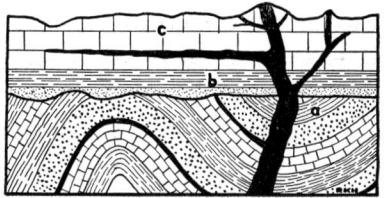 that occurred in this region. Of course, we cannot be sure from what we have thus far learned from the canyon wall how long the area has been undergoing erosion. Nor have we determined anything but relative ages of the various rocks studied. If, however, the sedimentary beds a, b, and c contain fossils, we will be able to reconstruct the geological history of the canyon area and fill in the story with relation to the general geological calendar.

OBVIOUSLY, the professional geologist is always busy reading the history of some region or other. But rarely has Earth made his work as simple as we have made it appear in this foreshortened outline. The successful field geologist requires a profound knowledge of all the principles of physical geology discussed in chapters 1–24. Further, in working out any relatively complete sequence of earth history it would be ideal if he could employ every ramification of all of the methods briefly discussed in the foregoing account. In actual practice, the unraveling of the tangled threads of the earth's ancient history is carried on by experts in each of the general methods. Thus, the complete story is the result of a synthesis of information which has accrued from the labors of many specialists working in every country of the globe. It is obvious that, as new information constantly comes to light, the details of our ancient history are modified; but it is unlikely that our interpretation of its major episodes, as we now know them, will be greatly altered by future investigations.

OUT of such studies there has slowly evolved the great geological calendar universally used by geologists today. The methods of determining the actual number of years that must be allotted to the major divisions of the calendar, and the calendar itself, are discussed in chapter 32. But before we are ready to study this geological time-table we must give some consideration to a number of subjects which either make possible the construction of the table or are important in the philosophy of its growth and use.

THE first of these topics properly concerns the study of fossils, the remains of the animals and plants of the past, for without the information obtained from this study the geological calendar could never have been constructed in its present form.

CHAPTER 26

UNIVERSAL CEMETERY

> The Earth is a vast cemetery—
> The rocks are tombstones on which the buried dead
>   have written their own epitaphs.
>
> —Louis Agassiz

THOSE who remember their Shakespeare may recall that in *King Henry IV* Owen Glendower boasted, "I can call spirits from the vasty deep." To which Hotspur, with proper scientific doubt, replied, "Why so can I, or so can any man, but will they come when you do call them?" The geologist is always conjuring up the life of the vasty past; and, despite the doubts of a multitude of modern Hotspurs, many of these ancient spirits do come when they are called.

The geologist is able to visualize such a complete picture of the forerunners of the modern animals and plants only because through-

The buried dead write their own epitaphs

out the past (as well as at present) everything in nature was busy writing its own history. It is true that only a small percentage of the once living things has written indelibly; further, it must be obvious that much of the written record that is available either has not yet been found or, if discovered, has not yet been deciphered. In addition there are doubtless a multitude of fascinating epitaphs which might be read were they not inscribed on rocks so deeply buried, either beneath the sea or under other rocks, that man will never be able to find them. Finally, the various agents of erosion have destroyed many of the tombstones, and with them their inscriptions.

FOSSILS, chief tools of the historical geologist, are the records of prehistoric life. They tell the story of the animals and plants ancestral to our modern faunas and floras. The word itself is derived from the Latin *fossilis*, which is in turn a derivative of *fodere*, "to dig." Consequently, in older works the word meant literally anything curious which had been dug up. Agricola, whom many suppose to have invented the word "fossil," and who wrote *De natura fossilium* (1541), was much more concerned with minerals than with organic remains. But a few years later (1565) Conrad Gesner, a Swiss of Zurich, published the first account of "fossils" essentially in the modern sense, and with illustrative plates which were the crude prototypes of present-day reproductions.

IT IS difficult to say when man first became interested in fossils, but doubtless the more curious had noticed them even before the dawn of written history. They have been found associated with cave-man remains in such a way that it seems likely that the objects were of some vague significance to our early ancestors. They have been discovered in ancient Greek temples, and in some cases they have been cemented into the walls of medieval structures such as French cathedrals. Some of the American aborigines also regarded fossils with veneration. To the Tusayan Indians of Arizona, a certain fossil (cephalopod) shell was a fetish highly revered by the medicine man. The Blackfoot Indians of Alberta even used another fossil, an iridescent and mysteriously marked cephalopod, in their hunting. A specimen was set up in front of a tepee and permitted to fall. As it came to rest, it pointed, so the Indians thought, toward the best area for game.

DISREGARDING some of these crude views concerning the meaning and value of fossils, a number of which are still entertained by the uninitiated of all lands, it is possible to construct a synoptic history of the various stages in the development of the study of organic remains. Such a history is readily divisible into four great eras of unequal length and importance:

    I. Ancient, *circa* 600 B.C. to 79 A.D.
   II. Medieval, 79 A.D. to 1758
  III. Renaissance, 1758 to 1859
  IV. Modern, 1859 to the present

The Ancient period was marked by relatively correct speculation on the part of a few Greek and Roman philosophers and Egyptian priests. The latter were familiar with fossil shells, and correctly reasoned not only that they had had a marine origin but that they indicated that much of Egypt had previously been beneath the sea. Anaximander (born about 611 B.C.) believed that human beings were first fishlike in form, and apparently he had some vague evolutionary concepts. In fact, both Cuvier and St. Hilaire centuries later ridiculed Lamarck because of his alleged following of the "silly transformation ideas of an ancient Greek." Xenophanes of Colophon, a contemporary of Anaximander, also pointed out that the organic remains found high on Pharos and Malta demonstrated that these islands had once been engulfed in the Mediterranean.

In contrast to these semi-enlightened ideas, most of the early philosophers of this period were given to abstract speculations. Even Aristotle (384–322 B.C.), who is regarded as the "Father of Natural History," believed that the earth was the center of the universe and that its development was comparable to that of an organism with periods of growth, maturity, and decay. Apparently from observing fossil shells in the rocks he reasoned that the lower animals originated in the muds of the earth and that they gave rise to the higher animals by sexual generation. Furthermore, he thought that all animals were related to one another through numerous transitional forms. Thus Aristotle had some glimmering of fundamental facts mixed in with his misconceptions; and, compared to his predecessor, Pythagoras (582–507 B.C.) of earth, air, fire, and water fame, he was accurate, indeed, in his reasoning. Aristotle's most famous student, Theophrastus (362–284 B.C.), actually wrote a treatise on fossils with which Pliny was familiar, though it has subsequently been lost. Theophrastus believed that there was a plastic force in the earth by which imitative forms (fossils) are produced, but that in reality these objects are just as inorganic as the plantlike forms made by frost.

Several centuries later, Strabo (54 B.C.—25 A.D.), the well-known geographer, noticed the great numbers of coinlike single-celled foraminiferal fossils in the rocks of the pyramids and in the débris around them. He says that legend has it that these objects are the petrified lentils dropped from the lunches of the workmen who made the pyramids. Scoffing at such a story, he is yet unable to offer a better one.

Pliny, the Elder (23–79 A.D.), who met his death because his curiosity got the better of him during the great eruption of Vesuvius, brings the first period of this brief history to a close. In his monumental thirty-seven-book *Historia Naturalis* he summarizes the concepts of natural sciences prevalent during his time, and leaves the subject of fossils in a state little better than it was in Aristotle's day.

THE second, or Medieval, period in the study of fossils was marked by a great deal of speculation and wild discussion mostly directed toward harmonizing the fossil record with biblical accounts. One notion which met with widespread approval was that fossils were simply the remains of organisms which perished in the Noachian deluge; but many held that they were not organic at all, but rather the result of a sort of fermentation within the rocks. Even the reasonably intelligent and ardent fossil-collector, Professor Johannes Beringer, of Würzburg, knew so little about these remains that he was duped by his students into describing as bona fide fossils all sorts of curious objects which they themselves had made and "planted" in the professor's favorite collecting spots. His *Lithographia würceburgensis*, published in 1726, contained the figures of many of the spurious fossils. Beringer a little later discovered the hoax himself when his tormentors overplayed their hand and planted the professor's name in the rocks. This proved too much of a strain even for Beringer's great credulity, for, when he found this crowning bit of humbuggery, he realized at last his earlier mistakes. In a desperate attempt to save his reputation he spent a fortune trying to buy up his publication. That he had good success is well indicated by the fact that his curious volume is today a collector's item of considerable value. But some

.... duped by his students ....

UNIVERSAL CEMETERY

of the actual faked fossil specimens still can be seen in the museum at Würzburg. And a crowning dénouement of this sad story came when, after Beringer's death, his family had his *Lithographia würceburgensis* reprinted as curio, to their great good fortune.

But throughout a time in which religious dogma made for less generally intelligent discussion of the topic of fossils than had been prevalent in the ancient period, there were, nevertheless, a few keen minds puzzling over the subject. Of these, three are sufficiently outstanding to merit individual mention. In addition to Gesner and Agricola, whose works we discussed earlier, Leonardo da Vinci (1452–1519), the Florentine of universal genius, as a by-product of his work in the construction of canals, had studied fossils, knew them for what they were, and appreciated their significance. Fracastoro (1483–1553), a contemporary of Da Vinci's, and a professor of philosophy at Padua, is perhaps best known for his distinguished work in medicine. But Fracastoro also was a profound student of geology, and as early as 1517 published some essentially modern views on fossils. Steno (1631–87) was a Dane who lived much of his life at Florence under the sponsorship of the De Medicis, whose Grand Duke Cosimo III he educated. Many of his enlightened views on fossils, and on geology in general, are almost embarrassingly similar to those presented in this book. He it was who first showed that a mere part of a fossil animal, such as a shark's tooth, was sufficient evidence for the former existence of the entire creature, and that the fossil tooth could be compared directly with a tooth in the jaw of a modern shark taken from Mediterranean waters.

The recent fossiliferous deposits of Italy contained the remains of animals so little changed from the ones these Italian investigators saw round about them that it is no wonder that they hit upon the correct idea that there had been a continuity of organic strains from the past to the present. On the other hand, those English writers of the period who had any correct notion as to fossils at all believed that they all represented life which had become extinct. Nor is this surprising, in view of the fact that most of the English fossils are very old, and consequently have a quite different appearance from most of their living descendants.

THE third period was made possible when Linnaeus devised the so-called "binomial system" of classification. Prior to 1735, when the first edition of his *Systema naturae* was published, only a relatively few animals and plants had been described; and these had been given either common names or long, unwieldy descriptive designations. But, since the time of the tenth edition of Linnaeus' famous work (for convenience considered as January 1, 1758), all animals and plants have been given two scientific names. The first of these, generally of Greek derivation, indicates the group, or genus, to which the organism belongs; the second, ordinarily of Latin origin, gives the species, that is, it points out with which of the one or more naturally breeding-strains making up the genus it is classified. For instance, *Homo sapiens* is the scientific or binomial name used to indicate the present species of man. Other species, such as *Homo heidelbergensis*, or the Heidelberg man, and *Homo neanderthalensis*, the Neanderthal man, have existed but are now extinct. Likewise other genera of ancient or fossil man are known, such as *Pithecanthropus*, the "ape man," and *Eoanthropus*, the "dawn man." All of these genera of man belong to the family Hominidae, in contradistinction to the family Simiidae, to which belong the genera of the so-called "great apes." But both of these families are classed under the order of Primates, one of the many still more comprehensive groups of the class Mammalia. The fish and the reptiles make up other classes of the great vertebrate phylum, which, in turn, is only one of the branches of the animal kingdom.

ONCE this relatively simple plan of Linnaeus' was put into use, there was a convenient pigeonhole for every organism, living or fossil. Naturally, from the first "authorities" have disagreed as to the validity of certain parts of the classification, but nevertheless the scheme has permitted them to speak a sort of universal language; and with this new tool the routine description of ancient faunas and floras went on apace.

OF THE many important workers during this third period in the history of the study of fossils we have space to mention only three, namely, Lamarck, Cuvier, and Smith. But these three are commonly regarded, respectively, as the Father of Invertebrate Paleontology, of Vertebrate Paleontology, and of **Stratigraphy, the study, chiefly of the time equivalency, of the bedded or stratified rocks. Paleontology, the science of ancient life**, received its very name in this third period (1834), and somewhat earlier, *circa* 1800, the first real geological maps were made, a point which we will discuss in a later chapter. In their construction William Smith, facetiously dubbed "Strata" because of his preoccupation with the rocks, made use of a certain principle in paleontology, namely, that **the fossils of a certain bed of rock are diagnostic of it and of the time in which it was formed.** Hence these ancient organic remains may be used to identify equivalent strata in whatever country they may be found.

The fourth, or Modern, period in the study of fossils began with the publication of Darwin's *Origin of Species*. Darwin's ideas gave a new significance to the entire subject of paleontology. No longer were the ancient faunas and floras merely described and catalogued as interesting records of the past. *The geologist now became interested in them not because they were old and dead but rather because they once were alive.* Scientists began to see that, after all, the only *direct* evidence for evolution was to be found in the fossils themselves. Entire groups were studied to determine their phylogeny, or evolutionary history; and the results were carefully checked against embryological studies of the living end-products of the supposed evolution. Such investigations are still being pursued with satisfactory results; and, although the full meaning of life is not yet understood, paradoxically enough, the study of fossils, or dead things, has brought us much closer to the answer to Nature's greatest riddle.

J. B. de Lamarck, The Father of Invertebrate Paleontology

WE HAVE described this historical development in some detail not only because it is interesting and significant but because, in so doing, it has been possible to bring out relatively painlessly a good many paleontological principles. We must now consider briefly a number of additional facts concerning fossils and their types of formation, for these data are essential even to an elementary knowledge of the subject.

IN ORDINARY cases **an organism, in order to become a fossil, must be buried at or soon after death; and it must possess some hard parts.** The media of burial are many and varied, such substances as resin, volcanic ash, loess, dust, peat, ice, and the common sediments such as sands, clays, and lime muds being the more usual ones. The actual sites of burial are obviously just those areas in which deposition ordinarily takes place. Consequently, one might expect that the best fossil record should be found in those common sediments which throughout the ages have been deposited near the shores of the seas. And this we actually find to be the case. The marine invertebrates, which have always swarmed in shallow coastal waters, live and die in a habitat in which the chances for burial at death are very good. Moreover, a great many of them secrete hard parts, usually in the form of an external skeleton. Thus it happens that the lowly invertebrates have written the most legible epitaphs, constitute the bulk of the fossil record, and are of the greatest use in developing the complete story of geological history.

CREATURES that are terrestrial in their habitat usually live in areas of erosion. Therefore, even if they possess hard parts, their bodies at death are easy prey for the larger scavengers and bacteria. The skeletal parts also are commonly widely scattered, weathered, and destroyed. For these reasons the ancient plant and animal life of the continents is not so well preserved as we should like. But in every area of continental deposition, the land life may be preserved, as, for instance, in swamps and bogs. Falls of volcanic ash also may engulf entire representative faunas; dust storms may actually bury complete herds of mammals; and some animals and plants have even been preserved in ice or in oil. Thus, although the continental life of the past is not nearly so well represented as the life of the oceans, some of the most interesting and perfectly preserved of all the fossils are those of land forms.

So far as the requisite of hard parts is concerned, it usually is about as important as rapid burial in some protective medium. But there are just enough instances of the preservation of entirely soft-bodied creatures that from these rare occurrences we can get some idea of how much of this sort of life of the past must not be represented at all in the fossil record.

Entirely irrespective of zoölogical affinities, it is possible to classify the various kinds of fossils as follows:

I. Actual remains of organisms
   a) Unaltered
      1. Soft parts preserved
      2. Hard parts preserved
   b) Altered or petrified
      1. Calcified
      2. Silicified
      3. Carbonized
      4. Pyritized

II. Objects indicative of former organisms
   a) Casts and molds
      1. Imprints of shells, leaves, and the like
      2. Tracks and trails of vertebrates
      3. Trails of invertebrates
   b) Gastroliths
   c) Coprolites
   d) Artificial structures

THE unaltered actual remains of organisms are the rarest of fossils, but naturally they are also among the most interesting. If the soft parts are preserved, the protecting medium is usually ice, frozen soil, amber, or oil. The hairy mammoth and the woolly rhinoceros are recently extinct arctic mammals. During the last great glacial period a number of these animals were frozen in the ice or soil, particularly of Siberia and Alaska; and they have remained in this excellent sort of cold storage for thousands of years. One of the mammoths thus preserved was found about the year 1800 in the delta of the Lena River. A few years later it was sent to the (now) Leningrad Museum, where the hairy hide may still be seen attached to parts of the mounted skeleton. The so-called Beresovka mammoth, in the same museum, is, however, much better known. Discovered in 1900 by a hunter who stumbled over the ancient carcass, the specimen was not collected until 1901. In the intervening year prowling wolves destroyed a part of the trunk, but nevertheless the animal was remarkably well preserved. The hide, hair, flesh, and internal organs were, for the most part, intact. A mass of clotted blood in the chest, broken bones in the fore and rear limbs, and the remnants of the last meal in the animal's mouth gave indication of the sudden death which overtook it. Apparently, in feeding, it had stumbled into a fissure and was unable to extricate itself, despite a violent effort to do so. Since the flesh was still firm and "fresh," it was possible to study this fossil as a zoölogical specimen. Even the contents of the stomach were examined!

Mammoths at the Leningrad Museum. From an old print

In all, something like two score similar finds have been made; and a fairly high percentage of the ivory of commerce has come from the isolated mammoth tusks found somewhere on the frozen Russian tundra. It also is of considerable interest that these extinct elephants, which lived several thousand miles north of the present proboscidean habitat, were also well known to the ancient cave men. We know this to be a fact, since the cave carvings and paintings of the hairy mammoth match perfectly with our present knowledge of these extinct creatures, gleaned from studying their unusual fossil remains.

The woolly rhinoceros has been found preserved not only in ice but in oil. The oil-saturated soils of eastern Galicia, in Poland, are particularly well known for this type of fossilization. Bat guano is another medium which helps to make possible the preservation of some soft parts. At Aden Crater, New Mexico, a skeleton of a ground sloth was found buried in guano in an old volcanic vent. It was held in complete articulation by its tendons and sinews, and the claws and some of the hide and hair were still attached.

UNIVERSAL CEMETERY

A LL of these instances of the preservation of soft parts, however, concern late Pleistocene creatures whose geological antiquity is to be measured in thousands, rather than millions, of years. Amber (fossil resin), on the other hand, commonly yields Oligocene (the table in chapter 32 will show age comparisons) insects and spiders, so perfectly preserved that in some instances the soft parts of the body have been dissected and the intestinal parasites studied. The gummy exudates from some of the ancient Oligocene pine trees, particularly in the Baltic region, was, on formation, sufficiently soft and sticky to trap and engulf these soft-bodied creatures. On hardening into amber, the ancient resin faithfully recorded the finest detail of anatomical structure.

HARD parts of organisms are fairly commonly preserved essentially unaltered. One of the best-known instances is the famous fossil bone beds at Rancho la Brea, California, where the skeletons of many Pleistocene birds and mammals are preserved in asphalt. But, in addition, many still older fossils are composed of hard parts so little altered after millions of years that, for instance, the pearly luster of a shell may still be preserved.

BUT ordinarily, if the specimen is of any very great geological antiquity, it will have been altered, or **petrified,** chiefly through the action of ground water. *"To petrify"* means to turn to *stone;* hence, although a great many fossils are petrifactions, it is obvious from the foregoing discussion that by no means all of them are. In all, some seventy mineralizing or petrifying substances have been recorded; but most petrifactions are either the result of **calcification,** of **silicification,** of **carbonization,** or of **pyritization.** Bone and shell, for instance, gradually lose their animal matter; and ground water, carrying, say, either calcium carbonate, silicon dioxide, or iron sulphide in solution, fills in the interspaces. Later even the original hard parts are gradually replaced with the result that either a *calcified,* a *silicified,* or a *pyritized* fossil remains. In the latter case the so-called "golden fossils" result, for the FeS_2, or pyrite, has the general appearance of gold, and is called "fools' gold." Thus it is apparent that the general shape of the original object may be preserved although the details of internal structure are lost. When such petrification takes place, a *pseudomorphic* fossil results, the name, of course, referring to the fact that it simulates the form of the original. But commonly the replacement is a slow molecular process. Molecule by molecule, the original structure is replaced by the mineralizing substance carried in ground-water solution. In such instances, although little or nothing of the original organism remains, its minute anatomical features are well recorded. The replacement in these cases must precede complete decay, otherwise the morphological structures would have been obscured. As a consequence this type of fossil is not as common in the animal kingdom as it is among the plants, where the decay of cellulose may be relatively slow.

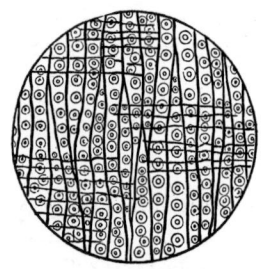

Above: Microscopic section of a silicified fossil conifer. *Below:* Microscopic section of a modern conifer. After Nicholson, 1876

The process of *carbonization* is considerably different from the other three processes which we have listed. Marsh gas (CH_4), water (H_2O), and carbon dioxide (CO_2) are given off from decaying cellulose $(C_6H_{10}O_5)_n$. In the process of formation of these products of decay there is required little carbon, as compared to hydrogen and oxygen; and, since there is little free oxygen dissolved in standing water, the original cellulose is gradually altered by a relative concentration of carbon, or *carbonized* until, if the process is continued long enough, an anthracite coal ($C_{24}H_8O$) results. Ideally, this process takes place when leaves, or even entire trees, fall into bodies of water and are buried under sediment. Thus, most fossils which belong to this type are plant remains. Animals with a skeleton of *chitin* ($C_{15}H_{26}N_2O_{10}$), however, are also subject to carbonization after death; and the fossil hydrozoans known as graptolites are almost invariably preserved in this fashion.

UNIVERSAL CEMETERY

CASTS and **molds** are the most common of the objects which are indicative of former organisms. The shell of a snail may be imbedded in sediment and subsequently be dissolved by percolating waters. The resulting cavity, however, may faithfully preserve both the internal and external surface features of the shell. Consequently, by making a wax impression of the natural mold of the exterior, an artificial cast of the exterior of the snail shell can be obtained. Or, as happens in many cases, natural processes may have resulted in the filling of the area once occupied by the shell itself by some sediment or mineral, so that a natural cast results. Leaves, trunks of trees, and even the skin of dinosaurs have been preserved as molds or casts. When Pompeii was destroyed by the famous Vesuvian eruption of 79 A.D., many men and animals were buried. When the city was excavated in relatively recent times, it was found possible to make excellent artificial casts of the natural molds of the bodies of the victims. The resulting casts showed such details as racial features, type of clothing worn, and the position in which death was met. Even the brain cases of some of the earliest vertebrates have been studied by using a similar method of reconstruction; and invertebrates as lacking in hard parts as the jelly fish have, in some cases, left enduring molds in the ancient sediments.

Cast of a Pompeian dog

The footprints or trails of all sorts of animals also may be preserved in the form of casts and molds. Indeed, this is a type of fossil which is not only unusually common but which, in the case of the invertebrates, is well represented even in the early Paleozoic rocks. The burrows of worms and rodents make up another type of fossil commonly recorded as casts or as molds.

GASTROLITHS are the stomach stones which characterize some of the Mesozoic marine reptiles, such as the plesiosaurs. They apparently were used as an aid in digesting the food which the animals gulped down so rapidly. *Coprolites* are fossil excrement. Since this type of record of the past can be thin-sectioned and analyzed, coprolites are of great importance in deciphering the feeding habits of long-extinct animals. The casts of the intestinal portions of worms, sometimes called *lumbricaria*, also belong to this category.

A coprolite

Artificial structures include the artifacts and cave paintings made by ancient man, as well as certain inorganic structures produced by some worms and some protozoans.

WE have now had a fairly complete, and possibly tedious, survey of the kinds of fossils. After we catch our breath, we must next inquire into the significance and use of these records of the past.

CHAPTER 27

CORRELATIONS

> Science is nothing but trained and organized common sense. . . . .
> —HUXLEY

LONG ago, Agassiz pointed out that a great scientific truth is likely to develop through three stages of criticism. First, people say that it conflicts with the Bible. Second, they decide that, after all, it has been discovered before. Finally, they complacently remark that they have always known it, anyway. To a certain extent the great concept of geological correlation passed through some such history. Steno, Fracastoro, and Da Vinci all had fairly decent ideas as to the possibility of determining the age equivalency of beds of rock in widely separated areas. But since such concepts involved a natural philosophy very different from that founded on the biblical description of Genesis or of the Garden of Eden, the general idea of correlation was not advocated to any great extent even by those who had a faint glimmering as to its possibilities. When, many years later, in the last decade of the eighteenth century, William Smith independently stumbled on the idea of correlation, the church, far from hampering him, came to his rescue; for, had it not been for Reverend Benjamin Richardson, rector of Farleigh, Hungerford, it is doubtful if the relatively untutored Smith would have appreciated the full significance of his discovery. Later, even though Smith finally was dubbed the "Father of English Geology," his critics often pointed

. . . . in widely separated areas . . . .

. . . . determining the age equivalency of rocks . . . .

out that there was nothing new in his ideas. Finally, as if to bear out Agassiz's contention completely, many other geologists of Smith's day were perfectly willing to minimize his importance by permitting it to be thought that they, too, had been using the principle of correlation for many years.

WILLIAM SMITH was a young surveyor's assistant when he began to notice that the fossils which could be found in a certain rock layer were apparently diagnostic of it. That is to say, any specific layer of rocks in Southern England with which he was familiar always yielded a definite assemblage of organic remains. Furthermore, the rocks above the specific layer, he noticed, contained a second and different fossil group; and, in addition, the bed below was characterized by still a third aggregation. Common-sense reasoning made it possible for Smith to figure out that he could, by using such information, work out some scheme showing the age relationships of all the strata which he had observed. To this end in about 1796 he began to make crude stratigraphical collections of fossils of the sort still used as a basis for correlation. Then one fateful day he called on the Reverend Richardson, who had a large collection of local organic remains. Immediately he amazed that worthy by picking out the unlabeled fossils in his curio cabinet and telling precisely the rock and area from which each had come. Such a bit of clairvoyancy, of course, called for an explanation. When it was forthcoming, Richardson was quick to see that out of Smith's common-sense reasoning and sound observation there was very likely to come a scientific development of far-reaching consequence. As a result of his encouragement, Smith continued in his studies; and by 1815, using the information he had gleaned from the study of the strata and their fossils, he had completed the world's first pretentious geological map, which depicted the various rock layers of England and Wales. Other geologists have followed Smith's example, so that today practically the entire surface of the earth has been mapped geologically, though naturally there are still large areas in which many details still must be added before the picture is complete.

CORRELATIONS

SINCE we now know that all animate nature represents the fruition of one great evolutionary process, it may seem obvious to many of us that the life of the past, ever increasing in its complexity, could be used as the basis for this world-wide mapping. What could be more reasonable than that the life of each succeeding geological period should be more advanced and thus different, or characteristic, when compared with the life of any other period? But it must be remembered in this connection that in Smith's time evolution had only been written about in circumlocutions by a few men like Buffon and Lamarck, and that even these writings were chiefly in a language which Smith could not read.

THE geologist has sometimes been accused of a gigantic hocus-pocus in this matter of correlation. The critic says that the earth-scientist determines the age of a fossil by finding the age of the strata in which it was found, and then has the audacity to turn around and give the age of the rock by making a pronouncement as to the age of the fossil. It is true that ever since Beringer's hoax the geologist has been a fairly wary chap and will generally ask for all possible information regarding a fossil before he names it. But, as a matter of fact, all fossil sequences used in correlation were discovered, not conjured up. Hence, in the first place, the age of all fossil assemblages has been determined on the basis of the strata in which they occur. The region in which the geological formations of a certain age are first studied is called the "type," and the rocks exposed there form the "type section." The fossils that are found in this type section are collected, described, and catalogued. In North America, for instance, the Devonian rocks of New York, whose position is shown in the table accompanying chapter 32, were the first of that age to be studied—hence all other Devonian rocks on this continent are compared with and referred to this New York type section. Putting it in another way, when fossils like those from the New York type Devonian section are found in some little-known area, the rocks of that region are therefore also assumed to be Devonian in age. Here, then, is one of the common cases wherein the age of sedimentary rocks is determined on the basis of their fossils.

ALTHOUGH almost any fossil has its own story to tell regarding its geological age, some, like people, speak more specifically than others on this point. The marine invertebrate *Lingula*, a veritable "Methuselah" of the seas, has persisted throughout geological eras so little changed that it is of almost no use in determining the age of the beds in which it occurs. Similarly, another long-lived, related form, called *Leptaena rhomboidalis*, is found with little variation in the rocks of three of the great geological periods. This type, therefore, indicates merely that the rocks in which it is found were deposited sometime during the great time range of the three periods. But yet, a third ancient invertebrate, called *Hypothyridina cuboides*, lived only at the beginning of late Devonian time. It consequently is a fine horizon-marker, or index fossil, for the sediments which were deposited during that restricted time interval. Since it occurs literally around the world in the Northern Hemisphere, its horizon is often conveniently referred to as the "cuboides" zone.

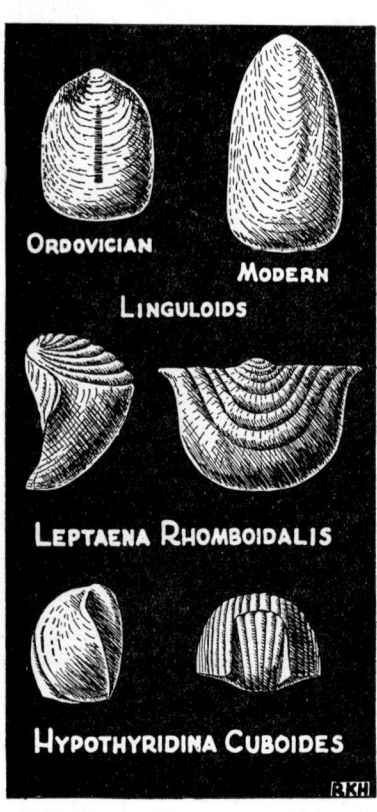

Although the professional geologist uses all of the fossils of a formation in determining its time equivalents in other areas, it has been learned that considerable reliance can be placed in index fossils of the *H. cuboides* type. Like this Devonian example, **all good index fossils have a wide geographic, or horizontal, range and a narrow geological, or vertical, distribution.** The most valuable index fossils are also numerous, and have some distinctive feature so that they can be identified readily.

THE extinct marine hydrozoans, the graptolites, answer these requisites relatively well. A rapidly evolving group, living chiefly in the surface waters of the open ocean, these ancient animals had a world-wide distribution during the early part of the geological era in which invertebrate animals were dominant. As a consequence, geologists are able to recognize some score of graptolite zones which make possible a convenient classification of the early fossiliferous sediments around the world.

EVEN a casual observer of nature knows that there are today a number of organic realms. That is, he knows from common experience that certain animals and plants are confined to definite physical areas. The animals of the mountains are different from those on the plateaus; and the latter, in turn, have characteristics which distinguish them from the creatures of the low plains. The animals of the strand are not the same as those inhabiting the open ocean, nor are the floras and faunas of fresh-water lakes the same as those found in the seas. This strong present manifestation of the dependence of the organic world on the physical environment apparently was nearly, if not quite, so defined throughout the geological past.

But in most of the recognized modern organic realms we notice that there may be considerable local differences in the plants and animals. For instance, the assemblage of invertebrate marine animals living today along the rocky coast of Maine is obviously somewhat different from that found along the low-lying shores of Chesapeake Bay, in spite of the fact that the two assemblages are, of course, contemporaneous. Thus we again see the inevitable result of differences in physical environment upon all organisms. Similar situations have existed for millions of years. We do not expect to find the same assemblage of fossils in sandstone that we look for in limestone, even though, by the tracing of two such beds to the place where they grade into each other, we know definitely that these two different kinds of rock were both deposited at exactly the same time. The faunas of these two kinds of rock, deposited under different physical conditions, are therefore also contemporaneous; and hence there generally are *some* fossils which are common to both sediments. Nevertheless, the faunal differences are sufficiently great to give the fauna of the sandstone a **facies,** or aspect different from that found in the limestone. Each gradation in type of deposit, therefore, makes for another facies of the same fauna. Here, then, are other situations which obviously complicate the problem of reading the record of the past. But fortunately there are almost invariably intermigrations between organic realms or between the various habitats of a single realm. Those hardy groups that have thus been able to live about as successfully in one habitat as in another are therefore of great service to the geologist. They commonly make it possible to determine the time equivalency of sediments of different lithology and different faunal facies.

WE HAVE now considered the problem of correlation in sufficient detail that it must be apparent that the process of working out the age equivalency of widely separated rock outcrops is generally a fairly complicated one. Let us now briefly consider the simplest type of correlation. This is a sort of work commonly carried on by the oil-company geologist who is examining a possible producing-area in a preliminary, or reconnaissance, fashion.

EVEN the beginner in geology soon learns that there is no single earth feature which can be regarded as a complete entity. Every geological phenomenon is inseparably linked with other earth features or processes. Thus, in order to get a true picture of the geology of any area, large or small, it is essential to discover the interdependency of the myriad earth phenomena. For example, as the oil geologist studies an isolated outcrop of rock, he knows that in all probability it is a part of a large body of rock invisible elsewhere in the immediate vicinity because of a covering of soil. He therefore attempts to find the equivalent of this bed in some adjacent area. In short, he attempts to make the simplest sort of correlation, which in this case may involve nothing more than comparing kinds of rock, or the kinds of surface features made by the exposed

CORRELATIONS

strata. He may also make other similar and rapid observations, involving field comparison of such fossils as can be quickly discovered. If he is lucky, he may be able actually to demonstrate the continuity of the bed in which he is interested by tracing the outcrops from one good exposure to another. Such correlation, involving areas of relatively small size, is called *local correlation*. But where the equivalency of rock strata is studied over regions involving several states, it is designated *regional correlation*. The latter type, which is so important in an understanding of the major aspects of earth history, is dependent on the fact that ancient seas were widespread over areas now making up large parts of all the continents, as we shall see in the next chapter.

THERE is another type of detailed, and relatively local, correlation which has great economic importance. It is the oil-field correlation made possible by modern methods of studying the information yielded by every well that penetrates the rocks of the earth. Thousands of these oil, gas, and water wells are drilled every year; and for most of them, the driller keeps a record to show what kinds of rock the drill encountered, and the thickness of each rock layer. But before this information can be employed in the manufacture of maps and sections to show the structure and extent of a layer of rock deep beneath the surface, there must be some methods devised so that all key beds can be recognized in all the wells which encounter them.

WITHOUT attempting to become technical, we may take the reader behind the scenes long enough to discover that five main methods are generally used in making the correlations. The first scheme is the simplest to employ, but the results are not always very accurate. In brief, a graphical record, or "log," of the driller's information is drawn up, with a definite scale to show the relative thicknesses of the kinds of rock penetrated. The types of rock are indicated in different colors, such as yellow for sandstone, blue for limestone, and so on. Such logs of wells drilled fairly close together can thus be matched by mere inspection. If conditions during the deposition of the sediments in the area were fairly uniform, the results of such correlation may be accurate; but inasmuch as most sediments change character from place to place, various sands may become confused in the interpretation. Thus, by this method many an underground structure suitable for oil accumulation has been "discovered," although none actually occurred in nature.

The second method is an elaboration of the first, and involves the chemical analysis of waters taken from the various sandstones penetrated. It has been found that the waters of any sandstone have the same chemical peculiarities over relatively large areas. Hence this method serves as a check on the first and may make certain the identity of sands in relatively widely separated wells.

The third scheme involves dissolving the calcareous rocks encountered by the drill in acid. The so-called "insoluble residue" left behind has certain definite characteristics for each limestone layer.

A fourth, and still more useful, method is very similar to the third. In it the "heavy minerals" are separated out of the oil-well sample. This is usually accomplished by pouring the cleaned and powdered material into bromoform, as has been described in chapter 14. When this is done, the heavy minerals sink and the lighter ones remain on the surface. Since many sedimentary rocks have their own characteristic heavy mineral associations, rock layers can be correlated on the basis of information yielded by such analyses.

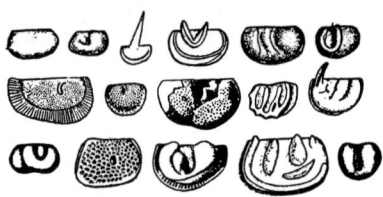

Ostracode crustaceans of the type used in micropaleontology. After Ulrich and Bassler. See Plate 62 for illustrations of foraminifera

Finally, and most important of all, the microscopic fossils taken from the well can be used in this oil-field correlation. These organic remains are so small that many escape breakage from contact with the drill, and thus all of their characteristics can be ascertained under the microscope. These small animals can therefore be employed as index fossils for the various buried rock layers, just as the larger fossils are commonly used in correlating exposed strata. Their study is the science of **micropaleontology,** which has grown and flourished because of its utilitarian value in the oil fields. The animals most commonly used as the tools of the science are the single-celled foraminifera and the ostracode crustaceans, about both of which we shall have more to say later.

This subject of oil-field correlation, which we have touched upon so briefly, is nonetheless a subject of utmost importance in one of the greatest of the modern industries. In fact, it is perhaps not too much to say that billions of dollars depend upon the correctness of the correlations which are daily arrived at by large staffs of trained workers.

THUS far in our discussion of correlations we have been talking about sedimentary beds; and this is only natural, since the very idea of correlation is dependent on the sedimentary strata. But, of course, it is also possible to correlate igneous and metamorphic rocks, though generally not nearly so successfully as sedimentary deposits. For instance, the age of an igneous rock is fixed by the time it was intruded or extruded, and thus came into a definite relationship with the adjacent, and pre-existing, rock. It must be obvious, therefore, that **every igneous rock is younger than the rock which it cuts or intrudes,** or that it is younger than the rock on which it rests, providing it is an extrusive such as a lava flow. Thus, igneous rocks are commonly correlated on the basis of some structural relationship to sedimentary beds of known age.

METAMORPHIC rocks commonly can be traced into unmetamorphosed sediments of known age or even into contact with igneous rocks whose age has been ascertained. Nevertheless, the correlation of either igneous or metamorphic rock masses is not yet on a very satisfactory basis. On the other hand, most of the criteria for correlation which apply to the sediments can be used with these two other great classes of rock, with the notable exception, of course, that they do not ordinarily contain fossils.

These criteria, a number of which we have hinted at in chapter 25, may be summarized as follows:

1. **Lithologic character.**—Rocks with the same lithological characteristics may be of the same age. This is particularly true if the lithological properties are very peculiar. Obviously, there are thousands of limestones of many different ages, but two limestones both deep red in color and highly crystalline in texture may well be suspected of being contemporaneous.

2. **Topographic expression.**—Many rocks have a characteristic expression when they are weathered. In the Mississippi Valley area a sandstone known as the St. Peter usually makes the same kind of bluff wherever it crops out. On the other hand, the Kimmswick limestone generally makes low benches where it is exposed, and its surface is beset with small holes, so that it seems to be honeycombed. The experienced geologist soon learns to recognize these peculiarities and to use them in his field correlation.

3. **Structural position.**—The position of rock layers with reference to any structure such as an unconformity or an igneous intrusion may also be used in correlation. For instance, if, in two fairly widely separated areas, a limestone rests with unconformity upon a sandstone and is in both areas cut by the same kind of intrusion, there is then some evidence that the limestone may be the same in both regions.

4. **Stratigraphic sequence.**—Peculiar sequences of beds found in different regions may be of the same age. Thus, if a sequence of black shale, greenish limestone, a lava flow, and a coarse conglomerate occurs at one locality, the finding of an identical sequence in another area might well be regarded as evidence that the beds of the two sequences were contemporaneous in their origin.

5. **Derived soils.**—The soils formed as the result of weathering of a certain rock under similar climatic conditions are likely to be the same over reasonably wide areas. Similarly, the nature of the vegetation may be quite the same wherever the rock crops out and weathers to soil.

6. **Mineral content.**—Igneous, metamorphic, and sedimentary rocks alike, if they are characterized by mineral assemblages somewhat out of the ordinary, may be correlated on the basis of these peculiarities.

CORRELATIONS

7. **Fossil content.**—This is the one positive criterion, and it has been elaborated on earlier in this chapter. In most cases of detailed work in correlation, however, all of the criteria available are employed.

ALTHOUGH we have placed a great deal of reliance on the sedimentary rock sequences themselves, together with the all-important organic remains which they contain, we shall later learn that the *breaks* between layers of sediments are perhaps of even greater importance than the actual sediments in the fundamental theory of correlation. In fact, this point is of such consequence that it has given rise to the geological dictum, **"Diastrophism is the ultimate basis of correlation,"** a statement which we shall have occasion to analyze in subsequent chapters.

CHAPTER 28

OLDER FLOODS

> There rolls the deep where grew the tree.
> Oh Earth, what changes hast thou seen!
> There where the long street roars hath been
> The stillness of the central sea!
> —TENNYSON

IN SPITE of the fact that most religions have their flood legends, some people find it difficult to believe that large areas of the continents have been repeatedly submerged beneath the sea. A catastrophic Noachian-type deluge is widely accepted as a reasonable event; yet, the far-reaching effects of the slow, diastrophic movements described in chapter 16 to many people seem as fantastic as the land of Oz. Nevertheless, the ocean has flooded the land not once but scores of times, for these ancient inundations have written plain and indelible records of their existence in the great sedimentary deposits which they have left behind them.

.... waves rule Britannia ....

Tennyson, in penning the quotation which stands at the head of this chapter, therefore was writing not only good poetry but sound geology. Today's busy thoroughfares of Riverside Drive, Michigan Avenue, the Strand, Unter den Linden, and the Champs Elysée have all been beneath the "central sea," and at repeated intervals. Even now, Great Britain is sinking at a rate which, if uninterrupted, is sufficiently rapid to effect complete submergence within the next 40,000 yr. Picture, if you can, a map of Europe without the identification tag of the British Isles! Or, imagine the consternation of the Englishman 400 centuries removed who is at last forced to admit that the "waves rule Britannia," instead of the traditional converse!

OLDER FLOODS

THE fluctuations in the outlines of continents, and consequently of ocean basins as well, are indeed ancient, as well as modern, earth phenomena of profound importance. The many theories which have been advanced dealing with the actual origin of the continents and ocean basins themselves are all inevitably linked with theories put forth to account for the origin of the earth itself. As a consequence, some phases of the subject of continental origin will be discussed further in chapter 31. But in the meantime we may inquire into some of the features which characterize these major earth structures.

FOR many years scientists have thought of continents as basin-shaped land masses with mountain chains near their borders, and low, relatively flat central plains. This is accurate enough, especially for the present, because we are now in a general period of emergence, despite the downward movements of certain areas which we have already noted. But throughout the *ancient periods of submergence*, which *have been much longer than the times of elevation*, the continents consisted of partially, or even nearly completely, flooded basins. These basins have generally been surrounded by high lands, called **borderlands.** The borderlands commonly have been nearly or quite continuous on three sides of the central basin, so that the ancient floods have in many cases been spoon-shaped extensions of the open ocean. Thus the geographical aspect of the typical continent of today is a far cry from its appearance during most of geological time.

THESE so-called basins, which throughout the past have witnessed so many floodings, are in reality great platforms of ancient igneous and metamorphic rock. Although they have been shallowly inundated time and again and have never stood very high above sea level, they, nonetheless, are the only relatively stable portions or bucklers of the continents. These platforms are readily divisible into two major elements: (1) **a positive, or emergent portion;** and (2) **a negative, or submergent area.** Although these two major divisions of the platforms are to be seen in other continents, let us turn our attention to North America, in which all the continental elements are especially well developed.

Neither of the two North American platform areas stands ruggedly above the sea today, nor have they during the past. Both have been warped repeatedly during periods of mountain-building, and repeatedly have suffered slight uplift or relatively minor depression. Pressures from the direction of the borderlands, however, have resulted in close folding and faulting at or near the edges of the platform, rather than in the platform itself, which is relatively free from major structural features of this sort.

Old rock platform of North America with its subdivisions, and the principal geosynclinal areas. After Moore

The positive element has been a relatively permanent land mass throughout the geological past. The amount of sediment that has been derived from it, however, is apparently small. North Central Canada and Greenland, commonly called the "Canadian Shield," is North America's positive area, the district which, throughout the ages, has most persistently kept its head above the older floods.

THE submergent portion of the ancient platform encircles the Canadian Shield on the west, south, and east. This is the part of North America which has been covered some score of times by shallow epicontinental waters whose flood area has varied from minor seas the size of the British Isles to great inundations with a total extent equal to the entire area of Europe. Of course, the largest of these floods were not entirely confined to the submergent or negative area. But this region is marked today by relatively thin layers of the sedimentary beds deposited during each submergence. In few places are these strata as much as 5,000 ft. thick. In fact, over much of the negative area wells can be drilled through the sedi-

OLDER FLOODS

mentary cover and into the old crystalline rocks in less than 3,500 ft. One reason for this thinness of sediment is that this negative area was never depressed much below sea level. Another reason may be seen in the fact that, during every one of the ancient periods of emergence, all sedimentary beds laid down during earlier floods were, of course, subjected to erosion. Consequently, there is today no one place on the platform in which there is a complete record of the ancient seas.

THE borderlands are elevated with respect to the old rock platform; and, although erosion is always removing countless tons of débris from them, they are constantly being pushed up, for reasons which will be discussed later. Thus they stand relatively high during periods of submergence and are only moderately depressed during times of general emergence. Like the Canadian Shield, these borderlands have been positive elements in the North American continent; but, unlike that nuclear element, they have been the great sources of supply for the sediments now found on the interior of the continent. In fact, they have contributed so much of their substance to the formation of younger beds that they are now greatly reduced in size.

THE third portion of the typical continent is the trough, or **geosyncline.** This structural element is a narrow belt along the inner edge of the borderlands; and it roughly encircles the old rock platform, though commonly several distinct parts of the belt can be recognized. The location of the geosyncline between the slowly rising borderland and the generally slowly sinking central platform places it in a position of crustal stress. Moreover, the geosyncline, being located at the edge of the high borderland, is continually receiving sediment, which is dumped into the basin of deposition by the inward-flowing borderland streams whose gradient is abruptly changed at the foot of the highlands. As a consequence, the geosynclinal belts, already weakened by crustal stresses, tend to sink under the weight of sediment poured into them. Sinking commonly keeps pace fairly well with the accumulation of sediment.

Hence the surface of the geosynclines remains near sea level. At times, however, owing in part to extremely rapid sedimentation, they may stand slightly above sea level; and at such times their sediments are, of course, eroded. But, in the main, the geosynclines are sites of relatively permanent deposition. In them sediments have accumulated as though in colossal dump heaps, in some cases possibly making up thicknesses as great as 20 mi. **Thus, the sediments deposited during any one period of submergence are typically thickest and coarse in the geosynclinal areas, and thinnest and finer-grained in the submergent platform area.** Thus, also, the scattered and fragmentary records of older floods in the submergent platform area can be filled out by the nearly complete account of deposition to be found in the adjacent geosyncline.

We have been talking rather glibly about so many different factors of these ancient floods, which no human eye ever witnessed, that we should be chagrined, indeed, if the critical reader did not raise some "how's" and "why's." For instance, how do we know that in a geosyncline sinking generally keeps pace with deposition? The answer is found in the fact that the rock layers in a geosyncline are commonly marked by ripples and mud cracks. Since these features persist through thousands of feet of strata, shallow water conditions must have continued also. Similarly, every statement which we make has been arrived at after some such critical analysis, but space does not permit us to stop and authenticate them all.

THUS far we have seen that during every ancient flood the continent is composed of, first, a relatively stable central basin, or ancient rock platform, which, however, has a general negative or submergent tendency. But, a part of this central basin remains weakly positive, for the nuclear element, or shield, still stands low above the central sea. Although being eroded, the amount of sediment derived from this nuclear mass is small. Second, there are the borderlands, or actively rising positive areas, which are being eroded and which supply the sediments which are deposited in the geosynclines and on the negative platform area. Third, there are the

OLDER FLOODS

geosynclines themselves, which are actively sinking troughs receiving the greatest quantities of sediment.

During the shorter periods of emergence, however, the movements of these three main continental elements are surprisingly altered. The most active sites of movement and uplift are the once-sinking geosynclines, which are now dominantly positive. The ancient-rock platform, which previously has been slightly negative in its movement, is now slightly positive; and the once positive borderlands are now somewhat negative in tendency. We shall inquire into some of the possible causes of these cyclical earth movements in succeeding chapters. For the present it will suffice to point out that, since the time of deposition of the first abundantly fossiliferous rocks, some half-billion years ago, there have been at least three of these major cycles. These major cycles have superimposed upon them a number of minor cycles.

A complete, or major, cycle involves (1) mountain-building and an interval of continental uplift, with the attendant restricted epicontinental seas, extreme climatic conditions, and the marked resultant changes in the organic world; (2) a long period of erosion and reduction in the height of the continent, with the resulting inundation of both geosynclines and central basin, followed by uniform and mild climatic conditions, with the organic world prospering greatly, though changing slowly; and (3) a return to another period of mountain-building, withdrawal of seas, climatic refrigeration, and rapid evolution of such animals and plants as are able to withstand the rigors of the rapidly changing conditions.

.... if North America were to be depressed..... The white continent would result if the sea rose 500 ft.; the black one if the land were elevated, about 600 ft.

AT THE present, according to some scientists, we are probably entering a new cycle. The sea has already begun to spread over the central portions of the continents, as is evidenced by such embayments as the North and Baltic seas and Hudson Bay, as well as by the submerged continental platforms. As a matter of fact, if North America were to be depressed a mere 500 ft., the Gulf of Mexico would extend so far northward that Louisville would be inundated and St. Louis and Cincinnati would both be seaports. A further sinking amounting to an additional 500 ft. would result in the central sea extending completely across the continent from the Gulf to the Arctic regions. In Europe the results of such a slight depression would be even more startling, for, should that continent sink even so little as 500 ft., the Ural Mountains would become islands and the waters of the Baltic would mingle with those of the Caspian and Black seas.

WITH such far-reaching and entirely possible results of slight modern depressions in mind, it is not so difficult to imagine the great fluctuations in land and sea during the past. From detailed studies of the sediments and their organic remains left behind by each of these ancient floods, it has been possible to construct the general table of geological chronology discussed in chapter 32. If we reflect a moment, we can readily see that, since all of the oceans are connected, any real change of ocean level should be recorded on all the continents alike. That is, if the seas withdraw from North America, they should also recede from Europe; if they inundate parts of Asia, they should also flood the correspondingly low portions of South America. Hence, the great erosional breaks, as well as the sedimentary records, should coincide the world around. In so

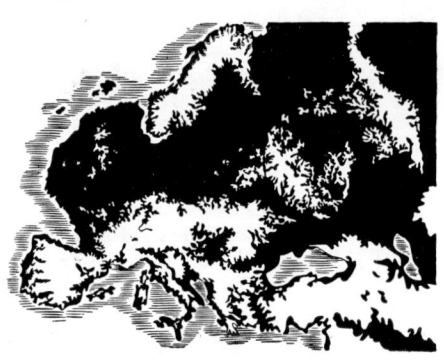

.... the Urals would become islands ....

far as this is true, our geological chronology not only is a natural one but is readily applicable on a world-wide basis.

Unfortunately, however, in order that the changes of sea level mentioned above should have the same effect everywhere, all the continental masses must remain absolutely stable, something they never have been able to do. In fact, we know definitely from present-day measurements that, not only may one continent sink with relation to the sea while another rises, but different portions of the same continent are undergoing different movements at the same time. Hence our general scheme of world-wide correlation runs into all sorts of difficulties. But by utilizing the fact that even on an otherwise emerging continent small negative portions will invariably be beneath marine waters, the geologist is able to use the telltale sedimentary record of the ancient sea, and of the animals that inhabited its waters, in piecing together the record of the past. Thus, by examining every sedimentary deposit, however limited in extent, and recording all of the information it can be made to yield, and by using all of the principles of correlation outlined in earlier chapters, it has been possible to make fairly uniform subdivisions of earth history which can be used on all the continents.

ONE of the major divisions of this universally recognized scheme of subdivision is known as the **period.** Periods are sufficiently large sections of the great geological calendar that they have the same name round the world. Smaller divisions of geological time, such as **epochs** and **stages,** however, have different names in different areas, as we shall later see. A typical period may be considered to have one major flood time commonly at or near its middle. This flooding commonly is both preceded and followed by smaller invasions of the sea. Thus, the period, like the greater geological cycle of which it is a part, begins with highlands and ends with them. A typical period characteristically is comprised of three epochs. The first of these is the time of slight submergence of the still dominantly emergent land. Then, after a relatively short time of withdrawal of the seas, there comes the middle epoch of widespread marine waters. The third and last epoch is one of retreat of seas as the result of relatively rapid emergence. The early and late epochs of the period therefore are times of large land areas with seas largely confined to the geosynclinal troughs. The mid-epoch is one of great reduction of lands, only the nuclear platform area and the borderlands standing high.

THE rocks deposited during any single period of this sort are said to comprise a **system;** those laid down during an epoch comprise a **series;** and those that are deposited during any one of the many stages making up an epoch are called **formations.** Most formations are marked by a characteristic assemblage of fossils; in general they also constitute rock units of sufficient distinction that they may be traced in the field, and their areal extent indicated on a map. Some writers also consider several formations, laid down during a single advance of the sea, as a **group.** Each group is said to have been deposited during an **age.** And just as several stages make an age, and an epoch is made up of two or more ages, so several epochs (usually, but not always, three) comprise a period; and there are a number of periods in the largest division of geological time, known as an **era.** Since we will be using these terms often in the remaining chapters, their relationships are summarized here in the accompanying table.

Time	*Rocks*	*Examples*
Era	Sequence	Paleozoic
Period	System	Silurian
Epoch	Series	Niagaran
Age	Group	Clinton
Stage	Formation	Rochester

WE SAY, for instance, that the rocks of the Niagaran series were deposited during the Niagaran epoch of the Silurian period. They form a part of the Silurian system of rocks which was laid down during the Paleozoic era. This particular era name is a coined one meaning "ancient life." As we shall later see, all of the other era designations have a similar origin. But the names of the smaller geological divisions are obviously geographical in their derivations. In other words, they usually refer to the *type area* in which they were first studied. Thus the Silurian system was so designated by Sir Roderick Murchison because he first studied it in the land of the ancient Celtic tribe of the Silures, on the borderland between England and Wales. In the century which has elapsed since Murchison first investigated these beds in the type area, they have been recognized as occurring on every continent. The Niagaran series takes its name from the falls area, where it was first studied in this country. Further, the Rochester formation was named from exposures at and near that city, and it is a part of the Clinton group, so designated because of its early study near the New York town of that name. At its type locality the Rochester formation is a shale, but it is nevertheless correlated with the Osgood limestone formation of Kentucky and Tennessee because the fossil content of the two different kinds of rock is similar and thus shows that they were both deposited during the Rochester stage.

It should be obvious from the foregoing account that the records left by older floods in large measure form the real basis for historical geology. But as we earlier pointed out, the breaks, or erosional and mountain-building intervals, between periods of deposition are of equal, or even greater, importance in marking off the major divisions of earth history. And this is a point to which we will return many times in our discussion of geological periods. For the present let us turn our attention briefly to the use of the ancient rock formations in determining the extent and configuration of the older floods responsible for them.

THE study of the distribution of the sediments of any one series gives a clew to the extent of the sea in which it was deposited. Such studies have now been carried on intensively for many years, with the result that maps have been prepared to show the configuration of land and sea for most of the geological epochs. Such studies form the basis for the science of **paleogeography.**

The paleogeography of the world in fairly recent geological times can be deciphered without a great deal of difficulty. But the outline of the very ancient seas can be reconstructed only after the most detailed of investigations. For example, the deposits of the older epochs commonly are deeply buried beneath younger sediments. The information we can get regarding them all comes from deep-well records. Where not deeply buried, the sediments may have been partially eroded; or, for large areas, the beds may have been entirely removed. But it is possible, despite these ever present difficulties, to determine, for instance, the position of ancient shores by ascertaining the direction in which the sediments become coarser grained. Occasionally, also, we find down-faulted blocks in areas of ancient rocks. In such structures the sediments we have been studying have been preserved from erosion, and thus they afford certain information concerning the extent of the sea in which they were deposited. Thus, by using the various principles developed in the earlier chapters of this book, the trained paleogeographer evolves his ancient seas and older continents.

IN LATER chapters we shall see that the results of his researches have yielded relatively precise information regarding lost land masses quite as interesting, and far more real, than the fabled Atlantic continent of Atlantis and the mythical Pacific land masses of Pan and Mu.

CHAPTER 29

MOUNTAINS AND CLIMATES

> Hills, rock-ribbed, and ancient as the sun.
> —BRYANT

NOT so long ago there was a fairly well-founded historical and sociological axiom to the effect that "mountains are museums of antiquity." But with our new methods of transportation and communication, customs are no longer changing nearly so slowly as they once did even in the most remote "hill countries." Thus the sociological validity of the saying is today thrown into considerable doubt. But in a very real sense mountains always will be museums to the geologists. Sharply serrate, little-eroded ranges,

.... mountains are museums of antiquity ....

standing stark against the sky, show, however, that mountains need not be earth features of any great antiquity, at least so far as geological time is concerned.

The fact is there are many generations of mountain ranges. The older mountains have literally crumbled to dust; and newer ranges have sprung up, phoenix-like, out of their ancient débris. The Laurentians, for instance, are older than the Appalachians, which are ancient, indeed, as compared to the Rockies, which in turn are the Cascades' elders. But regardless of their age these geological museums are likely to have their specimens well displayed. In them rocks of many types, which elsewhere are deeply buried, have been folded upward, so that the geologist is able to examine them at the earth's surface. Thus the highest points of the continents, as well as their deepest gorges, are both equally revealing areas in which to decipher the more complete records of historical geology.

Since we can recognize many generations of mountains, and since earth tremors occur daily, we know not only that mountain-building is an ancient process on the earth but that the causes responsible for it are still operative today. But the search for the answer to the question, "What *causes* mountains?" although an old one, does not yet have an entirely satisfactory answer. In the following paragraphs, however, we propose to outline briefly some of the theories which have been advanced to account for this major earth phenomenon.

Old Mrs. Appalachian says to young Miss Rockies, "You won't cut such a fine figure either when you're as old as I am"

In the preceding chapter we have seen that earth history is marked by geologically long periods of relative crustal stability alternating with shorter periods of rapid movement in which continents are elevated with respect to sea level, and mountains are folded out of former areas of geosynclinal deposition. This folding (and faulting), which we have studied in some detail in chapters 17 and 18 naturally involves crustal shortening. That this surface shortening exists, or that it is of considerable magnitude, none deny. But just how much shortening has taken place is difficult to determine. For instance, the shortening represented by the Alpine folds alone has been variously estimated from as little as less than a hundred miles to more than a thousand miles. Even if we accept the most conservative estimate, and remember that the folding has been more intense in the Alps than in most other ranges, we must still admit that in all the globe's mountain chains there is conclusive evidence of a most notable shrinkage of the earth.

If the earth, in dwindling to a smaller size, has witnessed a wrinkling of its surface material in much the manner that the skin of a withering apple is wrinkled to fit its diminished interior, then we are taxed with the responsibility of finding out why our planet has contracted. Perhaps it is because it was, and still is, a cooling body. Now, if the radius of the earth is shortened by as much as a mile, then its circumference must inevitably shrink a little more than 6 mi. Thus, if in all the periods of mountain-building the radius of our globe has decreased by a mere 200 mi., all of the ancient and modern foldings could be accounted for.

But from the observed conductivity of rocks, and the earth's known thermal gradient, it has been variously calculated that its temperature could only be lowered by radiation of heat outward somewhere between 10° C. and 45° C. in 100,000,000 yr. Using the higher figure and allotting to it the maximum possible radial shortening due to this cooling, we can achieve only about 10 mi. of diminution of circumference during 100,000,000 yr. Neglecting, therefore, the patent fact that radioactive disintegration is a great source of heat within the earth, which hence may not be cooling at all, **we see that mountains cannot be accounted for on the simple basis of a cooling globe.** Consequently, those who believe in a once molten earth have to find some other explanation of mountain-building.

ON THE other hand, those who consider that the earth grew by the accretion of solid particles point out that this method of growth would result in an interior somewhat heterogeneous in character. Thus, under the great pressures of millions of pounds per square inch in the earth's deep interior there would naturally be much molecular rearrangement and compaction. Hence the interior would gradually come to be made up of denser molecules, and consequently have a diminished volume.

As we shall see in the next chapter, the continental segments of the earth are made of lighter rocks than the intervening oceanic areas. Therefore, according to this concept, during the process of earth compaction the stresses generated cause all the wedge-shaped segments of the earth to move inward; and the oceanic segments, being heavier, move inward first. The continental areas therefore sink less, and hence relatively seem to rise. As the oceanic wedges move farther inward, the continental wedges at last are actually lifted, relative to the former; and thus the shallow seas are drained from the continental shelves. This is a matter which we discussed on page 135, to which the reader should refer for an explanatory diagram.

More and more crowding results from continued sinking, and the consequences of it are felt most strongly at the boundaries between segments. And since the oceanic wedges are most actively sinking, stresses are greatest at the continental borders. Thus,

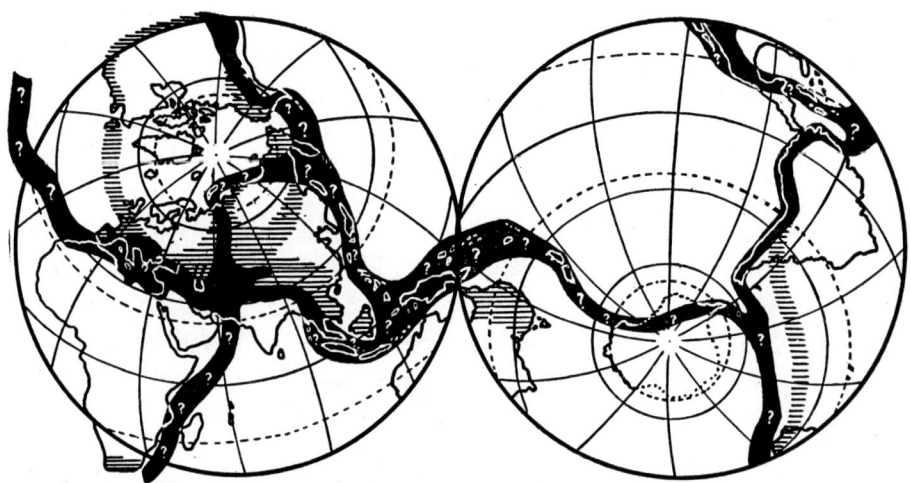

Geosynclinal belts of the Mesozoic and early Tertiary, shown in black. Late Paleozoic troughs indicated by horizontal lines. After Haug and Bucher

horizontal pressures are developed whose relief causes folding of the rocks into mountain ranges. The location of the folding is determined by the areas in which the compressional forces are localized and by the position of the geosynclinal belts of weaker sediments. These mobile belts generally are located near the margins of the continents, as we have earlier pointed out and as is well shown by the accompanying figure. According to this wedge concept of diastrophism and mountain-building, the uplift of the continents should precede the great periods of folding, though, naturally, the two phenomena would accompany each other during most of the duration of crustal unrest.

Other students of geology think that mountains are formed because different parts of the earth's crust are in *isostatic*, or equal-pressure, adjustment. In fact, precise determinations of the

MOUNTAINS AND CLIMATES

force of gravity suggest that **the rocks which make up mountain masses are lighter than those which underlie the plains,** a fact to which we will have occasion to refer again. According to the adherents of the *theory of isostasy,* the rocks in the "zone of flow" yield to slight differences in pressure, and thus flow to areas of reduced pressure. Above this "zone of flow" the earth may be considered as divisible into any number of columns of equal surface area. If there is no yielding in the "zone of flow," the columns must have the same weight. If this be so, then the rocks under the sea must be much heavier than those under the mountain ranges. Stated in another fashion, a column of rock a mile on a side which extends down to the "zone of flow," located at Chicago, should weigh as much as a column of similar cross section measured at Mount Everest, or as one studied at the deepest part of the ocean, say at the Tuscarora Deep. True, the column will be nearly 12 mi. longer at Everest than at the Tuscarora locality, and more than 5 mi. longer than the Chicago column; but in each case the difference in rock density will compensate for the disparity in column length.

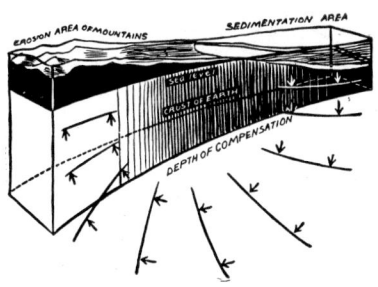

Suggested movement in the "zone of flow" as mountain area is eroded. After U.S. Coast and Geodetic Survey

Now, when a mountain is eroded, the weight of its rock column decreases; but the eroded material is deposited on another rock column, whose weight is correspondingly increased. For a time the outer shell of the earth stands this unbalanced condition, but eventually there is a transferal of material in the zone of flow from the area of deposition toward the area of reduced pressure under the rock column being eroded. Since the sides of the columns cannot be precisely parallel but must con-

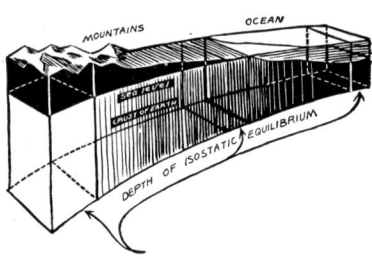

The three imaginary columns under sea plain, and mountain theoretically have the same weight. After U.S. Coast and Geodetic Survey

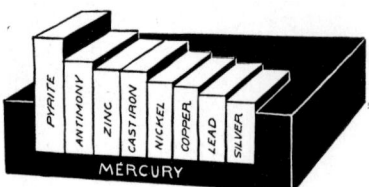

Substances of equal mass and cross section, but of different height, stand in mercury as shown above. Why? After U.S. Coast and Geodetic Survey

verge inward toward the earth's center, the wedge effect again comes into play. The heavier and shorter columns of the oceanic areas squeeze the continental columns and thereby cause them to rise. Lateral compression results, and the continental borders are folded. Thus the earth is kept in a condition of isostatic equilibrium, except in those areas in which crustal movement has just occurred, is now going on, or will soon take place.

THIS theory fits the fact of continual erosion rather well. It also does no violence to the notable fact that generation after generation of mountain ranges spring up along the same mobile belts, or to the concept that the continents and oceans are permanent facial features of the earth. But the great overthrusts with lateral displacements measured in tens of miles are difficult to explain on the basis of isostasy. Similarly, it is hard to account not only for widespread epicontinental seas on the basis of this theory but for some of the great peneplains of the past as well. If isostatic adjustment is nearly perfect today, and has been at all other times when the lands have stood correspondingly high, it must have been singularly poor during those long periods of crustal quiescence when the lands were standing nearly at sea level or were shallowly submerged.

THE hypotheses of continental migration and their relationship to mountain-building are touched upon in chapter 31, and they need not be reviewed here. The *theory of periodic melting* of the lower part of the crust, however, may be mentioned. According to this concept, radioactive disintegration generates heat in the rocks of the subcrust. This heat accumulates during the long periods of crustal rest, because at such times the rocks, being solid and stable, do not *readily* permit its escape to the surface. When the rocks are near or at their melting-point, they at last yield readily to the stresses which have been accumulating along with the heat of radioactivity. Since some of these stresses, at least, have resulted from the horizontal expansion of the heated rock, compressional forces are also generated and folding results. Of course, during the mountain-building great quantities of the subcrust are brought to or near the surface as a result of volcanic activity. This is a point at which the theory fits observed facts rather well, for **igneous rocks commonly are found at the core of folded mountains.** Furthermore, this theory of the periodic melting of the subcrust can be fitted with profit into most of the other concepts of mountain-building.

IT IS plain from the foregoing discussion that the mechanics of mountain-building are not yet thoroughly understood. On the other hand, it must be apparent to all that whenever mountains are formed, or whenever they are brought low by erosion, important climatic sequels attend these events. For this reason it may not be inappropriate at this point to consider briefly some of the aspects of climate throughout geological time.

FROM evidence which we will later examine, it is clear that life has existed on this planet of ours for many hundreds of millions of years. Since most plants and animals cannot stand long-continued temperatures above the boiling-point of water or below freezing, we can assume that much greater temperature ranges than this have not been common on the earth, at least during the past billion or two years. When you stop to think about it, this is a remarkable, and fortunate, lack of variability in our geological climate.

But nevertheless, there have been some notable climatic fluctuations which have written their record into geological history. In the period just antedating the present, great ice sheets covered much of northern North America and Europe. During other, and more ancient, glacial episodes ice sheets have even formed in what is now the Torrid Zone. Corals once thrived in areas now bathed by arctic waters, and palms grew in Greenland. The area which we now call Ohio and New York was once a desert, as was the district which now is industrial Germany. All these are but passing events in the story of the past which we will study in later chapters. For the present we may well inquire as to what are the physical controls of these past and present climates.

THESE climatic controls may be grouped into two main categories: (1) *astronomical* and (2) *terrestrial*. Some of the astronomical factors are (a) **possible changes in the eccentricity of the earth's orbit,** thus affecting the length of the seasons; (b) **possible changes in the position of the poles,** thus altering zonal arrangements; (c) **variations in sun-spot activity,** thereby modifying the energy output of the sun; and (d) **possibility of the solar system passing through clouds of dark matter** in space, which might reduce the amount of heat the earth receives from the sun. Certain features of ancient glaciation have been explained by some scientists on the basis of changes in the position of the poles, but the explanation has not met with great favor. Other astronomical factors, with the exception of sun-spots, cannot be mentioned here. But there is a growing conviction that sun-spots have, and have had, considerable influence upon cyclonic movements of our atmosphere.

Some of the terrestrial controls are (a) **latitude,** (b) **altitude,** (c) **atmospheric circulation,** (d) **oceanic currents,** (e) **carbon-dioxide and water-vapor content of the atmosphere,** (f) **volcanic activity,** and possibly (g) **escape of terrestial heat generated by radioactive decay.** The importance of latitude and altitude is too commonplace to need comment; but it may be pointed out that mountains apparently always have been dry on their lee, and well watered on their windward side. The very positions of land masses and bodies of water are climatic factors of consequence, and the oceanic currents presumably have been as important in influencing climate in the past as they are today. Carbon dioxide and water vapor are effective thermal blankets. They let the sun's rays in to the earth's surface but retard the radiation of heat out into space.

MOUNTAINS AND CLIMATES

DURING periods of submergence carbon dioxide was added to the earth's atmosphere by the precipitation of limestone from sea water, according to the following equation:

$$Ca(HCO_3)_2 \rightarrow CaCO_3 + H_2O + CO_2$$

Moreover, at such times rock-weathering would be slow so that little carbon dioxide would be consumed by the process. This all should result in large amounts of the gas in the atmosphere, with its tendency to produce the warm climates which we actually associate with periods of submergence. But during mountain-building erosion goes on apace, few limestones are deposited in the sea, and the carbon-dioxide content of the atmosphere is depleted. Hence, periods of mountain-building presumably have always been times of refrigeration not only because of the attendant general elevation of the land but because of the diminished thermal blanket effect of the reduced carbon-dioxide content of the atmosphere. Of course, critics have been quick to point out that a small amount of carbon dioxide in the atmosphere seems to be just as effective in reducing radiation as larger quantities. Moreover, water vapor may be even more efficient in this regard than carbon dioxide, so that here again we are groping with a situation we do not entirely understand.

BUT there seems little reason for doubting the efficiency of great clouds of volcanic dust in shutting out the sun's radiant energy from the earth. In fact, during such historically recorded explosive eruptions as those of Krakatoa and Katmai earth temperatures were apparently lowered by a measurable amount. Since, as we have earlier pointed out, volcanic activity commonly attends mountain-building, here is another possible factor in accounting for the climatic refrigeration which has so often attended the periods of crustal instability.

We have now examined some of the climatic controls; let us next consider briefly a few of the climatic indicators.

GROWTH rings in trees plainly indicate alternation of wet and dry, or warm and cold, seasons. Cell growth is retarded during periods of drought or cold, but warm or wet seasons accelerate growth and consequently cause ring structure. Such rings have been found in some of the earliest of land plants from the Devonian period. They likewise occur in a number of plants of the later geological periods. But these structures become more common in ancient plants toward the present poles; thus they have been interpreted as being due to the same alternation of warm and cold weather which we experience today.

Limestone at present most characteristically forms in warm waters. Most deposits of ancient limestones also thin toward the poles, thus tending to confirm the evidence afforded by tree rings. Similarly, most of the ancient glacial deposits, which are, of course, excellent climatic indicators, are found in what are now temperate or polar areas, though exceptions to this generalization are known.

PLANTS and animals which are characteristic of certain areas today are often used as climatic indicators for the time in which the sediments carrying their fossilized ancestors were deposited. These organisms, however, may occasionally give us the wrong information. This results when such organisms have changed their characteristic habitat during the past. For instance, we think of corals as warm-water animals today, and so they are. But we are accustomed to use their Paleozoic ancestors as indicators of the very conditions which the modern types require for life. We forget that the ancient forms are all extinct and had a different structure from the modern corals. They may, therefore, have been able to live in the cooler seas.

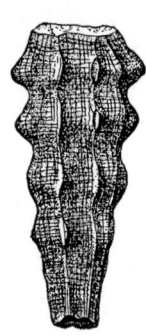

Hydnoceras, a Devonian shallow-water sponge whose descendents today inhabit the abyssal parts of the ocean

Coal beds also have been cited as climatic indicators, since conditions requisite for the lush plant growth required for coal formation at once suggest warm, wet climate. There seems to be little doubt that there was plenty

MOUNTAINS AND CLIMATES

of moisture at the times in the past when coal was forming, but unfortunately the present conditions which most closely approximate the ancient coal swamps are found where the climate is relatively cool, not warm.

Red beds have been taken to imply desert conditions, though we know that some very deep-red soils form under tropical vegetation. But red beds in association with salt and gypsum, which have formed from the evaporation of large bodies of water, are pretty certainly indicative of aridity, though not necessarily of great warmth.

FROM all these and many other lines of inquiry into the ancient climates of the earth comes evidence of wet and dry seasons and records of cold and warm weather. But the present zones of the earth were quite apparently not commonly so well defined as now, and there were times of fairly universal warmth. These have not been adequately explained. But there is evidence that they occurred chiefly during periods of submergence and quiescence; whereas the **times of cold, or at least of variable climatic conditions, coincide rather perfectly with periods of continental uplift and mountain-building.**

IT IS unfortunate in some respects that this chapter on such fundamentally interesting subjects could not be made a more definitive one, that it must bristle with "ifs" and "buts." On the other hand, much of the attractiveness of geology as a subject for mental gymnastics results from the fact that, although the results of geological processes are plain, their causes are still obscure—their complete history not yet written.

CHAPTER 30

VESTIGES OF CREATION

Before the beginning of years.
—SWINBURNE

WHEN Robert Chambers published his *Vestiges of Creation* in 1844, he was astute enough to anticipate the storm of criticism and protest it would arouse; so he prudently neglected to affix his name. Although he was more interested apparently in the creation of life itself than in the origin of the earth, he discussed the cosmogony of Herschel and Laplace and affirmed that "the work [of creation] may be said to have been done by the will of God, expressed in the form of the law of gravitation."

.... work of creation. From the *Nuremberg Chronicle*, 1493

However our planet came into existence, it must be evident that any explanation which seeks to account for the actual event must give due consideration to the observable vestiges of creation. These vestiges are of two types, the first being astronomical, and the second geological, in character. Let us first briefly consider the astronomical vestiges.

When we spectroscopically examine the sun's light, we find that we can identify spectral lines which reveal that many of the chemical elements making up the earth also occur in the sun. As a matter of fact, nearly sixty of the earth's ninety-two elements occur in the sun's atmosphere; but what its deep interior is made of we do not certainly know. It is of some interest to note, however, that we cannot be certain that *any* of our elements are missing from the sun; and, on the contrary, we have the example of helium recognized in the sun in 1868, twenty years before it was identified on our planet.

Despite this chemical unity of the sun and the earth, the former has a specific gravity of only 1.4, as compared to the earth's 5.5+. This results from the fact that the sun's surface temperature is about 6,000° C, or sufficiently great to vaporize all our known substances. The temperatures deep within its interior are, of course, unmeasurably greater still. This enormous heat keeps the sun gaseous, self-luminous, and greatly expanded in spite of the fact that gravitative forces 27.6 times as great as the earth's are tending to compress it into a sphere of higher specific gravity. Without the benefits of the sun's heat the earth would have had a far different geological history, for in some way or other the sun's energy motivates all its erosional processes and makes possible all the organic activities of its countless living things.

THE source of the sun's great energy, of which, of course, the earth receives a mere pittance, is not yet completely understood. Even supposing it were made of carbon and oxygen alone, the amount of heat now being radiated would require the complete burning of all the material in a scant two thousand years. But, of course, it has long been known that oxidation, or burning, is a process not taking place on the sun. Long ago Helmholtz got around the difficulty by demonstrating that if the sun were contracting even as little as 120 ft. a year, a sufficient amount of kinetic energy would be supplied to account for solar radiation. On this basis Lord Kelvin made his famous calculations as to the age of the earth, a point which will be mentioned again in chapter 32. For, although the shrinkage required would not be marked within historic times, on such a shrinkage program, the sun would have contracted from a positively tremendous heavenly body in a few million years. Accordingly, Lord Kelvin placed the upper limit of the age of earth, the sun's offspring, at 40,000,000 yr., but he really thought that it was less than half that old.

Until the discovery of radioactivity, at the turn of the past century, Lord Kelvin's estimate carried great weight, in spite of the fact that most geologists felt it represented far too small a lapse of years to account for the known events of geological history. But, once it was known that radioactive elements (such as uranium, thorium, and radium) undergo a slow atomic disintegration, with the liberation of helium and heat, an entirely new aspect was placed on the situation. As such elements are widely distributed in the earth, there is little doubt of their presence in the sun. Furthermore, many of the earth's elements can be made artificially radioactive by certain laboratory procedures; thus it is probably not too much to expect that in the sun some of the elements may be naturally radioactive. Although radioactive disintegration is slow, the amount of heat liberated is enormous, so that even minor amounts of radioactive materials in the sun might account for its radiant energy. And in addition to these data and inferences, we must remember the observable fact that one of the products of atomic disintegration, helium, is common in the sun. Thus, it seems at least logical to assume that radioactivity may in part account for solar radiation.*

THE surface of the sun, called the *photosphere*, is composed of white-hot gases which telescopic examination shows are in turbulent motion. The *chromosphere* is only a relatively thin shell of brilliant red cooler gases; yet it is thicker than the diameter of the earth. From this rarer atmosphere of the sun leap out the fountainlike *solar prominences*. These rise to heights of a mere 20,000 mi., or belch out into space twice the distance from the earth to the moon. Comprised of light gases, such as hydrogen and helium, direct observation shows that in some cases they attain a velocity of 100 mi. a second. This is a fact of considerable importance in examining theories of earth origin, since the velocity of escape of the

A solar prominence, showing the chromosphere

---

*See, however, in this connection, Bartky, *Highlights of Astronomy*, pp. 223–24.

VESTIGES OF CREATION

sun,* 385 mi. a second, is not tremendously greater than the velocities observed in some of the prominences. It is also important to observe that the solar eruptions are most common along two highly unstable belts in the sun. These belts are 25° of sun's latitude in width, and lie from 5° to 30° on each side of the solar equator.

WE HAVE now briefly examined a few of the vestiges of creation to be observed in the sun itself. The vestiges to be seen in the solar system as a whole have been partially outlined in chapter 2, and they do not require a great deal of further elaboration. Some of the peculiarities of the system which must be accounted for in any plausible theory of earth origin may, however, be summarized here.

The company of tiny wanderers as compared with the sun's great size

The earth is one of a company of wanderers about the sun. There are probably nine planets and about thirteen hundred planetoids, all of which revolve in the same direction and lie in nearly a common plane. The sun also rotates, though relatively slowly; and its equator is inclined only 7° from the common plane of the system. The orbits of the planets are nearly circular, but those of the planetoids are distinctly elliptical, and some are not in the general plane of the system.

The four inner planets have a much greater density (3.9 to 5.5+) than the sun (1.41), whereas all of the larger, more-distant planets are less dense than the hub of the system (0.7 to 1.3).

All of the planets, except Mercury and Venus (and Pluto), have satellites which, in the main, revolve in nearly circular orbits in the direction of the planets' rotation, and nearly in their equatorial plane. The eighth and ninth satellites of Jupiter, and the ninth of Saturn, however, revolve in the opposite direction, and thus are *retrograde* in their motion. Moreover, the satellites of both Neptune and Uranus have orbits which cross the plane of planetary revolution at high angles. The earth's moon also is exceptional in its great

* See Lemon, *From Galileo to Cosmic Rays*, chap. 5.

size as compared to its parent body, and the moons of Mars revolve at a rate more rapid than that of the planet.

All of the hundreds of solar fragments make up only about 0.14 per cent of the mass of the parent body; yet most of the systems' moment of momentum* belongs to Jupiter and the other outer planets.

The importance of all these features of the solar system becomes more readily apparent when we recall that the earth's mode of origin is not peculiar to it alone but to all its solar associates as well. We must, therefore, critically examine all the features of regularity in the system, as well as every eccentricity. Each and every vestige tells its story, however involved, regarding the birth of the system. At present we are still puzzled by some of the accounts, but no doubt we shall someday learn to read them all accurately.

THERE is another astronomical vestige which can be examined directly here on the earth. We refer to "shooting stars," or meteors, which are so familiar to all that they may be thought of as one of the earliest of the natural phenomena observed by children. Most of these meteors, which, of course, have nothing to do with stars, are merely small pieces of stony or metallic material which become heated to incandescence by the friction encountered as they enter our upper atmosphere, and thus are dissipated into dust. But a few of the larger fragments reach the earth and are then designated as *meteorites*. About thirty of the known elements have been found in these extra-terrestrial bodies, oxygen, silicon, iron, nickel, aluminum, and carbon being the most common. Most scientists regard these meteorites as fragments which are actual relics of the creation of the solar system. With this in mind, German investigators have determined the minimum age of such bodies as 2,600,-000,000 yr., on the basis of their helium content. This is a figure not

---

* The moment of momentum, or the quantity of rotation of a body, is the sum of the rotations of all of its parts. It has been shown that for a rotating solid body the rotation is proportional to the product of the square of the radius and the angular velocity of rotation, the latter being the angle through which the body turns in a unit of time. Thus, if the sun should shrink, its angular velocity of rotation would increase, since the product of the square of the radius and the rate of rotation must be a constant.

markedly inconsistent with the other estimates which have been made as to the date of origin of the system, as we shall later see.

Although most meteorites are found at or near the earth's surface, there are a few cases on record in which the impact of the meteorite was responsible for the formation of a "crater." Meteor Crater in central Arizona, probably formed in this fashion, is approximately $\frac{3}{4}$ mi. in diameter and about 600 ft. deep. Other "craters" occur at Henbury in Australia and at Wabar in Arabia. The great Siberian meteor of June 30, 1908, which is supposed to have generated air waves as far from the impact as 300 mi., has not left very spectacular evidence of its fall, nor has any undoubted meteoritic material been found. But all such vestiges of creation are nonetheless important and should be carefully investigated, since they throw light on a process which once must have been very common, if the Planetesimal hypothesis, which we will examine in the next chapter, gives a correct picture of the earliest stages in earth history.

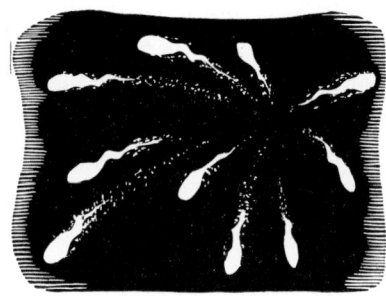

A meteorite shower

THE geological vestiges we can dismiss in short order. The first part of this volume has been devoted largely to their enumeration and explanation. It is obvious, therefore, that no theory of earth origin can be seriously considered if it does not logically account for all the facts we have previously studied regarding the earth's composition, structure, and physical processes. Moreover, before any theory can gain supporters, it must also square with the evidence afforded by the known facts of geological history which we are to consider in the succeeding chapters.

At the risk of a little repetition, a few of the most important of the geological vestiges may, with profit, be summarized briefly at this point. The earth reacts to stresses of short duration, such as tides, like a rigid solid with about twice the strength of steel. To long-continued stresses, such as those common in mountain-building, however, it yields relatively readily; and the structure of mountains indicates that the earth probably is a contracting planet.

The earth is comprised of a number of shells, some of which have more or less sharp boundaries. These shells become progressively heavier with depth.

The rocks of the continental areas are relatively light, having the specific gravity of granite, or about 2.6. Moreover, the mountain areas are, on the average, made of lighter material than the lowlands. Carrying this point a step further, we find that the oceanic segments of the earth are made up of still heavier material, for the rocks from oceanic areas have a specific gravity of about 3.3, which is the density of the average basalt.

Despite much theorizing to the contrary, there is still good evidence that there has been no general interchange of continental and oceanic areas.

The "original crust," so-called, has never been discovered; but the record of early diastrophism and volcanism is clear. Periods of glaciation also occurred in very ancient geological times. Nevertheless, conditions favorable to life must have existed even earlier still.

SINCE these are only a few of the extraordinary geological features to be explained, it is small wonder that no theory of earth origin which attempts to account for these, and the astronomical facts as well, has met with universal acceptance. In the following chapter, however, we will examine some of the better-known theories and see for ourselves how they fit the facts.

CHAPTER 31

EARTH'S BIRTH

> Earths round each sun with quick explosions burst,
> And second planets issue from the first;
> Bend, as they journey with projectile force,
> In bright ellipses their reluctant course.
> —Erasmus Darwin

ROBERT WARING DARWIN, although himself a man of very considerable note, in his later years had the grace and wit to remark that he was famous solely because he was the father of Charles and the son of Erasmus Darwin. For, in many respects, Erasmus was as outstanding in the field of science as his more widely known grandson. The quotation heading this chapter was written by this Erasmus Darwin in 1791 as a part of one of his scientific poems, and it is one of the saner of the many almost forgotten statements of cosmic evolution which appeared in the eighteenth century.

Earth's birth as a puny planet

BUT a mere five years later, in 1796, the famous French astronomer and mathematician, Laplace, proposed the hypothesis of earth origin which was to become so influential and to be so widely known as the *Nebular hypothesis*. Inasmuch as a number of other students of cosmogony, notably Buffon, Kant, Herschel, and Lockyer, also have speculated on possible modes of earth origin, out of some sort of nebulous material, it is perhaps better to designate this scientific guess as the **Laplacian hypothesis.** Although Laplace himself first proposed his statement of earth's birth as *a mere suggestive footnote to an appendix of his 1796 textbook*, it was not destined to remain in this lowly position. People apparently read footnotes in those days, for the hypothesis immediately became attractive to the scientists of the period. Consequently, to Laplace's great amazement, for he had never placed any great faith in his own suggestion, it was necessary to allot more and more space to the hypothesis in subsequent editions of his text.

For approximately a century the Laplacian idea of planetary origin completely colored the thought in many of the fields of science, and its influence was also great in the realms of philosophy and literature as well. Therefore, although the hypothesis is now outmoded, we will briefly examine its major features and point out a few of the weaknesses which ultimately caused its downfall.

LAPLACE did not attempt to account for the origin of matter (nor, as far as this point is concerned, can modern scientists offer much enlightenment), but he assumed as already extant a great gaseous nebula. This nebula he considered to be so diffuse that its outer limits extended beyond the present orbit of the outermost planet. Since he postulated a slight rotational movement and high temperature for this nebula, it was a mass of gas inevitably radiating its heat out into space. Thus it was a cooling and, therefore, shrinking nebula. Today we have information from physics indicating that the radiation of heat and light may result in diminution of both mass and energy, but in Laplace's day it was considered certain that the mass-energy total of a nebula was a constant. Consequently, Laplace thought that in order to conserve the moment of momentum in his postulated cooling nebula, its speed of rotation would surely increase as shrinking continued.

Laplace

As its speed of rotation did increase, according to the hypothesis, it bulged at the equator. Finally a ring of nebular material was

EARTH'S BIRTH

left behind when centrifugal force just exceeded the inward gravitational pull. This process of ring formation was repeated again and again until ten rings had been left behind—one for each of the nine planets, and one for the planetoids, though, of course, in Laplace's time not all of these parts of the solar system were known.

EACH of the successive rings collapsed; but while the planets thus formed were still gaseous, they, too, developed rings out of which the satellites formed in precisely the same way that their parents had originated before them. The planetoids resulted from the disruption of one of the rings into many bodies instead of a single one, and the sun is the residual portion of the extraordinarily shunken nebula. According to such a hypothesis, the earth originally was a starlike body shining in the heavens. As it cooled, it liquefied and then later solidified. The original crust was bathed in rain as soon as its primitive atmosphere had cooled sufficiently to permit condensation.

Here, obviously, is a hypothesis which at first sight does little violence to the "vestiges of creation" examined in the last chapter. And in the light of scientific knowledge at the close of the eighteenth century, it must be regarded as one of the shrewdest of all scientific gropings. But Laplace himself was well aware of certain difficulties in the way of his hypothesis; and many astronomical, physical, and geological discoveries since his time have finally caused the most ardent of the supporters of the hypothesis at last to abandon it.

MANY of the difficulties regarding the Laplacian hypothesis were determined and studied at the turn of the century by Thomas Chrowder Chamberlin and Forest Ray Moulton, respectively geologist and astronomer at the University of Chicago. Chamberlin had had his attention turned to the problem by the discovery of glacial deposits in some of the most ancient rocks. Since, according to Laplace's idea, the earth's atmosphere has been progressively diminishing since the time of its origin, the large amount of carbon dioxide in the air in those early days should have formed an effective thermal blanket and prevented early periods of glaciation. The whole story seemed to Chamberlin to lose strength in the light of this discrepancy, and consequently he began to look into the problem with greater care. It is unnecessary to list the difficulties which he and other workers have eventually discovered, but a few can be mentioned as typical.

IF LAPLACE were correct in his hypothesis, the sun should be ready to give off another ring, it should be rotating rapidly, should bulge greatly at its equator, and should possess practically all of the moment of momentum of the solar system. What do we actually find to be the case? Well, unfortunately for the hypothesis, the sun is rotating relatively slowly, it has no appreciable equatorial bulge or polar flattening, and most of the moment of momentum of the system is in the outer planets. This is a particularly strong objection, since these bodies cannot possess much more than approximately one-tenth of 1 per cent of the original mass of the nebula. Other objections are found in the fact that, as shown in the preceding chapter, the motions of some of the satellites demonstrate that the system is not nearly so simple as it was regarded in Laplace's time. His hypothesis, famed for its simplicity, is in difficulties with these very irregularities. Furthermore, according to the ordinary behavior of gases, it is very doubtful whether material leaving the nebula would ever have formed a ring. And rings, even granting, for argument's sake, their formation, could not break up into planets.

WITH these and other difficulties in mind, Chamberlin and Moulton suggested a radically new mode of origin for the solar system. In brief, their hypothesis, which is usually designated by the names of the authors, involves the partial disruption of our sun as a result of the gravitational effect of the close approach of another star. It is entirely fair to say that this hypothesis has had as profound an effect on the scientific thought of the twentieth century as the Laplacian hypothesis had during the previous hundred years. It was so attractive from the start that, paradoxically, many scientists have tried to modify or improve it; but since, as originally stated, all of its details were worked out with an unsurpassed elaborateness, these attempts have not been particularly successful.

IT HAS been demonstrated mathematically that the chance for collision among all the myriad stars of the heavens is today scarcely a remote possibility. But the same sort of analysis will also demonstrate that, given enough time, the possibilities of a reasonably close approach are fairly good. Now, we are all more or less familiar with the oceanic tides on the earth due to the proximity of the moon; and, although we may not understand this special tidal phenomenon completely, it helps us to comprehend the general tidal effects which would have been generated in the sun by the hypothecated close approach of a passing star. We know, for instance, that, neglecting the lag, there are two high tides—one on that side of the earth which is nearest the moon, the other on the opposite side of the globe. And we realize that if the earth were gaseous or liquid, the tides would, of course, be much larger. But most of us do not know that even the surface of the solid earth is periodically raised and lowered by the moon's gravitational effect, as is well shown by detailed geophysical observations. At Pittsburgh, for instance, the maximum recorded rise is just slightly under 2 ft.!

Thus it seems obvious that any star which passed close to our sun would pull sun-material out of the side nearest it, and also in effect pull the sun away from matter on the opposite side. The differential attraction on opposite sides of a star, which, after all, is the important factor in determining the size of its tides, would have been particularly great in a body as large as the sun, especially if another star came within a mere few million miles of colliding with it. Under such circumstances, therefore, the sun, even were it a quiescent heavenly body, would develop great tidal bulges. But, according to Chamberlin and Moulton, explosive forces in the sun also aided in its tidal disruption.

TODAY, as we pointed out in the preceding chapter, the sun has two great unstable belts which encircle it, one on either side of the solar equator. Each of these eruptive bands has a width of about 25° of solar latitude and extends from latitude 5 outward to latitude 30°. We have seen that it is in these two great belts that the solar prominences are at present the most pronounced. Let us

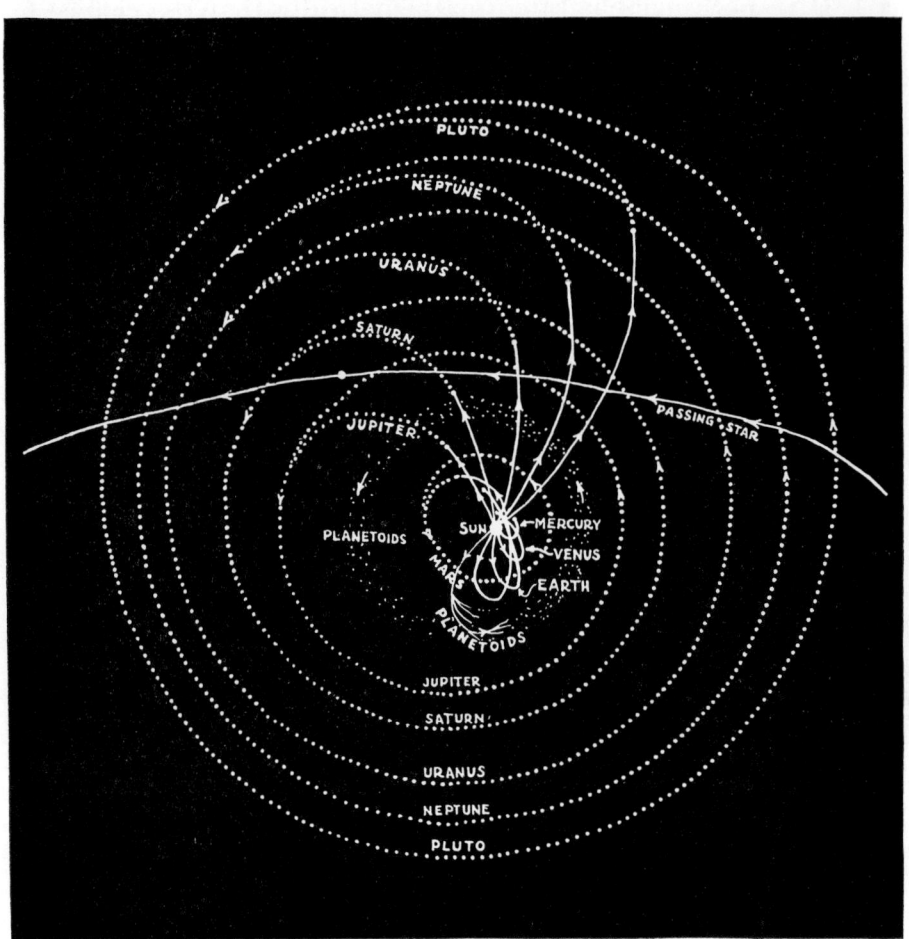

Diagram to show how the pairs of bolts from the sun may have given rise to the large and small planets, as well as Pluto and the planetoids. The passing star is small and the approach is close. The orbits are not shown to scale. Modified from R. T. Chamberlin

therefore imagine that these belts were also in existence some billions of years ago, before the Sun gave birth to her family of ill-assorted children.

Thus, as the disturbing star approached the sun, two gaseous jets, or bolts, were shot out from the sun's nearest eruptive belt. The bolt moving toward the star was to become Neptune; the one moving away from it on the other side of the sun was to become Mars. Then a second pair of bolts from the first eruptive zone gave rise (ultimately) to Uranus and Earth. Likewise, Saturn and Venus, and then Jupiter and Mercury, developed from two pairs of bolts issuing from the sun's second eruptive bolt as the disrupting star moved over that area. The planetoids and Pluto (with its possible associated bodies) are thought to represent a very early pair of bolts that originated even before the Mars-Neptune pair. Thus it is argued that the five great bolts on the side of the sun toward the passing star have developed, as might be expected, into the five larger, lighter, and more distant solar bodies. The bolts on the opposite side of the sun became the inner, smaller planets, and planetoids. Here is a hypothesis, therefore, that attempts to account for the difference in the densities, in the size, and in the spacing of the planets.

ALTHOUGH each bolt was gaseous when it left the sun, it soon condensed, cooled, and thus formed into liquid and finally solid particles. The coalescing mid-portion of the swarm of particles formed the nucleus of the planet. This central part of each bolt, and each little particle as well, were drawn around the sun in elliptical orbits of their own. The nucleus of the planets, in making many revolutions about the sun, gradually met and added to its mass countless particles of sun matter appropriately called **planetesimals**. Since the planets eventually grew to their present size by constantly adding these planetesimals, the Chamberlin-Moulton concept of cosmic evolution has commonly been called the **Planetesimal hypothesis**.

THE more planetesimals a planet gathered to its nucleus, the more their impacts caused nearly circular orbits to be formed out of their originally elliptical ones. Moreover, these planetesimal impacts are also considered to be responsible for the rotary motions of the elements of the solar systems. Satellites are regarded as masses of sun-substance which, from the start, have been under the gravitational dominance of the associated planet nucleus. Thus we see that the Planetesimal hypothesis well accounts for the very irregularities of the solar system which prove embarrassing for the Laplacean hypothesis. For instance, the larger planets have the most circular orbits, just as would be expected, since they presumably had more encounters with planetesimals. The retrograde movement of some satellites is also readily explained by those exceptional cases in which planetesimal impacts were so located as to give a backward spin to these growing heavenly bodies.

The chief objection to the Planetesimal hypothesis is found in the density stratification of the earth, previously alluded to, which seems to some to be the natural result of the cooling of a once molten globe. There are also astronomical data which are cited as evidence of a tidal retardation in the period of rotation of the Moon and Mercury; and these facts, according to critics of the Planetesimal hypothesis, suggest that all members of the solar system were once molten. It is also stated that, if the oceans had been in existence since long before the earth achieved its present size, they would now be much more saline than they actually are.

The criticisms of the Planetesimal hypothesis seem to us far from insurmountable, but because of them a number of modifications of the Chamberlin-Moulton hypothesis have been suggested, all with a view toward somehow accounting for a once molten earth. Joseph Barrell, of Yale, for instance, suggested that the planetesimals were the size of the modern planetoids and that their impacts generated sufficient heat to liquefy the growing earth. Sir James Jeans and Harold Jeffreys, British scientists, have proposed a far more elaborate modification of the Planetesimal hypothesis, which is called the *Gaseous-Tidal hypothesis*. In brief, this concept of planetary origin begins with the dynamic encounter of a passing star and our sun at a time when the latter was a gaseous mass with a diameter much greater than at present. The explosive character of the sun plays no rôle in its disruption. But a great tidal cone was lifted up and pulled away from the sun as the passing star reached its point of closest approach. The vast drawn-out tidal filament was made up of lighter material at its outer portion, of heavier material near the sun. The outer portion broke up into the outer, and lighter, planets; and the minor, but heavier, planets were formed from the near part of the filament. The satellites (except the moon) originated from the embryonic planets when, in their early gaseous condition, they first wheeled passed the sun. From these youthful planets filaments of gas likewise were drawn out because of the sun's gravitational influence upon them, and out of these smaller filaments developed the satellites.

EARTH'S BIRTH

Circularity of orbits is obtained by calling into play a "resisting medium" of escaping gas through which the planets would have to push their way. The efficacy of such a medium in causing the various irregularities now seen in the orbits of planets, even granting the possibility of its stability, which is remote, is open to serious question. Moreover, critics point out that earth's moon, with its relatively large size, cannot be accounted for very satisfactorily under this hypothesis. And, in addition, whatever criticisms having to do with a molten earth are applied to the Laplacian hypothesis apply with equal cogency to the Gaseous-Tidal hypothesis. Thus, although this is a concept now in style in parts of Europe and in a few sections of the United States, it seems to us very unlikely to stand up under the additional scrutiny time is bound to bring to its critical examination.

NO MATTER how the earth originated, it seems probable that either it was once much smaller than at present, and its principal growth was accomplished as a solid planet, or it was once a molten body whose size was not so very different from its present dimensions. If our planet actually originated after the fashion postulated by the Planetesimal hypothesis, the original earth nucleus was relatively small in comparison to its present mass. In the very early stage the nucleus may have been either molten or gaseous; but in any case, it is considered to have cooled to a solid body probably before there was a great deal of accretionary growth. According to Chamberlin, general conditions favored an early accumulation of the heaviest planetesimals, and thus the core of the earth came to have its present great weight and density.

Earth has not yet answered the question of her origin

During this early period of earth growth the nucleus probably was too small to hold an atmosphere. Like our present satellite, the moon, it was baked by day, frozen by night—a barren, lifeless, uneroded sphere. But once it achieved a diameter of something in excess of 2,000 mi., it began to retain a few gas molecules. With every planetesimal infall its mass increased; and, of course, with this growth, its capacity to hold an atmosphere also was enlarged. It must be presumed that a part of this original atmosphere was acquired through the capture of planetesimals made up of individual gas atoms or molecules. Some of it was also obtained at the time of collision and disruption of large gas-containing planetesimals. Or it may have been acquired by the later melting of the planetesimals and the transferal of the gases to the surface due to early volcanism. Whether free oxygen has existed from this early atmosphere stage on down to the present is a debated point. Some scientists feel that primitive plant life was established long before the earth achieved its present size, and thus that plants, through the ages, have been contributing oxygen to the atmosphere by the liberation of this gas from carbon dioxide during the process of photosynthesis. The recent discovery of O_2 in the sun, however, strongly suggests that free oxygen has been an integral part of the atmosphere since before the origin of life itself.

THE early earth had no more than acquired the beginning of the most tenuous atmosphere when the process of weathering commenced. But the process of volcanism had even a more ancient origin. If the earth grew by planetesimal infall, there was a heterogeneous arrangement and composition of the various earth layers added to the original nucleus. As material was added, compression of earlier layers increased and heat was generated. In addition, planetesimal impact raised the temperature at the surface. Radioactive substances also produced heat during decomposition; and with these sources augmenting the original internal heat there was enough carried outward by conduction to melt some of the relatively light silicate rock. Thus volcanic conditions had their beginnings in the early transferal of molten material from depths upward toward the surface. Since, as a result of this process, the heavy metallic substances would be left behind in residual concentration as the lighter substance worked toward the surface, here is a process by which the earth shells of different densities could have come into being.

In this connection it is interesting to note that the only other member of our solar system whose surface we can examine in some detail, that is, the moon, has a surface beset with great craters, presumably of extinct volcanoes (although it has been suggested that they represent meteorite scars). We may assume, therefore, that even if, or when, the earth nucleus was as small as its satellite, it was doubtless wracked by volcanic outbursts of great violence.

AS THE earth's atmosphere increased in size with the growing of the planet itself and through the increase of volcanism, it finally became able to hold water vapor. Finally, when locally the saturation-point was reached, came that dramatic moment when water first fell to the earth as rain. The first of these early downpours evaporated so rapidly that there was no great accumulation of water, but soon the natural depressions began to have permanent pools and the early work of running water had its crude and ineffectual beginnings. Soon sediments of the water-carried, as well as the wind-borne, sorts began to accumulate. And as the hydrosphere increased in size, conditions became favorable for the origin of life, in spite of the fact that our planet was not yet full grown.

Now let us review briefly what the early sequence of events must have been if the earth, in originating in accordance with any of the molten-earth hypotheses, was once a liquid globe.

If once molten, the earth-materials arranged themselves according to their densities, just as relatively light slag floats on the heavier iron in a blast furnace. Thus, the surface of the globe is made up of its lighter materials and its interior of the heavier substances. But since convection currents, liquid immiscibility, and other modifying factors all played a rôle in this gravity arrangement, the earth is not made up of perfectly delimited shells of different density.

During the molten stage the temperatures must have been sufficiently high that many earth-materials existed in a gaseous state. Some of these gases undoubtedly were dissolved in the molten earth-material, but there also must have been a hot gaseous atmosphere. This atmosphere and the gas dissolved in the liquid earth must have contained all the carbon dioxide now incorporated in our carbonaceous rocks, all the oxygen in our oxidized rocks, all the gases of our present atmosphere, and all the great quantity of water in the earth's hydrosphere.

As heat was radiated off into space, the temperature of the globe was reduced until silicates crystallized out at the surface. Because of their high specific gravity they sank, and with a long continuation of this process there was a tendency for the lighter granitic materials to be concentrated at the surface. With further cooling a thin crust of granite tended to form. But because of the great tidal effect of the sun upon the still rather incompletely shelled, and otherwise liquid, globe, the crust was disrupted again and again. As it was broken, the solid fragments, being denser than the liquid out of which they crystallized, tended to sink. Melting and remelting, crystallization and recrystallization, probably followed in monotonous succession; but eventually there came a time when the globe completely crusted over. At this period of earth history there should have been an original continuous, relatively smooth crust of granite.

After the crust had formed, cooling of the earth should have been much slower, inasmuch as convection currents no longer could play a major rôle in the dissipation of internal heat. But as the temperature of the crust diminished, there at last came a time when the first rains fell. After many downpourings the hydrosphere was initiated, probably as a shallow, essentially universal, freshwater ocean burying the universal granitic crust. Or, if sufficient cooling of the interior took place prior to the formation of the hydrosphere, the crust may have been wrinkled as it shrunk to fit this cooling interior; and thus the oceans would early have been confined to the natural depressions. Under the molten-globe hypothesis the earth becomes suited for the development of life at a much later date than under the solid-earth hypothesis.

WE HAVE now seen in outline form the different paths followed by early earth history under the two alternate hypotheses of solid earth and liquid earth. Let us next point out some of the strong points and weak features of each of the two hypotheses.

SOME consider it simpler to achieve the layering of the earth by cooling from a liquid globe than by growth by accretion of planetesimals, but the observed density stratification could be the result of either method. Furthermore, the rather indefinite boundaries between some of the zones are much more easily explained by the solid-earth hypothesis than by the alternate view of earth origin.

IT IS true that a cooling molten globe would develop a wrinkled surface, but the crustal shortening required by the earth's great mountain ranges (amounting to scores of miles for a single system) is greater than a cooling molten globe is able to supply. Nor can this cooling conveniently account for the origin of continental masses

EARTH'S BIRTH

and oceanic depressions, particularly when these major earth features exhibit the density differences previously commented on.

An original crust of the earth, which should be found somewhere if it ever existed, as the liquid-earth hypothesis requires, has never been discovered. This is a fact which favors the planetesimal hypothesis but does not by any means establish it.

THE amount of salt in the sea, according to some critics, is too great for the planetesimal concept to account for; but essentially it is equally difficult of explanation on the basis of a once molten earth, and it can be accounted for if all pertinent factors are properly evaluated. Early climatic conditions, as we now know them, favor planetesimal accretion and a growing atmosphere, rather than a molten globe and a diminishing atmospheric envelope.

VOLCANISM, at least during the later geological periods, is explained more readily on the basis of an original heterogeneous earth structure than on the basis of molten planet. But if we rely on the heat generated by radioactive disintegration, as indeed we must, volcanism can be accounted for by either hypothesis.

There are many other points; but we have already shown that the topic of earth origin, fascinating as it has always been to thinking man, is not yet an entirely closed subject.

ON THE basis of what we have now learned regarding hypotheses of earth origin, we may examine some of the ideas regarding the origin of continents and ocean basins which really depend on concepts of earth origin itself. In this connection there have been many explanations advanced, but we can mention briefly only five of the general postulates here. The first two of these hypotheses are predicated on an earth growing after the manner suggested by the Planetesimal hypothesis; the last three, on the basis of a once molten earth. The first of the five concepts is the **Segregation hypothesis** of T. C. Chamberlin. This idea involves an initial segregation of light planetesimal material in the continental areas and heavy material in the oceanic districts. Weathering, volcanism, and diastrophism are all considered to have increased this early differentiation into the observed light rocks of the continental earth segments and the heavy oceanic rock materials. The second, which is, to a certain extent, a modification of the first, is the **Asthenolith hypothesis** of Bailey Willis. This is a concept of differentiation of the granite rock of continents and the basaltic rocks of ocean basins from large blisters, or *asthenoliths*, of magma at or near the base of the earth's crust. The lighter, or granitic, magma is thought to erupt near the edges of the asthenolith, and thus the granite is concentrated on the continents. The magma then left behind in the great blister is relatively denser; thus the crust sags, and an oceanic depression is formed.

The third concept is Dana's **Differential-Contraction hypothesis.** According to this postulate, the continents are those parts of the earth which first solidified from the molten globe, and consequently they have contracted less than the oceanic areas. Joseph Barrell offered a fourth, or **Basic-Eruption hypothesis,** which is also based on a once molten globe. After this globe had crusted over in a solid granitic shell, remelting of basic material at depth caused the eruption of this basaltic magma through the original crust. Great areas of the crust were depressed by the weight of the basaltic material, and in these depressions accumulated the waters of the early oceans.

The late German geologist and meteorologist, Alfred Wegener, who lost his life in the exploration of Greenland hoping to find confirmation of his hypothesis,

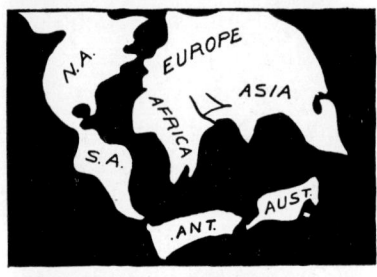

How the continents drifted apart, according to the Wegener hypothesis. Pennsylvanian, Eocene, Pleistocene, and present stages are shown. After Wegener

is generally considered responsible for the modern concept of **floating continents,** although the idea did not originate with him. This hypothesis, which is stylish in Europe today, postulates that originally there was but a single continent made up of the light granitic rocks which had solidified out of the molten globe. Relatively late in geological time this great continent broke up, and the Americas slid or floated westward on a base of yielding, heavy rock. The great mountain ranges along the Pacific border of the Americas are thus considered as wrinkles resulting from the lateral pressures developed by the westward continental movement.

Here is a fascinating and attractive concept which permits one to fit the continents back together like the parts of a gigantic jig-saw puzzle—and, surprisingly enough, the parts fit pretty well, as one can see from the accompanying diagrams. Moreover, some of the great mountain ranges coincide rather well when the various continental parts are brought back into contact; and some features of the distribution of ancient plants and animals are easily accounted for. Naturally, there are strong objections. There are as many poor fits in our puzzle as good ones. We have only seemed to put the parts together. There has been no adequate mechanism found which would cause the continents to migrate. Further, even if a force adequate to move the continents could be found, and the substratum were mobile enough to permit movement over its surface, it would, by the same token, be too mobile a substratum to permit the formation of great mountain ranges.

Obviously, here is a fine subject for a geological tempest, for the implications of the hypothesis are great. For instance, if it could be proved that the continents did float like granite icebergs on a sea of basalt, then it might be argued that the earth probably was once molten. In fact, so much controversy has raged over the Wegener hypothesis that as recently as 1926 the International Astronomical Union thought it wise to check up on it in a detailed fashion. The longitude of certain cities both in Europe and North America was determined more accurately than ever before, and their positions have been redetermined periodically. As a result of these observations it seems that Ottawa and Vancouver, Canada, in 1935 were 19 ft. nearer each other than they

Geologists with parts of a gigantic jig-saw puzzle

had been in 1926. During the same nine years, however, Washington, D.C., seems to have increased its distance from San Diego, California, by some 40 ft. To complicate matters further, the measurements show that the American continent is slowly moving toward Europe, rather than the reverse, as the hypothesis demands!

IT IS plain from this long chapter that we do not yet know for certain how the earth originated or how the continents and oceans came into being. But speculation on these matters has whetted man's mind and produced some of his greatest masterpieces of scientific reasoning. Even those hypotheses which we now know to be certainly false have aided man in determining more and more accurately the great sequence of early earth events before man or any other of the Earth's children were present to serve as witnesses.

CHAPTER 32

GARGANTUAN CALENDAR

> Some drill and bore
> The solid earth, and from the strata there
> Extract a register, by which we learn
> That he who made it and revealed its date
> To Moses, was mistaken in its age.
> —Cowper

VISITORS at the 1933 and 1934 Chicago Century of Progress Exposition saw in the Hall of Science a giant geological time clock. On this device the age of the earth was indicated as 2,000,000,000 yr. Priests and politicians, pedagogues and just plain people, calmly watched this clock mark off 10,000,000 yr. at a tick. Surely some of the millions of observers must have been skeptical regarding the figure given as the age of this terrestrial ball; but, surprisingly enough, no one was sufficiently disturbed about it to write a letter to the management. Within the past few years, even most laymen have come to know that the earth is "old even beyond tradition's breath."

Naturally, since the dawn of written history there have always been a few inquisitive-minded men speculating as to the age of the planet on which they lived. But prior to 1650 the common man had been given no very definite instructions as to what he should believe regarding this matter. Then, about the middle of the seventeenth century, Archbishop Ussher, of Ireland, declared that the exact date of Creation was October 26, 4004 B.C. Since he also was sufficiently omniscient to set the hour as precisely 9:00 A.M., it

.... a calendar fit for Gargantua himself

is small wonder that his dictum received church sanction, and not surprising that a few still stubbornly cling to his figures.

It is a far cry from Ussher's scant 6,000 yr. for the age of the globe to the Brahmin's view that the earth is eternal; the geologist's answer lies somewhere between these two extremes. Let us briefly examine some of the procedures the geologist uses in arriving at his estimate of the age of the earth.

APPROXIMATELY forty different scientific methods for evaluating geological time have been employed. These fortunately fall into four main groups, as follows:

1. By estimating the rate of loss of heat by the sun or earth.
2. By determining the rates of erosion throughout the past.
3. By calculating the rates of accumulation of salts and sedimentary beds.
4. By studying the rates of atomic disintegration.

THE first method must employ data which are almost entirely hypothetical. But, based on the assumption that both the sun and the earth were "simply cooling" bodies, Lord Kelvin, only a few short decades ago, determined that geological time could not be longer than 40,000,000 yr. This conclusion was arrived at since the earth, in simply cooling from a molten globe, should have, in all that time, become cooler than it now is. Since Lord Kelvin was a physicist of tremendous reputation, some geologists tried their best to trim their figures to fit his. This was difficult, for, since the time of the Scottish geologist, James Hutton, who, Ussher to the contrary, proclaimed (1785) that he could "find no vestige of a beginning,—no prospect of an end," the earth-scientists had been insisting that the earth was *very* old. By "very old" most of them meant approximately twice as old as Lord Kelvin's figures would permit. Fortunately, the physicists at last joined the geologists in pointing out the unreliable features of this heat-loss method of age determination, and it has gradually fallen into the discard.

GARGANTUAN CALENDAR

THE second method of estimating geological time, which is based on evaluations of erosion rates, is in general little more accurate than the first. It is, however, a method which can be employed in determining the age of such recent geological features as Niagara Falls. Since the recession of these falls has been studied for many years, an average annual erosional rate can be established. The determined yearly rate of approximately 5 ft. makes it possible to ascertain the time involved in the erosion of the original brink of the falls back to its present position. The various answers of from 18,000 to 25,000 yr. are all of the same order of magnitude. They give a reasonable idea as to the duration of geological time since the retreat of the last great ice sheet from the Niagara area, a retreat which permitted the waters of the upper Great Lakes to plunge over the Niagaran escarpment.

Retreat of Niagara Falls
See also Plate 37

THE manifold complicating factors involved in determining the age of Niagara are magnified many times in any attempt to ascertain the rate of erosion of an entire continent. Nevertheless, the amount of material annually carried in solution and suspension by the rivers of the United States has been determined by the Hydrographic Branch of the United States Geological Survey. The figures, which have been arrived at only after the most careful of calculations, are 270,000,000 tons of dissolved matter and 513,000,000 tons of material in suspension. Since the average weight of the earth's surface rocks is approximately 160 lb. per cubic foot, there is annually removed from our forty-eight states alone something like 10,000,000,000 cu. ft. of the stuff continents are made of. (Figure this out for yourself.) If you remember the area of the United States, you can easily determine that, at this rate, approximately 10,000 yr. will go by before an average of a solid foot of material is removed from the country's entire surface. Or, to lower the average height of the land by a mere 500 ft. will, at the present rate, require about 5,000,000 yr. Now, we seem to be getting somewhere!

But when we recall that the continents stand much higher now than throughout much of the past, we see that today's rate of erosion is far too high to be taken as the ancient average. Then, when we go still more deeply into the problem, we are confronted with the fact that several generations of lofty mountains have, in succession, been reduced to sea level by erosion. Moreover, for every change in the average elevation of the continents, there has been a change in the average rate of erosion. So that in the final analysis we can only safely say that this indefinite method of determining geological time indicates merely that the earth is prodigiously old.

THE third general method of time determination is closely related to the second. It is based on the rate of deposition of the eroded materials. But, unfortunately, there is no general agreement as to what the rate of deposition is. Herodotus, in the fifth century before Christ, observing the annual overflow of the Nile and the thin veneer of sediment the river deposited on the flood plain, realized that the building of the Nile Delta must have required thousands of years. A more definite idea as to the Nile's rate of deposition was obtained when in 1854 the 3,000-yr.-old statue of Rameses II was found about 9 ft. under river sediments. Hence the rate of deposition at Memphis, Egypt, is about 3.5 in. a century. But other investigations have indicated that the landward portion of the Nile has been elevated at a rate nearly ten times as fast, namely, nearly 3 ft. every hundred years. Since the total thickness of the sedimentary rocks of the world has been determined by Schuchert and others to be nearly 500,000 ft., or about 100 mi., the time required for the deposition of this gigantic column is something like 16,000,000 yr. on the basis of the more rapid rate of Nile sedimentation, and 160,000,000 yr. on the assumption that the slower deposition is more nearly the average. The rates of deposition of materials in the sea show similar wide variations, depending on a number of modifying factors. Furthermore, we know that the 100 mi. of sediment were not deposited continuously; the periods of erosion also have to be entered into the ledger, and no earth accountant has yet been able to evaluate them properly. Thus we can only say that the depositional method also indicates that the earth is millions of years old.

A VARIATION of this third method of age determination is based on calculations of the amount of salt in the sea. According to investigations sponsored by the National Research Council the total amount of salt in ocean water would, on precipitation, amount to nearly 5,000,000 cu. mi. Since the amount of salt annually contributed to the oceans by the rivers also has been pretty accurately estimated as about 160,000,000 tons, it is possible to ascertain roughly how long it has taken the seas to achieve their present degree of salinity. Most calculations result in a figure something in excess of 100,000,000 yr. Yet again, this does not take into consideration a number of complicating factors, and there are many indications that the annual rate used is far higher than the average of the past, so that again we are thwarted in our attempt to achieve great accuracy for our age determination.

Every "Old Man River" contributes his share of salt to the sea

BUT there are some deposits which are seasonally banded. These sediments can be examined in much the same way as tree rings are studied, and their age can thus be accurately determined directly. For instance, on this basis, the 2,600 ft. of banded Green River shales in Wyoming and Colorado probably required about 6,500,000 yr. for their deposition. Had all the 100 mi. of sediment we have previously mentioned been deposited at the same rate as these seasonally banded shales, well over a billion years would be involved. Recent glacial lake clays are also seasonally banded, the summer and winter layers being easily separable. Each double layer, representing a year's deposition, is called a **varve;** and since they can be counted, a direct determination of the time involved in their formation can be made. In addition a study of varves makes possible some significant correlations in connection with the most recent ice age. But these are points which will be discussed more fully in chapter 48.

THE fourth method of measurement of geological time is based on radioactive disintegration. This measurement is possible since radioactive substances, such as uranium and thorium, undergo a slow disintegration. During this radioactive decay the nuclei of the atoms lose alpha-particles and electrons, with the result that new elements are formed. An atom of uranium, for example, releases 8 alpha-particles and 6 electrons and is transformed into lead; but it is lead with certain peculiarities. That is, unlike ordinary lead, which has an atomic weight of 208, this lead has an atomic weight of 206. It is thus an isotope of the ordinary metal. Simultaneously the alpha-particles become helium atoms by the annexation of electrons.

Some radioactive elements disintegrate rather rapidly, but others change so slowly that billions of years may elapse before they are even half converted to lead. Uranium is one of the latter, and thus it is particularly well adapted to use for determining the great age of the earth. Usually the life of the radioactive element is expressed in terms of its "half-life," which is the time required for half of any original amount of the material to be transformed. In the case of uranium, this is of the order of 5×10^9 yr. for transformation to lead—a span of time inconceivably long. Thus, even if the earth were literally billions of years old, some of the original uranium would still be left.

If a curve is drawn of the proportion of undisintegrated uranium that remains from some original amount at any time, it will be like the following figure. Such curves are called "exponential"; and as they descend to the right, they flatten out and stretch along indefinitely. The half-life point lies on the curve at the value 0.5 along the vertical axis and is expressed in years along the horizontal scale.

As the uranium disintegrates, lead and helium result, so that we may picture the gradual diminution of uranium going on simultaneously with an accumulation of these decomposition products. The actual rate of decay has been found to be entirely independent of environmental factors, which is another point in favor of the method.

Now the problem faced by the geologist is to find a rock with uranium in it, and then, by a study of the amount of undecomposed uranium, in contrast with the amount of lead present or with the amount of helium present, to find out where along the curve the age of the rock lies. Igneous rocks, such as granites, are best adapted to this analysis, because they commonly contain small amounts of uranium in some of their minerals. Since we know that granite formed from a liquid state, any minerals present must date their origin from the time of crystallization. Thus, by taking samples of granite, which bear known relations to the sediments they intrude, it is possible to obtain not only a figure in years for the age of the granites but also some notion of the age of the associated sediments as well.

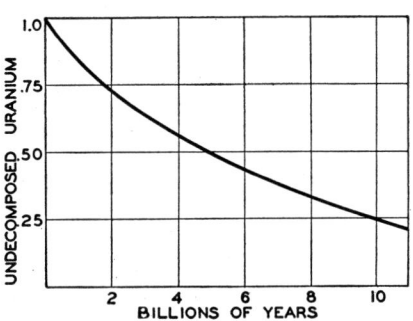

An exponential curve of uranium decay. Note that although the half life of uranium is 5 billion years, one-fourth of a given quantity remains undecomposed after 10 billion years

AS IN many other situations, numerous extraneous factors may enter the work. For example, primary lead may be present, and it may easily be confused with the radioactively formed lead of slightly different atomic weight. It is necessary also to have entirely unweathered samples for analysis. Moreover, if thorium is present with the uranium, as it commonly is, the mixture must be separately determined, since the disintegration rates of uranium and thorium are not the same. Furthermore, the method is applicable directly only to igneous rocks. The age of sedimentary beds in most cases can be interpolated only with difficulty from the determined age of the igneous rocks with which they are associated. Nevertheless, an

GARGANTUAN CALENDAR

WHEREIN ALL GEOLOGICAL TIME IS COMPARED WITH A YEAR IN WHICH EACH SECOND MARKS THE LAPSE OF 100 YR., EACH MINUTE, 6,000 YR.

Eras*	Periods and Their Duration in Millions of Years		Important Physical Events	Important Organic Events	Equivalent Calendar Dates if Ordinary Year = 3,162,240,000 Yr. of Geological Time
CENOZOIC 38— 60— 60 M.Y. ago	Quaternary	Holocene or Recent	Youthful land forms, high relief. Great Valley of California is a present geosyncline which originated as early as the Jurassic	Man makes iron implements beginning about 1,350 B.C. Man makes bronze implements beginning about 3,500 B.C. Man makes polished (Neolithic) implements beginning about 18,000 B.C.	About 11:57 P.M. Dec. 31
		Pleistocene Epoch 2 M.Y.	Glacial climate. Four great ice advances, separated by warmer interglacial intervals	Crô-Magnon man—first of present species Neanderthal man—Paleolithic culture Heidelberg man—Paleolithic culture	
	Tertiary	Pliocene Epoch 10 M.Y.	World-wide elevation continues; seas restricted	Intermigration of North and South American mammals. Horses and elephants become almost modern in appearance	7—days
		Miocene Epoch 18 M.Y.	Cascadian revolution begins: Cordilleras, Alps, Himalayas formed. Widespread volcanism: basalt flows in northwestern United States. Uplift results in cooler climatic conditions	Notable advances in the horse and elephant families. Apes appear in the Old World	
		Oligocene Epoch 10 M.Y.	Seas marginal; climate equable	Carnivores and ungulates become well established. Early ancestral elephants	
		Eocene Epoch 20 M.Y.	Seas marginal; extensive terrestrial sedimentation	Dawn of mammalian dominance. Reptiles in subordinate position. Coming of modern types of marine invertebrates	Dec. 26
MESOZOIC 200 M.Y. ago		Cretaceous 70 M.Y.	Widespread epicontinental seas; Cordilleran geosyncline the main site of deposition. Extensive chalk deposits here and abroad. At close of period occurred the Laramide revolution, during which the Cordilleran geosyncline was folded, and the Rocky Mountains formed	Introduction of modern flowering plants. Toothed birds and flying reptiles known. Marine pelecypods very abundant Extinction of all Mesozoic types of reptiles and birds at close of period. Climax of cephalopods. Armored and horned dinosaurs reach climax just before extinction	16+days
		Jurassic 38 M.Y.	Continent still largely emergent, but shallow seas encroach on northwestern and western positions of North America	First birds appear. Crustaceans take on modern aspect. Cephalopods very numerous. Gigantic reptiles common. Marine and flying reptiles abundant	
		Triassic 32 M.Y.	Continent emergent; seas marginal. Climate arid as shown by red beds, salt, gypsum. Extensive terrestrial deposition	Appearance of primitive mammals. Reptiles dominate. Cycads prominent. Marine invertebrates reduced in number	Dec. 9

*Figures at the right indicate age in millions of years and geological position of some rocks which have been dated on the basis of the radioactive disintegration of their uranium minerals.

GARGANTUAN CALENDAR—Continued

Eras*	Period and Their Duration in Millions of Years	Important Physical Events	Important Organic Events	Equivalent Calendar Dates if Ordinary Year = 3,162,240,000 Yr. of Geological Time
PALEOZOIC 550 M.Y. ago — 207— 310— 374— 380— 443— 450—	Permian 35 M.Y.	World-wide continental uplift and mountain-building: the Appalachian revolution. The Appalachian geosyncline folded; the Ouachita geosyncline also folded in early stages of the revolution. Widespread aridity as shown by red beds. Widespread glaciation	Extinction of most Paleozoic types of invertebrates and many Paleozoic plants. Reduction in all types of life. Primitive reptiles fairly numerous locally	
	Pennsylvanian 45 M.Y.	Continent alternately sinking and rising relative to sea level, forming marine deposits and coal swamps. World's greatest coal deposits formed	Great coal-forming forests. Earliest-known reptiles. Trilobites become rare. Brachiopods, pelecypods, and gastropods common	
	Mississippian 35 M.Y.	Low lands and widespread submergence. Climate warm	Great abundance of crinoids. Sharks numerous. Amphibians become well established	
	Devonian 35 M.Y.	Widespread submergence. Local volcanism and terrestrial deposits	First amphibian record. Armored fish numerous. Brachiopods reach climax. First forests known. Coral reefs important	40+days
	Silurian 25 M.Y.	Widespread submergence. Local aridity as shown by salt, gypsum	First air-breathers in the form of scorpions. First great abundance of corals and coral reefs. Eurypterids reach climax. Trilobites diversified. Land floras appear	
	Ordovician 70 M.Y.	Greatest submergence of North America: 60+ per cent of continent beneath epicontinental seas	First vertebrate record in the ostracoderms. Climax of invertebrate dominance. Brachiopods, trilobites, and cephalopods most numerous and characteristic	
	Cambrian 105 M.Y.	Transgressing seas develop widespread submergence. Appalachian, Cordilleran, Ouachita geosynclines main sites of deposition. Climate mild	Most invertebrates present. Trilobites most abundant and characteristic. Brachiopods and gastropods fairly numerous	Oct. 29
PROTEROZOIC 1,200 M.Y. ago 573— 587— 898— 913— 966— 1024— 1087—	Keweenawan 250 M.Y.	Great volcanism in central and eastern North America. Important copper, silver, gold, nickel deposits. At the close of the Proterozoic occurred the widespread Killarney revolution, during which the Ontarian geosyncline was folded into mountains. These mountains were completely eroded before the Paleozoic era opened	Most of the invertebrate phyla probably represented, but skeletons poorly developed, and fossils rare	75+days
	Huronian 250 M.Y.	Sedimentary processes dominant. Three main geosynclines: Ontarian, Appalachian, Cordilleran. Formation of world's largest iron-ore deposits. Earliest known glacial episode.	Bacteria and seaweeds present. Sponges and protozoa probably present	
	Temiskaming 150 M.Y.	Algoman disturbance and granite intrusion, after first period of sedimentation		Aug. 14
ARCHEOZOIC 2,000 M.Y. ago 1257— 1465— 1852—	Keewatin 800 M.Y.	Laurentian revolution. World-wide intrusive igneous activity. Development of extensive mountain ranges, all subsequently eroded before the Proterozoic era opened Igneous activity dominant; some sediments. Rocks highly metamorphosed. Records largely lost	Blue-green algae present Primitive one-celled plants and animals probably present	92+days May 14
AZOIC 3,162 M.Y. ago. Arbitrary date†	Not subdivided 1,162 M.Y.?	Birth of earth from the sun. Later stages marked by the formation of the rocks from which the earliest Keewatin sediments were derived	Colloids and organic solutions present but no actual life	134+days Jan. 1

† For purposes of calendar comparison in righthand column only.

ever increasing store of apparently reliable data procured by the radioactive method is available to us. These data indicate that the age of the earth is *at the very least* 2,000,000,000 yr., since the oldest rock thus far analyzed, from Carelia, Russia, has an age of 1,850,-000,000 yr.; and there is little reason for thinking that it is anyways near the oldest rock on the face of the earth.

There are many ramifications to the study of radioactive disintegration, one of the most recent being the investigation of pleochroic halos in mica. This and many other studies quite beyond the scope of this volume nevertheless show that the fourth method of age determination is, for measurements of great lengths of time, the most accurate yet devised.

THE principal events of earth history are indicated in the preceding table, which forms a calendar fit for Gargantua himself. The dates for major divisions are given as the best estimates available today from the studies of radioactive disintegration. Although obviously tentative, there is now little reason to believe that they are not of the right order of magnitude. The dates given in the right-hand column are, of course, fictitious. They are based on a year of 365 days in which a minute is equivalent to 6,000 yr. Although used for the purpose of comparison only, it is interesting to note that many modern estimates of the age of the earth place the date of origin at least 3,000,000,000 yr. ago. In other words, our Gargantuan calendar, which makes it possible to express any geological date in terms of the days and months of an ordinary year, is not entirely of the wrong order of magnitude.

CHAPTER 33

PRE-CAMBRIAN

>   Oh, I am so old, meseems
>   I am next of kin to Time.
>   —"A. E."

EVERY good old-fashioned home used to have an attic that served as a storehouse for all the odds and ends that were too good to be thrown out, yet not quite good enough to be used downstairs. Most sciences, too, have one or more "attics" to which are relegated those phenomena which do not lend themselves to ready explanation, or those organisms and objects which do not conveniently fit into any of the ordinary classifications.

GEOLOGISTS had such a catchall for many years in the sequence of rocks the older scientists called *"Primary,"* but which we today designate as **pre-Cambrian,** much in the manner that the historian uses the expression "before Christ." The beginning of the Cambrian is such an important geological date because from that time on our records are reasonably clear and complete. Moreover, an unbroken thread of life runs through the later periods, knitting their story together and giving it chronological continuity. The pre-Cambrian rocks, essentially lacking this record of life, cannot even be correlated very accurately. In addition, because they have suffered deformation in all of

. . . . Sir William Logan made detailed studies . . . .

the earth's great periods of orogeny since the time of their origin, it is obvious that they must in general be more highly metamorphosed than any of the younger strata. In fact, the pre-Cambrian rocks are so badly altered, and show such wide variations in composition and structure, that it has been well said that they are only "homogeneous in their heterogeneity." This may be a gilt-edged scientific expression, but it nonetheless conveys the right impression. Little wonder that the early geologists called these beds of rock "Primary," or "Primitive," and dismissed them as too complicated to be deciphered.

IN 1843, however, Sir William Logan made detailed studies of these "Primary" rocks along the north side of the St. Lawrence River, and later investigated the strata north of Lake Huron. To the granite and its metamorphosed gneissic equivalents in the St. Lawrence area he gave the appropriate name **Laurentian.** Then he discovered another sequence of rocks which he knew were younger than the Laurentian, since they rested unconformably upon the massive, crystalline Laurentian strata and contained boulders of granite and gneiss derived from them. Since these younger beds were typical of the country northeast of Lake Huron, they were named **Huronian.** Then at last another set of still younger "Primary" beds, the copper-bearing sequence of Michigan, was found to rest unconformably upon the Huronian sediments. This system of rocks came to be known as **Keweenawan** from its typical exposure on the great Keweenaw peninsula which juts out into Lake Superior.

Thus three great divisions of the pre-Cambrian were recognized in type areas. Subsequently, rocks in other districts were correlated with those in the typical exposures. Thus, by using most of the criteria for correlation outlined in chapter 27, geologists were at last able to bring some order out of the original "Primary" chaos. In unraveling the sequence of events for this ancient period, organic remains, of course, were of no assistance. Reliance had to be placed in such criteria as (1) composition and texture of the rocks, (2) order of superposition, (3) relative importance of the unconformities, (4) general or regional structure, and (5) dating of times of igneous intrusion and extrusion. Finally, as late as 1885, A. C. Lawson, by

PRE-CAMBRIAN

using the fifth criterion, found that the Laurentian granites were far from the oldest rocks, since they were intruded into the ancient **Keewatin** sediments of the Lake of the Woods district.

This was a most iconoclastic discovery since it laid low forever the fond belief of many geologists that the Laurentian granites represented the "original crust" of a once molten globe. Moreover, it not only added a new and major division to the pre-Cambrian, but it greatly extended the geologists' concept of the age limits of the time. This was inevitably so, since the new concept required the Keewatin *sediments* to have been derived from some still older set of rocks not yet discovered.

NEW discoveries of importance are still being made in the field of pre-Cambrian geology almost every year. The present state of knowledge, which is far from definitive, nevertheless enables us

General Time	Era	Period	Major Events
Pre-Cambrian	Proterozoic	Keweenawan	16. Lipalian erosional interval 15. Killarnean mountain-building and granite intrusion 14. Keweenawan volcanic activity and sedimentation
		Huronian	13. Post-Animikiean erosion 12. Animikiean sedimentation 11. Cobalt sedimentation and glaciation 10. Bruce sedimentation
	Archeozoic	Temiskaming	9. Post-Algoman erosional interval 8. Algoman mountain-building and granite intrusion 7. Temiskaming sedimentation
		Keewatin	6. Post-Laurentian erosional interval 5. Laurentian mountain-building and granite intrusion 4. Keewatin volcanic activity 3. Coutchiching sedimentation
	Azoic	?	2. Period of erosion 1. Formation of pre-Keewatin rocks

to give a rough outline of the major events for the pre-Cambrian of the southern Canadian Shield. Some of the highlights of this history are presented in the foregoing table, in which the events are listed from 1 to 16, the oldest being No. 1. This table may look formidable, but it actually does little justice to, and makes short shift of, dramatic events that required well over a billion years for their transpiring. So for those who prefer additional information, and for those who dislike tables, we may further summarize as follows:

THE pre-Cambrian rocks belong to two great eras, the **Archeozoic** and **Proterozoic,** and at least four periods are represented, although there is still no universal agreement as to classification. But regardless of such academic questions, we are *certain* of a number of facts. These rocks do indeed comprise the "basement complex," since all younger rocks rest on them. In other words, no matter where one were to drill, if the hole could be carried deep enough, the pre-Cambrian rocks finally would be encountered. The pre-Cambrian strata, however, appear at the surface at many places, notably (*a*) **in the cores of great mountain systems** where major uplift and profound erosion have combined to uncover them, and (*b*) **in the stable, positive continental areas known as "shields."** The largest of the latter is the Canadian Shield, in which about 2,000,000 sq. mi. of these ancient rocks are exposed. But the surface of much of Scandinavia is also made up of these rocks, as is about 300,000 sq. mi. of Australia, much of northeastern South America, and a goodly part of Africa south of the Sahara. In North America they are by no means confined to the Great Lakes area and the Canadian districts. They occur in the Grand Canyon, where the *Vishnu schists* are probably the equivalent of the Keewatin formations, and the *Unkar-Chuar* rocks are doubtless later Proterozoic representatives. They are also found in the Northern Rockies, where the thick *Belt* series is notable for its relatively slight metamorphism; and they occur in such mountain areas as the Appalachians, the Adirondacks, the St. Francis Mountains of Missouri, in the Black Hills and the Arbuckles, and in the very core of the Colorado Rocky Mountains, as at Pike's Peak. Some of these areas are illustrated in Plate 38.

PRE-CAMBRIAN

IN THE Great Lakes area it may be noted that there were three great periods of mountain-building and granitic intrusion followed by profound erosion. These represent events 5 and 6, 8 and 9, and 15 and 16 in our table. Only the "roots" of these ancient mountains are discernible now, but their ancient folding has given rise to an easily deciphered "grain" in the country. This grain, observable

ERA	PERIOD AND EPOCH	GROUP AND FORMATION	KIND OF ROCK	PREVAILING COLOR	THICKNESS IN FEET	DIAGRAMMATIC PROFILE
MESOZOIC	TRIASSIC	Shinarump congl.	Conglomerate	Brown	25	CEDAR MOUNTAIN
		Moenkopi fm	Shale and sandstone	Red	480	SURFACE OF KAIBAB AND COCONINO PLATEAUS — RIM OF GRAND CANYON
		—UNCONFORMITY—				
PALEOZOIC	PERMIAN	Kaibab limestone	Limestone and sandstone	Gray, buff and red	525	Sea deposits with marine shells, etc.
		Coconino sandstone	Sandstone	Light buff	350	Probably dune sands with tracks of primitive reptiles and amphibians
		Hermit shale	Sandy shale	Red	225	Foot-prints; primitive "evergreens"; fern-like plants; insects; and sun-cracked silts — ESPLANADE
		Supai formation	Sandstone and shale with some limestone	Red and gray	825	Red flood-plain deposits with land animals and plants
		—UNCONFORMITY—				← Old land surface
	CARBONIFEROUS (MISS.)	Redwall limestone	Limestone	Gray stained red	450 to 500	Sea deposits with shells, corals, etc.
	DEVONIAN	Temple Butte ls.	Limestone and sandstone	Pale purplish	0-36	← Fish scales
		GREAT UNCONFORMITY				
	CAMBRIAN (Tonto group)	Muav limestone	Sandy shale and limestone	Gray	100	Sea deposits with shells and seaweeds
		Bright Angel shale	Sandy shale	Greenish gray	450 to 650	Trilobites TONTO PLATFORM
		Tapeats sandstone	Sandstone	Brown	225	Seaweeds
PROTEROZOIC	ALGONKIAN	GREAT UNCONFORMITY Unkar and Chuar groups of Grand Canyon series	Sandstone and shale with some limestone; contains sheets and dikes of lava	Mostly red	0 to 12000	Shinumo quartzite "GRANITE" GORGE Hakatai shale Bass limestone Vishnu schist Pegmatite
ARCHEAN		GREAT UNCONFORMITY Vishnu schist	Schist, granite, and gneiss	Dark gray	Not known	

Generalized columnar section of the rocks in the walls of the Grand Canyon. Note the relationship of the Vishnu schist to the Unkar-Chuar rocks, and these in turn to the Paleozoic sediments. After L. F. Noble

both in the sediments and in the granitic or gneissic masses, makes it possible to determine the size and orientation of some of these eroded mountains. For instance, the old *Killarney* Range, which doubtless was of Alpine grandeur, ran somewhat north of east-west across the Great Lakes district, and probably had a length of well over 1,000 mi., although only 700 mi. of its eroded roots are exposed. Unfortunately, there is still some disagreement as to whether the three granites forming the cores of the three ancient mountain systems are actually distinct. When at last determinations of their ages are made by using the data obtainable from the radioactive disintegration of their uranium-lead minerals, this old question should finally be settled. Similarly, the great *Grenville* limestone of eastern Canada does not even appear in our condensed table, because, although it is one of the thickest limestones known, we are not yet certain whether it is Huronian or Keewatin in age. This mystery also will be solved when the age of the granites intruded into the limestone is accurately determined by geochemical methods.

BUT, whatever else we know about the pre-Cambrian, we are certain that even in this remote past most of the physical events were going on much as today. The most ancient rocks we know are *sediments* not greatly different from the modern ones except in their lack of organic remains. In fact, many of these age-old deposits were laid down in geosynclines whose alinement and extent have been determined. Moreover, mountains were folded out of these basins then much in the same fashion as in later periods or as at present. Truly, the pattern for more recent physical events in earth history has been modeled on an ancient and well-worn style.

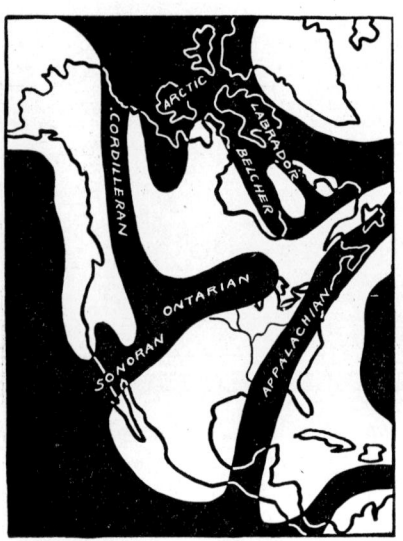

The location of the ancient geosynclines shown in black. After C. Schuchert

The pre-Cambrian rocks of the Canadian Shield have yielded mineral wealth in enormous amounts. The great "iron ranges" of the Lake Superior region have produced more than 50,000,000 tons of ore, valued at well over $100,000,000, in a single year. The ore occurs chiefly in the Huronian rocks. Originally it was an iron carbonate with some iron silicate. Owing to weathering, the primary ore was oxidized near or at the surface, and then concentrated in part by descending solutions, in part by the removal of siliceous materials by ground-water action. The enriched ore thus re-

PRE-CAMBRIAN

mains in a *residual concentration*. In the famous Mesabi Range it is sufficiently soft and fragmental to be mined by steam shovels.

In addition to yielding these Lake Superior ores, the Proterozoic rocks are the source of the iron at the great Kiruna, Sweden, mines; and the iron deposits of this age in Minas Geraes, Brazil, are the largest in the world. Here rich ores form entire mountains; and, although they are as yet little exploited, they promise a great supply of the metal in the future. Most of these Proterozoic iron ores have been formed through *sedimentary* processes. We will next consider the pre-Cambrian metals that owe their origin to, or were intimately associated with, Archeozoic or Proterozoic *igneous* activity.

NATIVE copper is found in the Keweenawan lavas of the peninsula of that name in northern Michigan. This is the site of the early Indian workings in the native metal. Mined actively for more than a century (since 1830) the peak annual production of the region was about 135,000 tons, valued at approximately $65,000,000 in 1916; and it has now yielded, as a district, nearly 4,000,000 tons of the copper. The metal, interestingly enough, not only serves as a cement in the conglomerate in which it occurs, but it also replaces in part the pebbles themselves. More than 90 per cent of the world's supply of nickel is also obtained from a single late pre-Cambrian gabro sill, which has a fabulously rich nickel-copper ore border. Located at Sudbury, Ontario, the production includes, besides nickel and copper, a number of the rarer metals.

SILVER and gold are other products of the late pre-Cambrian igneous activity. Perhaps the richest silver area in the world is located in the vicinity of Cobalt, Ontario, where the native metal occurs in veins associated with a large sill. Some of these veins have on occasion yielded several thousand dollars worth of silver from a single ton of ore! One of the most noteworthy gold districts is Porcupine, which is almost midway between Lake Huron and James Bay. The gold occurs chiefly as the native metal and in veins associated with a granite which is intruded into Keewatin lavas, and is unconformably overlain by mid-Huronian rocks. Not discovered until 1912, the district has nevertheless produced several hundred million dollars worth of gold, and its Hollinger mine has been considered one of the richest the world has known. The Kirkland Lake area is another, and more recently developed, spectacular gold district; and noteworthy new discoveries are made almost yearly.

So important are all these great mineral deposits to Canada that the government, in the summer of 1935 alone, spent $1,000,000 in organizing geological prospecting parties whose business it was to scour the brush of the southern Canadian Shield in the hope of finding new deposits.

We have already spoken of the rich iron-ore endowment of the pre-Cambrian in continents other than our own. Similarly, gold is by no means confined to the North American pre-Cambrian. Rocks of the same age carry the extraordinary gold-bearing conglomerates of the South African Transvaal, where the Witwatersrand, or "Rand," district has already produced about $6,000,000,000 worth of the yellow metal; and silver and gold occur bounteously in the pre-Cambrian of Australia. Thus it can be safely said that events which transpired nearly a billion years ago have been of the greatest importance in shaping the industrial and economic pattern of recent human history.

AMONG the more notable events of the pre-Cambrian were the two Proterozoic periods of glaciation. So far as we can now determine, the first of these occurred at sometime after the beginning of the Proterozoic era; the second, near or at its close. The basal conglomerate of the mid-Huronian Cobalt series of Ontario contains faceted and striated boulders of large size. They are overlain by what appears to be an indurated glacial boulder clay; and beneath them, in some places, is found a striated and grooved glacial pavement. Here is, therefore, unmistakable evidence of the oldest of the earth's definitely known glacial episodes; and the distribution of the tillite indicates

Areas in which there is evidence for Proterozoic glaciation

further that this ancient continental glacier was no mere patch of snow. As a matter of fact, some investigators feel that this ice sheet was more than a thousand miles across in the area just north of the Great Lakes.

Other evidences of glaciation at about the same time are found in the ancient tillites exposed on the Lufira River of Katanga, Africa, west of Lake Tanganyika; in the indurated tills of the Nama series of Southwest Africa; and in the thick older tillites of South Australia.

Later Proterozoic glacial deposits are now known from widely scattered localities. They occur in the Flinders Range of South Australia at least 1,500 ft. below beds which are known to be early Cambrian in age. Those of Norway, Greenland, China, northern Scotland, and the Wasatch Range of Utah are probably younger still. In fact, some geologists feel that they should be classed as early Cambrian; but, since the early Cambrian seas supported a limestone reef-building coral-like animal in abundance, warm-water oceans, not glacial climates, apparently are indicated for this epoch.

In spite of the essentially world-wide distribution of these glacial deposits, as is indicated by the accompanying map, the Proterozoic era is not to be considered solely or chiefly as a glacial period. The glacial episodes probably were of relatively short duration, and they were doubtless far from contemporaneous on the various continents. They do seem to coincide rather well with periods of continental uplift and mountain-building. In this respect the late Proterozoic glaciation was similar to the late Paleozoic period of refrigeration.

WE SAID that pre-Cambrian times were not all frigid. This is a statement that can be demonstrated in various ways. There are thick strata of limestone in the pre-Cambrian rock sequences, and many graphite beds as well. The former is a fairly certain indicator of warm seas, and the latter suggests the presence of life in considerable abundance. For the graphite was originally disseminated in black shales as carbonaceous material, which doubtless was derived in large part from primitive organisms such as seaweeds. In fact, the graphite beds of the Grenville limestone alone have been estimated by Sir William Dawson to carry as much carbon as our great Carboniferous coal seams!

Section of Grenville limestone showing irregular black masses of graphite—indirect evidence of life. After H. A. Nicholson

But these are *indirect* evidences of life. What *direct* proofs do we have of the existence of life in the pre-Cambrian? Unfortunately, there are not many. In fact, there are less than most scientists like to admit. For, after years of search for fossils in rocks of this age, and the publication of hundreds of papers on the supposed remains discovered, only a few of the "finds" bear the stamp of unquestioned authenticity. In this category belong the blue-green algae, brown algae, sponges, "worms," and a few arthropods, such as those discovered recently by Sir T. E. David in Australia. "Slim pickings," indeed, for the record of a billion years!

THE sudden appearance of the relatively diversified Cambrian fauna, therefore, has been rather difficult to explain. On theoretical grounds at least, this fauna suggests an enormously long period of evolution prior to its appearance. We shall examine this Cambrian fauna in chapter 36; but we may anticipate a bit to point out that most of the major invertebrate groups had a representation in it, and that highly organized representatives of the most advanced invertebrate phylum were dominant forms in its makeup. What, then, happened to the record of most of their pre-Cambrian ancestors? In attempting to answer this question, a number of theories have been advanced, of which the following are typical:

1. **Pre-Cambrian fossils were destroyed by metamorphism.**
2. **Pre-Cambrian organisms had no skeletons, since there was no available calcium in the seas.**
3. **Pre-Cambrian oceans were acid; thus calcareous skeletons could not be secreted.**
4. **Pre-Cambrian rocks which have been studied thus far were all deposited on the land in fresh water.**

5. **Organisms originated in the soil, and migrated by the rivers to the oceans, which were not reached until Cambrian times.**
6. **Pre-Cambrian organisms had no hard parts because they lived in the surface waters of the sea, where skeletons, because of their weight, would have been a detriment.**
7. **Pre-Cambrian organisms had no hard parts because the sessile or sluggish mode of existence had not yet appeared.**

Let us now take up these seven theories in order and briefly consider their merits. The first explanation doubtless is sound for all those pre-Cambrian rocks which have been altered by heat and pressure, for we know definitely that such alteration will obliterate fossils in beds normally replete with such remains. And we have already seen that most of these ancient rocks have been severely metamorphosed. But there are notable exceptions to the general rule. The Belt series of Montana, the Keweenawan of Michigan, parts of the Huronian of Ontario, and the *Jatulian* of Finland, among others, include pre-Cambrian formations which have escaped extreme deformation. It is true that such pre-Cambrian fossils as do come to light generally turn up in these same unmetamorphosed beds, but their general paucity of organic remains obviously must be explained on some ground other than metamorphism.

THE second theory has far-reaching implications which are entirely beyond the scope of this volume. But in summary form this concept holds that there were no scavengers in the pre-Cambrian oceans. Thus, organic matter, decomposed by bacteria, generated ammonia in these ancient seas. The reaction of the ammonium carbonate with the calcium which had been brought to the oceans by the rivers, in the form of calcium chloride or sulphate, would precipitate calcium carbonate. Thus the calcium salts would not be available for organisms, which, having no hard parts, had little chance to write their record indelibly. It is true that when known scavengers became numerous in Ordovician times invertebrate skeletons became much heavier. But this may well be a mere happenstance. Critics point out that, had these early oceans been ammonified, not only would lime ooze have been precipitated but life itself would have been annihilated. As a matter of fact, nitrifying bacteria at present limit the amount of ammonia in the sea by breaking it down to produce nitrites and nitrates, which in turn serve as food of larger marine plants. There is no reason for thinking the bacteria were not on the job a billion years ago.

THERE is some reason for believing that during the early history of the oceans they were slightly acid, as the third theory suggests. Under such conditions calcareous skeletons would be rather out of the question. Squaring with the theory is the fact that primitive radiolaria and sponges have siliceous, not calcareous, skeletons; and the skeletons of some other early animals are largely made of chitin. But the early blue-green algae somehow were not embarrassed by the supposed oceanic acidity, for they went merrily on secreting calcium carbonate in large quantities.

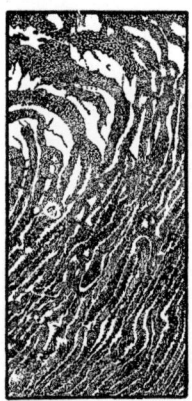

Fragment of *Eozoan*, $\times \frac{2}{3}$, one of the many so-called pre-Cambrian "fossils," which is probably inorganic in origin. After Sir William Dawson

C. D. Walcott, formerly director of the Smithsonian Institution, spent a score of disheartening years trying to discover pre-Cambrian fossils. It is not surprising, therefore, that he should be the author of the fourth theory. He believed that during the pre-Cambrian the continents were much larger than at present, and hence that all deposits which we examine were formed in shallow, non-marine bodies of water upon the continents. If we actually want to find pre-Cambrian fossils of marine type, we will have to look beneath the present oceans. But we have already learned that there were true *marine* embayments in these early days. We have even examined a map showing their extent and position. This is not the naïve statement it may seem at first sight, since one must know considerable about an ancient seaway before he has the temerity to abandon the generalities of a word picture of it in favor of the particulars of a map, however crude.

IF ONE accepts the Planetesimal hypothesis, outlined in chapter 31, then the suggestion of its senior author that life originated on the land, and in the soil, is a sweetly reasonable one. We shall have occasion to meet this latter concept again in the following chapter. For the moment we may merely point out that as the ancient waters of the soil finally found their way via springs, brooks, and rivers to the sea, the newly formed organisms were thereby transported to a new home. In the fresh waters of the land, there was no real reason to secrete hard parts, although in the salt waters of the sea such a process could readily have been initiated. In any case the period of transportation from site of life-origin to sea may have been so long that organisms may not have reached the oceans until Cambrian times. If so, it is futile to look for pre-Cambrian life in the marine deposits of that time.

THE last two theories merge into one upon close inspection. The pre-Cambrian animals lived in the surface waters and were floating or free-swimming creatures unencumbered by calcareous skeletons. At death they fell to the bottom or were eaten by other creatures, and in either case had little or no opportunity to be permanently incorporated into the rock record. But suppose they sank to the hospitable shallow oceanic bottom because they were too sluggish to keep afloat, or because they had been merely crowded out of the overpopulated surface waters, or because they were actively driven out by the advent of the first predacious carnivores. Then the active animals probably continued to swim, or at least crawl about in their new environment. But the more passive attached themselves to the bottom; and, because they were inactive, the secretion of calcareous skeletons began, since they were not able to get rid of the excess calcium carbonate. For it can be pretty well demonstrated that the formation of a calcareous skeleton is an involuntary chemical process inseparably linked with inactivity.

Probably some features of each of the majority of these theories combine to explain the poor record of pre-Cambrian life. But it seems fairly certain that most pre-Cambrian animals were motile and hence without calcareous skeletal parts. It has been suggested therefore that when additional fossils are found in these ancient strata, they will have either no skeletons at all or ones made of materials such as silica or chitin.

W E HAVE now been talking rather glibly, even if relatively briefly, about some aspects of life in the more ancient geological past throughout a number of chapters. Therefore, we may well pause in our general story in order that we may insert a chapter concerning the very origin of living things on the earth. Such a discussion is required at this point in order to serve as a background for the study of the following geological periods, all of which are marked by an abundant record of the life of their times.

CHAPTER 34

LIFE BEGINS

> Without parents by spontaneous birth
> Rise the first specks of animated earth.
> —Erasmus Darwin

TODAY the world teems with life—every little niche in nature, no matter how inhospitable it may appear to man, seems to be occupied. And little wonder, for there are no less than a quarter of a million species of plants and more than a half-million distinct kinds of animals! Since this vast number merely represents the kinds which scientists have described, and takes no account of the undiscovered forms of life, it is certainly erring on the side of conservatism to assert that there must be more than a million kinds of modern animals and plants. Moreover, paleontologists have described well over one hundred thousand kinds of fossils, and yet this great number represents probably no more than 5 per cent of the species of the earth's once living things. With this plethora of past and present organisms it would be a dull scientist, indeed, who failed to ask himself, "Where and how did all this profusion of life originate?" Even the uneducated must realize that it did not just spring up like Topsy.

.... animated earth ....

The creation of Adam, from the *Nuremberg Chronicle*, 1493

YET, after many centuries of crude speculation on this subject, and several hundred years of scientific, or at least semi-scientific, probing of the questions, it may as well be confessed that no one yet definitely knows the answers. How, when, and where life first began are still baffling queries. Answers have indeed been made, some of them with assurance; but many, especially those made by scientists, have been advanced with extreme diffidence. In the main these answers fall into three categories, as follows: (1) **Life originated as a result of creation through divine, or at least superhuman, motivation;** (2) **life was introduced on this planet after having originated on some other heavenly body;** and (3) **life originated as a result of the synthesis of inorganic elements into organic compounds through the action of natural laws.** It is almost unnecessary to say that many, if not most, laymen and a rather considerable number of scientists subscribe to the first answer in some form or another. The second solution deserves little consideration, since it merely removes the question from the terrestrial scene. Finally, it may be stated that most *students* of the subject believe the third answer to the puzzling question is the correct one. Let us therefore examine this solution in a little greater detail.

WHEN Strata Smith was laying the foundations for the modern science of geology at the beginning of the nineteenth century, there was little room for doubting that organic and inorganic matter were poles apart. But throughout the past century the demarcation between the living and the non-living has steadily grown less sharp, until today the gap has almost ceased to exist. Thousands of organic substances have now been synthesized. Hydrocyanic acid, standing in water, gives rise to urea and other substances found in living tissue. Formaldehyde, sugars, and nitrogenous organic substances can be generated under conditions similar to those that must have been common in pre-Cambrian times. The agents and conditions requisite for the synthesis of such organic materials are the simple things such as sunlight, carbon dioxide, and surface waters with which this mundane ball has been amply blessed ever since the appearance of the primitive oceans.

Even as this chapter is being written, there is new information concerning the borderland between the dead and the quick. For some years it has been evident that the viruses, which are smaller than any visible cell, nevertheless show the very *sine qua non* of life—the power of reproduction. These viruses are apparently intermediate in position between the smaller and lifeless organic molecules, which form spontaneously at the earth's surface, and the bacteria and cells comprising much larger molecular aggregates. Closer by far to typical life than the sugar and amino acids, they actually may well merit the designation "living."

SCIENTISTS have recently prepared a crystalline protein which possesses the properties of tobacco-mosaic virus. The protein has a larger molecule than egg albumen, is insoluble in water, and its properties remain unchanged after ten successive crystallizations. But on the living tissue supplied by a tobacco leaf the protein apparently becomes a virus which can reproduce itself indefinitely. With this remarkable discovery in mind, we can understand that many virus-like molecules which are not disease-producing must have, for that very reason, failed to attract our attention. Thus we must admit that we have not yet even gotten a good glimpse into that undoubtedly fascinating organic realm which lies somewhere betwixt the simplest bacteria and the spontaneously formed sugars and amino acids. But it can no longer be regarded as fantastic to suggest that in this realm creation of life is by no means a lost art.

IF THERE is even a possibility of the synthesis of new life today, how much easier the whole process must have been in that well-nigh lost limbo of earth history in the late Azoic or the early Archeozoic eras. How much greater must have been the chance of successful synthesis at a time when there was not an already existent swarm of hungry, and therefore hostile, living things. Pondering these questions as early as a score of years ago, T. C. Chamberlin outlined the geological setting he considered requisite for the best chances for natural organic synthesis of life out of inorganic substances. Since the great advance of biological information on this subject during the last twenty years has failed to alter the required geological setting materially, let us consider briefly some of the points in Chamberlin's thesis.

DOWN TO EARTH

Thomas Chrowder Chamberlin
1843–1928

"Aristotle, 322 B.C.; Copernicus, 1543, A.D.; Galileo, 1642; Newton, 1727; Laplace, 1827; Darwin, 1882; Chamberlin, 1928."

The foregoing succinct appraisal by Bailey Willis is suggestive of the greatness of the philosophic mind which developed, among many other fruitful theories, the Planetesimal hypothesis and the theory for the origin of life outlined in this chapter

The contacts of earth, air, and water have always been the sites of the greatest geological bustle and activity. Today, as throughout the ancient past, these contacts have not only been the scenes of deposition and of erosion but also the home of life itself. In such locations there is now, as always, (1) not only **an adequate supply of the sun's radiant energy** but some protection from an excess of it. Moreover, there has always been, at the contact of land with any body of water, (2) **a sufficiency of the materials which most commonly are requisite for life** as we know it. These include oxygen, nitrogen, carbon dioxide, water itself, and lesser amounts of phosphorus, sulphur, potassium, alkalies, and the alkaline earths. But in addition there must have been (3) **some mechanism for holding, protecting, and preserving the products of each synthetic advance** so that the next step forward could become a reality. There also must have existed (4) **a circulatory mechanism to bring suitable supplies to the new combination of materials,** to concentrate them there, and to carry away the excess and the waste.

We can find the required circulation, possibly ancestral to the typical organic sort of today, in the soils, for evaporation takes place at their surface and there is a capillary supply of moisture from below. Moreover, as we have said, the earth's crust, together with its atmosphere and hydrosphere, is an unfailing source of the elements required by life. The infall of meteorites in the ancient past, as today, brought a series of unstable carbides, phosphides, and nitrides. Spontaneously decomposing in the atmosphere or at the earth's surface, an entirely new set of compounds was formed. The opportunities for the chemical action and reaction, requisite for the first organic synthesis, thus were boundless.

IT HAS been well said that water is the birthplace of life, but by this statement ordinarily is meant sea water. The oldest-known fossils are indeed remnants of marine life; and when any life line is traced into its simplest ancestral types, the trail leads inevitably back to the sea. But the marine waters offer too great an opportunity for diffusion to make them seem the logical site of the origin of life. Concentration of the materials is required, not dilution. The early soils, and the waters of the land, not the primitive seas, are therefore pointed to as the more likely site for synthesis. In the zone of ground-water saturation, every imaginable relationship between solutions, air, and mineral matter must have existed. The slow circulation of the ground water tended to concentrate solutions in favored places where chemical reactions could be begun and continued. Many an inanimate trial and error must have taken place. Many a reaction must have occurred, many a synthesis must have been nearly completed, only to have the delicate balance between the reactions and their physical surroundings broken by a geological cataclysm. But we must realize that the place was only limited by the extent of the early continents themselves, and that the time available, although not illimitable, was nevertheless to be measured in hundreds of millions of years. Finally, at some place favored by all the short-term requisite conditions for synthesis, and marked also by comparative geological stability, the trick was turned.

HOW far back in geological history this dramatic event occurred none can say for certain. If the earth originated after the manner postulated under the Planetesimal hypothesis, it may well have taken place even before the earth achieved its present size. But in any case it probably occurred at a time more ancient than that recorded by the oldest of the earth's surface rocks. If the first important achievement in Life's long and varied career be regarded as the fabrication of the primitive viruses, we still must allot a great deal of time before the first appearance of bacteria. From bacteria to lime-secreting algae requires another long journey upward on the steep ascent of life; but on this climb the plants somehow acquired chlorophyl, their remarkable green coloring matter.

The importance of chlorophyl cannot be overstated. Though it is almost exclusively a property of plants, even animals could not exist without it, for this green coloring matter enables the former to utilize the sun's radiant energy to make starch from carbon dioxide and water. Since animals cannot manufacture either starch or protein out of inorganic matter, plants are the ultimate source of these substances in the animal world. Thus it is certain that plants long antedated animals, as we now know them, on the face of the globe. Moreover, through the long years of the well-recorded history of life, which we are to examine in later chapters, we will see again and again evidence of the fact that the appearance of new plants has always paved the way for new steps in animal evolution. Statements like the foregoing regarding the later manifestations of life on this second-rate planet we may make with some assurance—regarding life's earlier history we can only make what we hope are shrewd guesses based on limited data.

BUT no matter when, how, or why life first originated, there is little or no reason for doubting that every living thing today represents some twig-end on an ever expanding tree of life whose deepest roots go far back into the pre-Cambrian. Looking down this long vista of the past, the geologist, more than any other scientist, is able to visualize the great antiquity of life and the vast extent of its ramifications. In fact, the story of the various stages through which life has gone is a major part of geological history, beginning with the Cambrian period. And, since Man is merely an end-product of the most ancient successful life-synthesis, to the geologist he is, in a very literal sense, formed of the actual "dust of the ground."

CHAPTER 35

EARLY PALEOZOIC EVENTS

> This—all this—was in the olden
> Time long ago.
> —Poe

IF THE eager reader can be persuaded to turn back to the table on page 299, he will find that the date the curtain goes up on the early Paleozoic scene is October 29. By this we mean that, if all geological time since the birth of the earth is reckoned as 365 days, then nearly 10 months of our Gargantuan year will have elapsed before the beginning of the Paleozoic era. We have waded through the events of those scant 10 months in giant strides. There was nothing else to be done, since we do not know *with absolute certainty* a single happening before May 14; and for the succeeding months, up to October 29, there are several weeks whose records are essentially missing, and many a day for which the facts are entirely unknown.

Now, beginning with the Paleozoic era we must proceed more deliberately. It is true that many an hour from this time on is still unaccounted for, although the weeks and the days all have a relatively coherent story to tell. But when Roderick Murchison and Adam Sedgwick first started to decipher the history of the early Paleozoic periods over a century ago (1831), they ran into all sorts of difficulties. Murchison, working in the borderland between England and Wales, realized that the rocks he was studying represented a new system, which, as we have seen before, he designated the

Adam Sedgwick

Silurian. Sedgwick, studying the strata of the north of Wales, recognized a still older set of rocks carrying the most primitive fauna that had, as yet, been discovered. To this system he gave the name **Cambrian,** since the type area was in the ancient Roman province of Cambria. Unfortunately the rocks which Sedgwick called *Upper Cambrian* turned out to be the same as those designated *Lower Silurian* by Murchison. Out of this situation was brewed a fine scientific controversy, which was not settled until Professor Lapworth of Birmingham suggested (in 1879) that the middle set of strata of the three natural divisions of the early Paleozoic rocks be called **Ordovician.** Like the Silurian this system also was named after an ancient Celtic tribe, the Ordovici. Today all the arguments have ceased, and these three periods are universally designated by the three names just given.

WE HAVE seen in chapter 28 that the seas have commonly inundated the land and have written records of their former presence in the deposits left behind after their withdrawal. We have also learned that the geosynclines are *relatively permanent* sites of marine embayments, but even such basins of deposition characteristically give evidence of the universality of the alternation of marine invasions and widespread retreat of the oceanic waters. Such alternation, with the cyclical implications earlier touched upon, obviously results from the periodic deepening of the oceanic basins or the uplift of the continents, or both. But in any case, the ultimate cause is diastrophic; and thus we begin to see a point at which we have previously hinted, namely, that **major earth movements must form the ultimate basis for separating geological time into its logical historical compartments.** The largest of these earth movements naturally serve as the bases for the largest divisions of time, the eras. Thus, at the close of the Proterozoic, the great Killarney revolution, with its mountain-building, continental uplift, and widespread erosion, makes the separation of the Proterozoic era from the Paleozoic a relatively simple matter. Similarly, shorter and less intensive periods of diastrophism serve to separate periods such as the Cambrian, Ordovician, and Silurian. Disturbances of

EARLY PALEOZOIC EVENTS

still less consequence cause temporary withdrawals of the seas, so that the break in sedimentation and erosion resulting from such oceanic retreats can be used to separate the deposits of one epoch from the next succeeding one. The periods of the early Paleozoic are no different from any others in this respect, division into epochs being a perfectly natural result of the early Paleozoic oscillations of the sea.

WE HAVE learned in chapter 28 not only that the periods are commonly divided into epochs but that the division is usually a threefold one. The Cambrian is no exception to this general rule. In fact, the threefold division of this system of rocks can be recognized in many countries, where, of course, local names are given to the epochs. In North America the designations commonly employed have had the usual geographic origin based on areas in which there is typical development of the rock sequences. These names are:

Upper Cambrian = **Croixian,** named from the rocks at St. Croix Falls, Minnesota

Middle Cambrian = **Albertan,** named from the rocks of the Alberta Rockies

Lower Cambrian = **Waucobian,** named from the rocks at Waucoba Springs, Nevada

Let us now briefly examine the record of Cambrian physical events. At the beginning of the period all of the continents were land areas which had been subjected to deep weathering during the long Lipalian interval of erosion. Thus the first deposits laid down by the earliest encroaching Cambrian seas were the re-worked sands which had been millions of years in accumulating during Lipalian planation. The sites of these early seas are indicated on the accompanying map, which shows the alinement of the two great geosynclines which characterized the North American continent for much of the Paleozoic era. **These basins of deposition were the Appalachian trough on the east and the Cordilleran geosyncline on the west.** Their positions can be thought of as being essentially, but not quite, those of the Appalachian and Rocky Mountains of today. They had originated in a modified form as early as the late Proterozoic, as is shown by the map on page 306.

.... Appalachian trough on the east ....
Cordilleran geosyncline on the west ....

The early Cambrian, or Waucobian, seas were confined to these geosynclinal areas; and, although they were relatively large, they never inundated more than approximately 20 per cent of the continent. Thick sediments of the early Cambrian occur in both geosynclinal areas, the rocks being over a mile thick at Waucoba Springs, Nevada, and 3,000–4,000 ft. thick in the Appalachian trough. Since these rocks are predominantly of clastic material, with large amounts of quartzite, two conclusions may safely be drawn. First, the basins were progressively sinking, or else the shallow-water conglomerates and sandstones would not persist through such great thicknesses. Second, the lands bordering the basins must have stood relatively high or their streams would not have been able to carry such large quantities of clastic material into the basins. When the sediments are examined in detail, they show further that the bordering highlands lay to the east of the Appalachian trough and on the west of the Cordilleran geosyncline. This eastern land mass has been called **Appalachia,** and the western one is usually designated **Cascadia.** The great interior of the continent, or the old rock platform, however, was standing not far above sea level. Nevertheless, its contribution of sands to the Cambrian formations was notable. This is particularly true for the Cordilleran trough, since the sediments there indicate that the land mass *to the west*, although standing relatively high, was not being vigorously eroded.

The Middle Cambrian, or Albertan, sediments of the Cordilleran area are thick shales and shaly limestones, being particularly well represented in the southern Canadian Rockies of Alberta. They

EARLY PALEOZOIC EVENTS

indicate a widening of the basin of deposition and a reduction in the height of the adjacent lands. Not so much is known regarding the Middle Cambrian of the east; but here, again, shales are overlain by limestones in the southern Appalachian trough, indicating progressive reduction of the height of the mountainous borderland of Appalachia.

FINALLY, the late Cambrian, or Croixian, seas overflowed the boundaries of the geosynclines, and at a maximum flooded as much as 35 per cent of the continent. At this time sandstone was deposited over a considerable area in which wind-blown sands had previously been accumulating on the old Canadian Shield. The exposures of these re-worked sands at St. Croix Falls, Minnesota, where they are 1,700 ft. thick, are regarded as typical; but thick sandstones also occur on the flanks of the Adirondacks. As these seas spread over the negative element of the continent (see page 248), the waters became deeper in the geosynclines, where limestones and dolomites were being deposited.

A MINOR mountain-making period, which has been called the *Vermont disturbance*, separates the Cambrian and Ordovician periods. In the St. Albans, Vermont, area the evidence for this orogenic movement is seen in a basal Ordovician conglomerate, and in a minor angular unconformity between the Cambrian and the Ordovician formations. But on the whole, the Cambrian closed quietly in North America. The withdrawal of the late Cambrian sea from the continent, however, indicates that diastrophic movements must have taken place elsewhere in the world. Europe, for instance, was the scene of greater disturbances, volcanic action being prevalent in the type of Cambrian area of Wales.

THE Ordovician period is notable for its great inundations, the largest of which literally had the North American continent awash from stem to stern. Like the Cambrian, the Ordovician rocks also have a threefold division based directly on fluctuations of the Ordovician seas or, ultimately, on diastrophic movements. The names of these epochs, together with their more important formational types, are shown in the accompanying table.

System Divisions	Epoch Names	Appalachian Region Formations	Mississippi Valley Formations
Upper Ordovician	= **Cincinnatian** =	Juniata sandstones	= Richmond formation
		Lorraine sandstones	= Eden-Maysville shales
Middle Ordovician	= **Champlainian** =	Trenton limestones	= Galena dolomite
		Black River limestones	= Platteville limestone
		Chazy limestones	= St. Peter sandstone
Lower Ordovician	= **Canadian**	= Beekmantown dolomites	= Oneoto-Shakopee dolomites

As the Ordovician period opens, we find the Canadian seas moving into the Cordilleran, Appalachian, and Ouachita geosynclines,

Generalized paleogeographic map to show positions of Ordovician and Silurian troughs. After C. Schuchert.

as is indicated on the composite map for the Ordovician and Silurian which accompanies this chapter. At the beginning of this first Ordovician inundation both the Canadian Shield and the negative element of the old rock continental platform were emergent. But before this epoch was over, the negative element had been pretty much encroached on at a number of places. Certain small uplifts, such as the *Ozark Dome*, the *Adirondack Dome*, and the *Cincinnati Arch*, however, were minor positive structural elements which persisted through much of the Ordovician as emergent, or only slightly submerged, areas. They thus modified greatly the Ordovician paleogeography and, as we shall later see, played an important rôle in the geography of some of the subsequent Paleozoic periods.

The Canadian formations are of two very different lithologies, or facies. There is a dark shale facies rich in graptolite remains which accumulated in the northeastern portion of the old Appalachian trough, and a dolomitic facies which has a widespread distribution in the area of the southern Appalachian geosyncline and in a submergent area extending from Texas north into Minnesota. Similar facies occur in the Canadian rocks of the Cordilleran geosyncline, the dolomitic phases representing the deposits laid down near the middle of the basin, whereas the shale facies again occurs nearer the old land mass, as one would expect.

At the close of the Canadian epoch there was such a complete withdrawal of seas from North America that, largely on this basis, some scientists have attempted to elevate the Canadian epoch to period rank.

THE Champlainian formations are widespread, a goodly portion of the Canadian Shield having been inundated at this time. The rocks are commonly limestones, or shaly limestones, in marked contrast to the dolomites on which they commonly disconformably rest. The St. Peter sandstone, at the base of the Middle Ordovician series, however, represents residual sands re-worked by the encroaching Champlainian seas. It was during this epoch that the Cincinnati Arch was first elevated and the northern part of Appalachia began to rise. The result of this latter uplift is seen in the increased amount of clastic material in the eastern Champlainian formations. Another evidence of crustal unrest at this time is the record of volcanism in the eastern United States and northeastern Quebec. Ash falls are recorded in ash beds throughout the Appalachian geosyncline from New York to Alabama, and they also occur as far west as Wisconsin and Iowa.

THE Cincinnati seas spread over about 65 per cent of the continent as we now know it. The formations of this epoch are limestones and calcareous shales in the interior, grading into shales, sandstones, and finally conglomerates as one approaches the source of clastic materials in the slowly rising land mass of northern Appalachia. These coarse sediments stand as mute witnesses of a relatively important orogenic movement which is designated the *Taconic disturbance*. Other evidence for this mountain-building is found in the angular unconformities which occur between Ordovician and Silurian formations. The center of activity was in western New England and eastern New York, an area which has suffered so much in more recent disturbances that not all the details are as clear as we should like. We know, however, that, like most periods of mountain-building, its influence was felt far in the interior in the gentle bowing-up of such areas as the Cincinnati Arch and the Ozark Dome. We have already given evidence that the effects of the Taconic disturbance were felt long before mountain-building was at a climax. We shall later see that further effects will be felt until the ancient mountains are eroded to their very roots. Do not make the common mistake of considering that mountain-building is a physical event which took place with great rapidity. In the late Ordovician it was a slow process, just as it is today.

THE Silurian period, like its two predecessors in the early Paleozoic, is divisible into three epochs. These are listed, together with some of the important formations, in the accompanying table.

System Divisions	Epoch Names	Appalachian Formations	Mississippi and Ohio Valley Formations
Upper Silurian	= **Cayugan**	= { Waterlimes / Salina shales }	= Absent
Middle Silurian	= **Niagaran**	= { Lockport limestones / Clinton sandstones and shales }	= Louisville limestones
Lower Silurian	= **Alexandrian**	= Medina sandstone	= Brassfield limestones

At the beginning of the Alexandrian epoch North America was a low-lying continent except for the bounding land mass of Appalachia on the east. West of the Cincinnati Arch, clear arctic seas spread southward until they finally merged with southern interior

EARLY PALEOZOIC EVENTS

waters. In this area limestone deposition was the rule, but east of this low dome the early Silurian sediments were almost entirely clastic. The Medinan sandstone and its equivalents have a maximum thickness of 1,000 ft. and extend from New York to Alabama. Throughout this area these sediments represent the erosional products of the dwindling Taconic Mountains.

DURING early Niagaran time the area of Appalachia was sufficiently reduced that only fine clastics were poured into the geosynclinal trough. These sediments have long since consolidated to form the Clinton fine sandstones and shales. As these seas cleared, limestone deposition was prevalent throughout the interior of the continent in a great marine embayment almost as extensive as that of Middle Ordovician times. The formations are well exposed at Niagara, where the Lockport dolomite forms the very brink of the falls, and the Clinton sequence makes up the undercut lower part of the Niagaran escarpment.

At the close of the Niagaran epoch the Silurian seas were greatly restricted, being largely confined to the geosynclines and adjacent small embayments. In such a restricted basin of deposition were formed the salt and gypsum beds and the shaly limestones of the Cayugan epoch.

The Silurian rocks correlated between Niagara Falls and Rochester, New York, showing the typical detail into which we do not enter in the text. After Alling and Hoffmeister

There are extensive deposits of Silurian rocks in the Cordilleran area, the sequences being especially well developed in Nevada, California, Alberta, the Mackenzie River valley, and Alaska. Unfortunately, the precise ages of these rocks have not yet been completely determined.

No mountains were formed in North America at the close of the Silurian, deposition being essentially continuous from the late

Silurian into the early Devonian in the northern Appalachian trough. There is a great break between the Middle Silurian and the Middle Devonian rocks in certain parts of the central states, however, since neither late Silurian nor early Devonian waters markedly overflowed the ordinary confines of the geosynclinal basins. Although North American orogeny did not occur at this time, there was considerable volcanic activity localized in Maine, New Brunswick, and the Chaleur Bay district of Canada. And in western Europe the great **Caledonian Mountain disturbance** closed the Silurian period. This disturbance formed a great mid-Paleozoic range of folded and faulted mountains extending in a 4,000-mi. arc from Wales through Scandinavia and across the arctic islands into and along the north coast of Greenland. Their orogeny, and the resultant angular unconformity, serves as a perfect physical demarcation between the Silurian and Devonian formations in the British Isles, as we shall see in chapter 37.

THE climate of the early Paleozoic was mild and equitable throughout most of the 200,000,000 odd years involved in its long history. This is indicated by the distribution of marine animals

.... went dry ....

which seem not to have been influenced in their habitats by climatic zones of the present-day sharpness of definition. Widespread coral reefs in the Silurian seas and great limestone deposition throughout this and the other periods are further criteria suggestive at least of warm seas. But there are also evidences of minor episodes of glaciation in the Ordovician (possibly Silurian) glacial tills of northern Norway. Furthermore, additional climatic variation is seen in the late Silurian deserts of what are now the eastern Great Lakes states. Someone has facetiously said that this is the first time that North America went dry. Certain it is that this is the earliest *conclusive* record of aridity. In the district just outlined an arm of the Appalachian trough dwindled down to a very saline basin. As evaporation went on, and new supplies of sea water probably continued to enter the desiccating basin, enormous quantities of gypsum and salt were precipitated. Since both of these products occur over an area of some 100,000 sq. mi., arid conditions persisted for a long time, and a prodigious quantity of sea water must have been evaporated. On the southwestern shores of the ancient salt basin, in northwestern Ohio and southern Michigan, dune sands, now called the Sylvania sandstones, well over 100 ft. thick were accumulating, much as they do in the modern dunes on the southeastern shore of Lake Michigan today.

THE economic products of the early Paleozoic belong chiefly in the Ordovician and Silurian formations, the Cambrian being largely barren of important economic resources. The Silurian salt and gypsum just mentioned are extensively mined, the Salina deposits of New York State alone yielding about one-fourth of our domestic supply. The St. Peter sandstone of the Ordovician and the Sylvania of the Silurian, being nearly pure quartz sand, are largely used in the manufacture of glass. The Silurian Clinton iron ore is famous as an early source of the metal, although it is not so important since the Proterozoic ores of Lake Superior region have become available. But about 10 per cent of our annual production still comes from this ore mined chiefly in the Birmingham, Alabama, district.

Ordovician rocks are an important source of oil and gas in a large number of oil fields, many of them, like the Oklahoma City field, marked by spectacular gushers. Similarly, rocks of this age are extensively quarried for *slate* and for *marble*. The Ordovician slates have originated through the metamorphism of Ordovician shales laid down in the Appalachian geosyncline, and the marbles likewise resulted from the metamorphism of the limestones. Pennsylvania, Vermont, and New York are the great producers of Ordovician slate, which accounts for 90 per cent of the American total. Most of the better finishing marble used in the United States also is Ordovician in age, the most notable quarries being near Rutland, Vermont. Phosphate rock in Tennessee, iron ore in Newfoundland, and lead and zinc in a number of the Mississippi Valley states are further evidence of the rich and varied mineral endowment of the Ordovician formations.

WE HAVE now discussed the physical features of the early Paleozoic with almost no reference to the life of the times. There is no more interesting aspect of these early periods than the history of their faunas and floras. Furthermore, it is a great mistake not to realize that all of the events we have just described played an important part in determining the course of organic evolution during these times. For the purpose of clarity of discussion, however, we have separated out the life-story of the Early Paleozoic from its physical history, and we will discuss it in the following chapter.

CHAPTER 36

INVERTEBRATE HEYDAY

> The leader's deeds and hard won glory live;
> This remains; this alone survives. . . . .
> —Ovid

THE history of any period of human activity is often written in terms of its ruling families. Thus the story of Florence of the fifteenth century would certainly be divested of much of its glamor, and obviously of most of its accuracy, were the Sforza, the Borgia, and the Medici left out of the picture.

Our history of life in the early Paleozoic periods likewise is largely a story of the ruling groups of invertebrates. And, just as the historian commonly gives a genealogical record in order to clarify his account of the ramifications of the ruling families of the period he is discussing, we too must here include a table in order to indicate the relationships of the various invertebrate groups which make the early Paleozoic history a living document.

A most intelligent member of a ruling family: *Isotelus*, an Ordovician trilobite. ×½

IN THE accompanying table most of the groups which have had important rôles to play are listed, and their geological ages given. We must point out in this latter connection, however, that the actual ranges of many of the classes probably are greater than we have indicated, since the known geological range of any group merely means the duration of time throughout which there occurs direct fossil evidence of its presence. In the following paragraphs we will amplify the table by discussing the outstanding points in the geological history of the early Paleozoic invertebrates. And when new or unusual animal assemblages are first mentioned, we will pause momentarily to describe their salient features.

Phylum	Classes	Known Geologic Range	Common Name or Living Example
Arthropoda	Insecta Arachnida Myriopoda Crustacea	Devonian? to present Pre-Cambrian? to present Silurian to present Pre-Cambrian? to present	Insects Spiders Centipedes Trilobites, crayfish
Mollusca	Cephalopoda Gastropoda Pelecypoda	Cambrian? to present Cambrian to present Ordovician to present	Squids, nautilus Snails Clams and oysters
Molluscoidea	Brachiopoda Bryozoa	Pre-Cambrian? to present Ordovician to present	Lamp shells Moss animals
Echinoderma	Echinodea Asteroidea Crinoidea Blastoidea Cystoidea	Ordovician to present Ordovician to present Cambrian to present Ordovician into Pennsylvanian Cambrian through Mississippian	Sea urchins Starfish Sea lilies Sea buds Sea bladders
"Vermes"	Subdivisions omitted	Pre-Cambrian? to present	Worms
Coelenterata	Anthozoa Scyphozoa Hydrozoa	Cambrian? to present Cambrian to present Cambrian to present	Corals Jellyfish Hydrozoans
Porifera	Porifera	Proterozoic to present	Sponges
Protozoa	Classes omitted	Cambrian to present	Amoeba

GEOLOGIC DEVELOPMENT OF THE PRINCIPAL INVERTEBRATE ANIMALS

DURING the Cambrian period essentially all the living things of this second-rate planet were confined to the sea. The lands were barren of all plant life save possibly a few of the simple types that had ventured into the low swampy areas bordering the oceans and lakes. But in the oceanic waters algae were present in abundance, and all of the major divisions of the invertebrate animals had some representation. Truly, this first well-recorded fauna had made notable advances from the first crude organic synthesis hundreds of millions of years earlier.

The Cambrian fauna is now known from approximately two thousand species. On the basis of this species discrimination the **trilobites** made up 53 per cent of the total. In other words, here is a

INVERTEBRATE HEYDAY

ruling class which is far from exclusive. Picture, if you can, a medieval Florence in which every other Florentine belonged to the Medici family! The **brachiopods** also were a ruling class of great importance and little exclusiveness, for they made up 30 per cent of the Cambrian fauna. Arthropods other than trilobites account for 5 per cent of the remaining 17 per cent; gastropods, 3 per cent; the primitive, coral-like reef-formers, or *archaeocyathids*, another 4 per cent; and the remaining five species out of every hundred are distributed among the *protozoans*, the *sponges, coelenterates,* "*worms*," *echinoderms*, and *cephalopods*. So far as we know, the chordates were not present, the invertebrates being the unchallenged lords and masters of what little they could survey in the commonly murky Cambrian seas.

THE trilobites, which became extinct at the close of the Paleozoic era, are the index fossils on the bases of which Cambrian formations are zoned. For instance, to the expert the genus *Olenellus* connotes Lower Cambrian; *Ogygopsis*, a trilobite with a perfect crossword-puzzle name, Middle Cambrian; and so on. The body of these ancient crustaceans was trilobate in the longitudinal direction, since a pair of grooves separate the rounded axial lobe from the two lateral areas. This trilobation gave the name to the group. Trilobites were also composed of three parts from front to rear, there being a head, a thorax, and a tail. The head and tail were covered by solid shields of the skeletal material (of chitin and calcium carbonate), but the thorax was jointed and flexible.

The trilobites, in common with other crustaceans, generally bore a pair of anterior feelers and possessed compound eyes, although a few were blind. Each body segment carried a pair of jointed legs made up of two branches. The lower portion was a limb used in locomotion, whereas the upper and outer branch bore feather-like gills, which may also have been used in swimming. For an illustration of these features refer to Plate 42.

Although the trilobites were the most intelligent, as well as the largest, creatures of their day, few were more than 3 in. in length. On the other hand, some of these early crustaceans were over 2 ft.

long, and may have weighed as much as 15 lb. Furthermore, when we stop to remember that the trilobites were the ruling class for the entire Cambrian period, or approximately 100,000,000 yr., we may then appreciate the fact that man will have to continue his overlordship of the organic world for another 98 or 99 million years before he can point to a dynasty as time-honored as that of these early arthropods.

BRACHIOPODS were playing second fiddle to the trilobites in Cambrian times; but they have had the last laugh, for they have outlived their betters by 200,000,000 yr. Few groups have left such a long and varied fossil record of their presence as the brachiopods, and as such they have always been favorites of the geologists, who use them extensively in correlating the ancient rocks. The brachiopods have a body anatomically related to some of the worms, but it is enclosed in a bivalved shell and is attached throughout life by a fleshy extension of the body called the "stalk," or *pedicle*. The valve, or shell, through which the pedicle is extruded is the ventral shell, and is commonly designated the *pedicle valve;* the dorsal shell bears the support for the gills or *brachia*, and hence it is called the *brachial valve*. These shells are not of equal size, the pedicle valve commonly being the larger of the two; but the shells are equilateral. These are points which should be remembered, for they help to separate this class from the clam shells, which they may closely resemble.

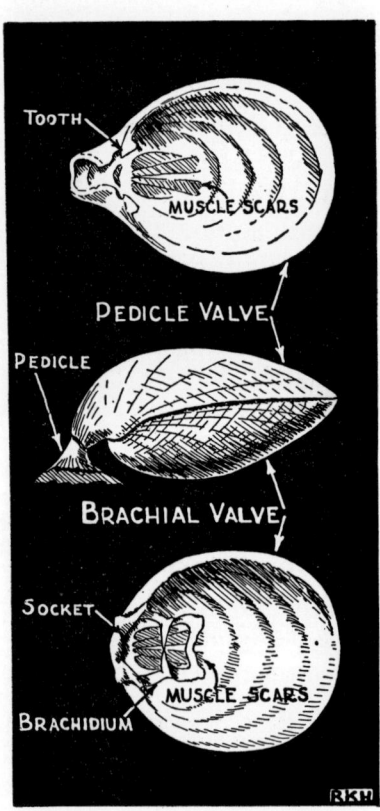

Terebratulina, a modern articulate brachiopod. ×2

THE classification of the brachiopods is a complicated one, more than seven hundred genera being involved. Roughly, however, they may be divided into *articulates* and *inarticulates*. The former are brachiopods with teeth and sockets for the articulation of the shells, which generally are composed of calcium carbonate. The latter lack definite hingement devices, and their shells commonly are chitinous or made of calcium phosphate. In the Cambrian it was the latter group of brachiopods which were the most abundant. Locally their shells may actually make up a goodly portion of the rock itself. Before the period was over, however, the articulate brachiopods began to become prominent, and the inarticulates never again were of any importance. In such unchanging forms as *Lingula*, however, they have clung tenaciously to life, and are still found in modern marine waters.

THE archaeocyathids were essentially confined to the early Cambrian seas, in which they built great reefs. These ancient structures are common in a great belt 400 mi. long in eastern Australia; and they also occur in Antarctica, Sardina, Spain, Newfoundland, New York, Quebec, Washington, and California. Such a world-wide distribution indicates uniformly warm seas at the time. Although their biological position is still in doubt, we may think of them as a sort of primitive coral.

THE gastropods were the only other relatively common members of the Cambrian fauna. Most of these ancient molluscs had simple, uncoiled, conical shells; but some possessed spiral shells with a few volutions. The *hyolithids* were early gastropods, related to the modern *pteropods*, or sea butterflies, whose long conical skeletons are common in some late Cambrian formations.

ALL of these classes of invertebrates to which we have allotted a little attention bore shells, but there must have been literally scores of important groups lacking in skeletal stiffening whose records have been lost to us. Fortunately, we can judge the general importance of these animals by the discovery of a great assemblage of them in the Middle Cambrian rocks called the Burgess shale, which crop out above the town of Field, British Columbia. All of these are preserved as mere films of carbon on a black slate, but in many specimens even details of the structure of the appendages and of the viscera are discernible. From this remarkable occurrence well over one hundred species of Cambrian life, totally unknown elsewhere, have been added to our knowledge of the fauna. Here, again, the inference is plain that we know but little regarding the totality of ancient life. The animals we do know rather well are commonly those turned up by some such happy chance as the discovery of deposits like that of the Middle Cambrian Burgess shale, part of whose fauna is portrayed on Plate 40.

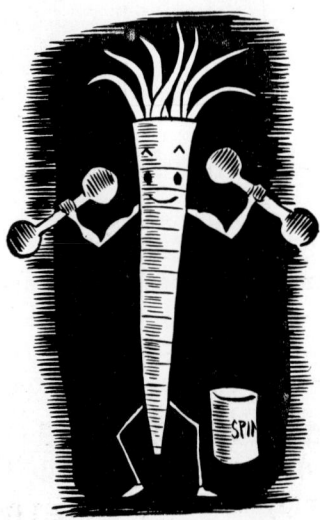

.... the cephalopods are the strongest ....

The ruling families of the Ordovician seas were the trilobites, the brachiopods, and the cephalopods. The first two of these we have met before and will meet again. The cephalopods, almost unrepresented in the Cambrian, are the largest and strongest animals of the Ordovician. They are such important and interesting marine invertebrates that we devote all of chapter 45 to their history. The interested reader is, therefore, asked to do something new, i.e., refer *ahead* for further details concerning the class. Sufficient for the moment to point out that the early cephalopods were externally shelled molluscs with either straight or coiled shells which were internally partitioned. Some of the Ordovician straight conical types attained a length of close to 20 ft. and had a maximum diameter of over 1 ft. Small wonder that the other denizens of the deep scurried before these predacious masters of the ancient seas.

THE trilobites of the Ordovician still constitute the dynasty of intelligence, and actually are more diversified than they were in the Cambrian. But so many other classes have improved their posi-

INVERTEBRATE HEYDAY

tion that relatively the trilobites are not so outstanding as they had been. Many of them have now learned to enroll, armadillo-wise, so that the soft parts of the ventral anatomy are protected. This may have been a response to the increasing menace from the carnivorous cephalopods, which not only were busy eating heavily at the trilobite's table but doubtless had begun to find the trilobites themselves rather palatable morsels.

THE brachiopods are the most numerous of the larger Ordovician invertebrates. Their deployment into a host of articulate types is as rapid as the decline of the inarticulates is precipitous. Most of the calcareous shelled articulate forms were striated or plicated; and the hinge line was long, thus making it more difficult for enemies to pry the valves apart to get at the soft bodies within.

IN ADDITION to the trilobites, brachiopods, and cephalopods, a number of classes poorly represented in, or absent from, the Cambrian assemblage have achieved positions of prominence in the Ordovician animal society. Among these are the true corals; the clams; the bryozoans, or moss animals; the crinoids, or sea lilies; and the *graptolites*. These latter extinct hydrozoans are so common in the darker Ordovician sediments that some scientists regard them as the most distinctive animals of the time. The graptolites, to which we have formerly referred in connection with index fossils, are coelenterates which are related to the modern colonial hydrozoans like *Obelia*. On many a dark shale their carbonized remains gleam like pencil marks when the specimen is turned so as to catch the light; hence the name *graptos* ("written") plus *lithos* ("stone"). But each mark is, in reality, a single or double row of chitinous cups on a rod-like central support. Some graptolites lived attached to seaweeds,

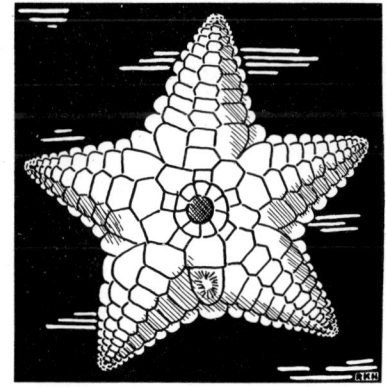

Ordovician starfish were very much like modern ones: *Hudsonaster*. ×2

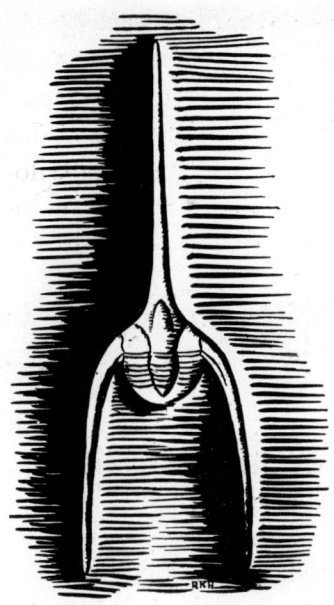

Ordovician trilobites were already becoming specialized: *Ampyx.* ×3/2

and some had floating organs of their own manufacture. In any case, this mode of existence in part accounts for their world-wide distribution and consequent use as index fossils.

At this point it may be well to repeat a statement which we have made before and, with your indulgence, will make again. Faunal realms and provinces were just as much a part of the Ordovician scene as they are of the modern organic world. The fauna of the Ordovician geosynclines, in which muds were accumulating, is markedly different from the life-assemblage of the clear central seas, just as differences in sedimentation greatly influence the ecology and distribution of modern invertebrate animals. Moreover, then, as now, there were areas of *provincial* life-development, as well as great regions of *cosmopolitanism*. In general the warm, clear seas of the middle epochs of Paleozoic periods were the sites of this cosmopolitanism in organic evolution, whereas the generally restricted early and late seas of a number of the periods harbored the provincial assemblages.

The Silurian invertebrates are descendants of the Ordovician stocks. In fact, we can consider the life of the early Paleozoic as representing one long essentially uninterrupted evolutionary trend. During this major earth cycle crustal unrest was at a minimum, continental submergence at a maximum. Consequently the times were ripe, and the

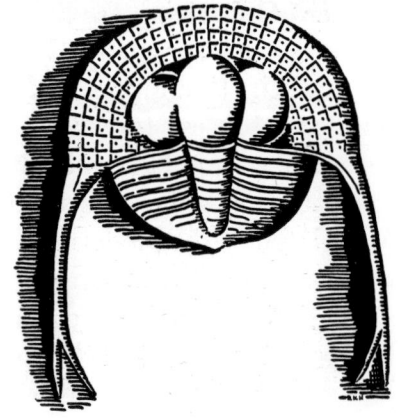

Trinucleus, a blind, burrowing Ordovician trilobite. ×1

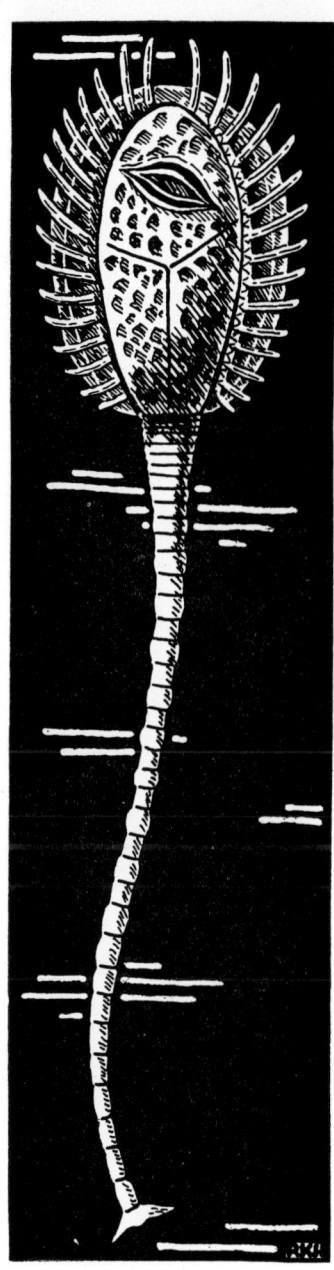 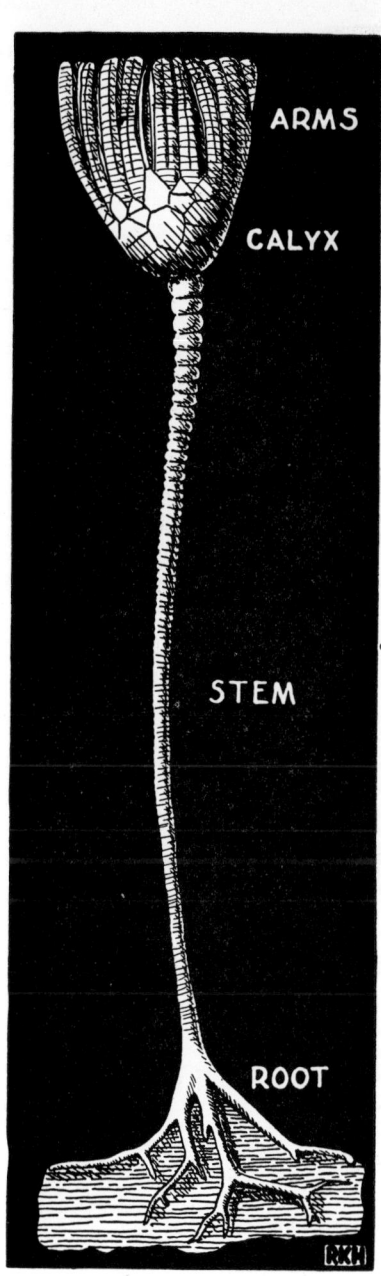

A Silurian cystoid: *Staurocystis*. ×3

A Silurian crinoid: *Eucalyptocrinus*. ×⅔

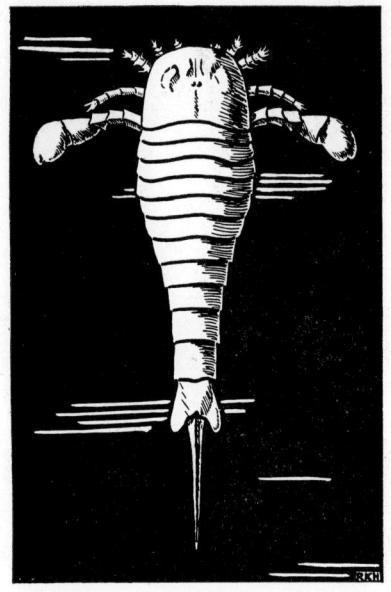

Eurypterus, a Silurian eurypterid. ×⅛

physical scene correctly set, to stage the greatest invertebrate show the earth has known. But, unfortunately, a minor part, created for the vertebrates as early as the Ordovician, is already assuming greater importance. New stars are to rise out of these obscure understudies.

Even as the show goes on we see the leads have changed since the last performance. The trilobites have already passed their climax and are going in for frills and furbelows in the form of spines and excrescences. These may have been protective devices against their old enemies, the cephalopods, or protection from their new foes, the fishes. But, in any case, it is interesting to note that here, as in so many other ancient groups, the appearance of spines and pustules is prophetic of the impending extinction of the animals which go in for such ornamentation.

The graptolites are already near the close of their short existence, and are greatly restricted in number, as compared with their Ordovician abundance. *Monograptus* types are the most typical. Bryozoa were still fairly common, but less so than in the Cincinnatian epoch. Corals were more abundant than ever before, and they built large reefs in the clear Niagaran seas. The echinoderms also had a marked rise in this mid-Silurian epoch, the *cystoids* being common and, locally at least, the *crinoids* wonderfully abundant. The starfish and sea urchins had put in their appearance but were still

Pterygotus, a Silurian eurypterid. ×1/40

INVERTEBRATE HEYDAY

rare. We shall have occasion to refer to the history of these corals and echinoderms in greater detail in chapter 39.

Possibly the most unusual element of the late Silurian restrictional, or provincial, faunas was the *eurypterid* assemblage. These aquatic arachnids have commonly been designated "sea scorpions" because of their close resemblance to the true, or land, scorpions, of which they probably were the direct ancestors. Originating in the pre-Cambrian, the group dies out by the close of the Paleozoic. During the Silurian, which was probably the time of their climax, one genus, *Pterygotus*, attained a length of 9 ft., apparently the all-time record for megalomania in the arthropod stock.

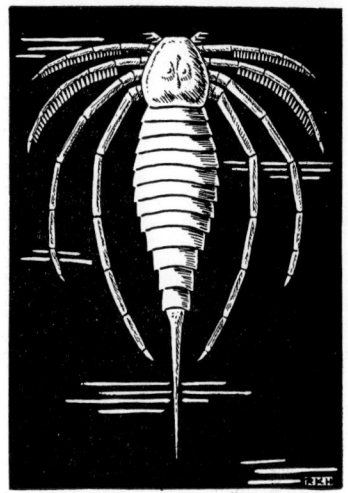

Stylonurus, a Silurian eurypterid. $\times \frac{1}{24}$

Scorpions and millipedes are also occasionally found in the late Silurian rocks, but whether they were air-breathers or not is still a moot question. We will refer to it again in chapter 39.

WE HAVE now completed a brief survey of a *few* of the invertebrates which were outstanding during a time when backbones had scarce put in their appearance in the organic world. In future chapters we will have to turn our attention more and more toward the vertebrates. Even though we must, therefore, change our emphasis to point out the new things in the history of the living, we hope the reader will not fail to remember that the invertebrates have continued on down the long road to the present in ever increasing diversity and vigor.

CHAPTER 37

LATE PALEOZOIC EVENTS

The little present must not be allowed wholly to elbow the great past out of view.—ANDREW LANG

HUGH MILLER was born in a small straw-thatched cottage in the village of Cromarty, Scotland, in the year 1802. As a boy he was fascinated by the picturesque shore features that had been carved out of the sandstone that cropped out along his native coast. This rock formation, he soon learned, had for years been affectionately called the "Old Red Sandstone." He had further opportunity to study these sandstones as, boy and man, he was taught his trade as a mason. In spite of these humble origins, he did a good job at self-education and learned to wield an extremely facile pen. Thus, finally, in 1841 he published a book; and, in view of his early interests, it is not surprising that he called it *The Old Red Sandstone*. By 1874 this volume was still a "best seller" in its nineteenth edition, and it seemed as if the entire educated world had read about this ancient rock formation. But even before the issuance of Hugh Miller's classic, our old friends, Murchison and Sedgwick, in 1837, had given the name **Devonian** to rocks of the same age which were exposed in the sea cliffs of Devonshire, on the south coast of England.

It is not remarkable that many of the Devonian deposits are clastic, like the Old Red Sandstone, for they are strata which ac-

Picturesque shore features that Hugh Miller knew. Old Man of Hoy, Orkneys, carved out of the Old Red Sandstone. After Geikie

LATE PALEOZOIC EVENTS

cumulated as the Caledonian Mountains, described in chapter 35, were being worn away. The events which took place between this Caledonian disturbance and the end of the Paleozoic era are sandwiched in between two mountain-building periods of prime importance. They thus fall naturally into a late Paleozoic cycle. This cycle begins with the earliest Devonian inundation of the continents which had been elevated during the Caledonian diastrophism. It continues through the *Devonian, Mississippian, Pennsylvanian,* and *Permian* periods, and ends with the continents again standing high at the close of the Permian *Appalachian revolution*. **It is a cycle of unrest when compared with the early Paleozoic cycle of stability.** It is marked by a number of periods of considerable orogenic disturbance, and by a gradual humping up of the backs of the continents.

The "Queen Mary" is a colossal Cunarder which is so long and draws so much water that she is difficult to handle even in the largest harbors. She is even too big for the Panama Canal. But during the mid-epochs of the early Paleozoic periods a decent skipper could have taken her across open seas where now lie most of the states of the Union. Similarly, the widespread mid-Devonian inundation would have made it possible for a smart pilot to have started from the position of New York and taken his leviathan to the Pacific. But from Devonian time on in the late Paleozoic, such feats of prehistoric navigation would have become increasingly difficult in North America. For, throughout the long expanse of years with which this chapter is concerned, the continents are gradually standing higher and higher, and typically marine embayments are becoming smaller and smaller.

Sir Roderick Impey Murchison, founder of the Silurian and Permian systems, and, with Sedgwick, founder of the Devonian system. From History of the Geological Society of London

WE HAVE already learned that, although the seas were withdrawn from North America during the close of the Silurian period, this retreat was not accompanied by an orogenic movement of any consequence. Thus there was almost complete sedimentation from the Silurian into the Devonian in that part of the Appalachian geosyncline which is now New York State. Since this latter area happens to be one in which the entire Devonian sedimentation is almost completely represented, it has gradually come to be regarded as the Devonian type district for North America. In this area the principal divisions of the Devonian rocks are as follows:

Series	Group	Chief Formations
Upper Devonian =	Chautauquan	= Bradford and Chemung sandstones
	Senecan	= Portage sandstones and shales
Middle Devonian =	Erian	= Hamilton shales and shaly limestones
	Ulsterian	= Onondaga limestones
Lower Devonian =	Oriskanian	= Oriskany sandstones
	Helderbergian	= Helderberg limestones

THE oldest Devonian rocks are fossiliferous **Helderberg** limestones which rest with slight disconformity on the Upper Silurian limestones. Above these limestones are sandstones known as the **Oriskany.** These two groups of formations, deposited in early Devonian time, are largely confined to the Appalachian geosyncline; but there was also sedimentation in the Cordilleran trough, especially south of the present Canadian boundary. In addition, deposits of this age in Tennessee, southern Illinois, Missouri, and Oklahoma indicate a considerable spread of the early Devonian seas into the south central states.

The middle Devonian **Onondagan** seas were evidently warm, clear marine embayments, for coral reefs of large size occur in the deposits which have been left as a result of this ancient flood. The new elements in the marine life of these seas came from the north, as they do not occur in the faunas of the same age in South America. But during **Hamilton** time there are several southern immigrants recognized in the faunas, so that these ancient seas had both northern and southern connections. During the Middle Devonian, Appalachia was again elevated, the results being seen at once in the change from pure Onondagan limestones to Hamilton shaly limestones or shales. Naturally the sediments become coarser eastward toward the source of supply.

With the beginning of late Devonian times the land mass of Appalachia was further elevated. To these earth movements, which were accompanied by considerable igneous activity, the name **Acadian disturbance** has been given, since the orogeny seems to have been centered in the Acadian area of eastern Canada. As erosion of the highlands continued, a great delta was formed in southeastern New York and northeastern Pennsylvania. This is generally described as the *Catskill delta*, since the mountains by that name represent a residual erosional pile of the deltaic sediments. When at its maximum, this delta must have been some 200 mi. wide, and it extended along the western shore of Appalachia for about 500 mi. Its deposits are thickest near Harrisburg, Pennsylvania, where about 13,000 ft. of sediment are exposed, of which about a third is made up of "red beds." Dark shales and sandstone "flags" make up much of the **Portage** formations, whereas the **Chemung** and **Bradford** beds are fossiliferous marine sandstones grading eastward into terrestrial deposits of what is generally called the "Catskill" type.

From the brief foregoing account of the Devonian we can see that the crustal unrest, manifesting itself in the *Acadian disturbance*, was by no means confined to the end of the period. We see the first effects of the uplift in the increased amount of muds brought into the middle Devonian seas by the quickened streams of Hamilton time.

THE igneous activity of southern Quebec, Gaspé, New Brunswick, and New England is now recorded by ancient lava flows, as in Gaspé; by stocks, or volcanic necks, as in the vicinity of Montreal; and by most of the granite batholiths of New England. In Europe also there was continued orogenic movement and volcanic activity throughout the period; and in Australia there occurred the greatest of all Devonian mountain-buildings, the *Kanimbla disturbance*, in which the entire east coast of the continent was folded and intruded by granite batholiths.

DEVONIAN economic resources are relatively of lesser importance than those of the other late Paleozoic periods. Oil and gas production from Devonian rocks is, however, of considerable consequence in the Appalachian states. The oil is notable especially for its high lubricating qualities. Glass sands, phosphate rock, and flagstones are other Devonian products of value to man.

IN SPITE of the Acadian disturbance, much of North America remained low-lying at the close of the Devonian. The **Mississippian** inundations, therefore, were relatively extensive in the central and western states. The records of these ancient seas are indelibly written in the limestones of this period which crop out extensively in the Mississippi Valley, from which the system takes its name. In Europe, however, this division of geological time is quite generally called the *early Carboniferous*. A simple subdivision of the actually complicated Mississippian formational record is given here:

General Series	Mississippi Valley	Appalachian Area
Upper Mississippian	= Chester series	= Mauch Chunk sandstones and shales
Middle Mississippian	= Valmeyer series	= ↕
Lower Mississippian	= Kinderhook series	= Pocono sandstones

The oldest Mississippian is a dark shale which is fairly widely distributed over the eastern Mississippi Valley area. During its formation Appalachia was still standing relatively high, and the sandstones of the *Pocono* group were deposited in the eastern portion of the Appalachian trough. A little later in the early Mississippian epoch the **Kinderhookian** seas were considerably extended across the south central interior of the United States, and one arm reached along the Cordilleran geosyncline northward to the Arctic. The presence of this sea over the Chicago area is indicated by fissure fillings in the Niagara limestone containing early Mississippian fossils.

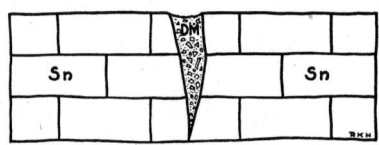

Kinderhook fissure filling in the Niagaran limestone in the Elmhurst quarry, Chicago area. After Chamberlin and Salisbury

The rocks of the **Valmeyer** series rest disconformably upon the Kinderhook formations, thus showing that the seas withdrew from much of the interior of North America at the close of the early Mississippian epoch. The Valmeyer rocks are essentially all limestones in the interior and western portion of the continent, indicating that the middle Mississippian seas were again clear and that Cascadia, at least, was not being actively eroded. Appalachia, however, was still supplying considerable quantities of clastic material to form the Upper Pocono sandstones. These coarse-grained deposits become finer-grained westward away from the source of supply, and in the Mississippi Valley the cherty limestones which characterize the middle Mississippian formations in that area are its western equivalents.

The Mississippian seas withdrew from the interior of the continent at the close of Valmeyer time. When they returned, the **Chester** sediments, comprising alternating sandstones and limestones, indicate that there were rhythmic oscillations of the marine waters prophetic of the Pennsylvanian cyclical sedimentation shortly to come. While Chester marine sandstones, shales, and limestones were being deposited in the interior, the largely terrestrial *Mauch Chunk* red sandstones and shales were laid down in the Appalachian district. A few limestones also occur in this latter area, showing that, momentarily at least, this eastern geosyncline was sufficiently deepened to permit the seas to clear.

At the close of the Mississippian period there was extensive mountain-building in western Europe, where ancient ranges known as the **Variscan Mountains** were formed. North America was elevated at the close of the period, and a series of crustal movements began which were merely the forerunners of greater orogenies

Generalized columnar section of the Mississippian rocks in Illinois and Missouri. The section shown involves a pile of sediments about ½ mi. thick. After R. C. Moore

which were to follow in the late Paleozoic. These movements apparently were strongest in the Ouachita and Appalachian areas. The mountain-building movements closing the Mississippian are generally grouped under the term **Culmide disturbances.**

MISSISSIPPIAN rocks yield a variety of important products, such as building-stones, lime, cement, lead and zinc, salt, coal, and oil and gas. Rock asphalt occurs in western Kentucky, and oil and gas are found in reasonably large quantities in rocks of this age in the Appalachian states, in Illinois, and in the mid-continent fields. Mississippian salt is mined in Michigan, and coal of this age is obtained in western Virginia and in southern Russia. The limestones of the Valmeyer series yield some of the most famous building-stones in the world. The "Bedford," or Indiana, stone, from south central Indiana, is the widest known in this respect; but the Carthage, Missouri, rock is also extensively used. The sandstone known as the Berea is actively quarried in Ohio, since it is popular for ornamental stone trim, and is considerably used in the form of grindstones. Much of the Mississippian limestone quarried in the Mississippi Valley is burned for lime. Lead and zinc also occur in limestones of Mississippian age near Picher, Oklahoma, and Joplin, Missouri; but these ores were deposited in post-Mississippian time.

THE **Pennsylvanian** rocks, named from their extensive exposures in the state of that name, correspond essentially to the *Upper Carboniferous*, or "Coal Measures," of Europe. The beds of this system are not divided into the three series which typify the earlier Paleozoic systems, nor are the terms which are used in the eastern states readily applied to the central and western sequences. The series recognized in the two areas are as follows:

Appalachian Region	Mid-continental Region
4. Monongahela	4. Virgil
3. Conemaugh	3. Missouri
2. Allegheny	2. Des Moines
1. Pottsville	1. Morrow

LATE PALEOZOIC EVENTS

In the Appalachian region the series have been arbitrarily established on the basis of coal beds. The rocks of this eastern area are largely of continental origin, thick conglomerates, sandstones, and shales being common. Fresh-water and marine limestones also occur, and coal seams are numerous.

The mid-continent region has a better record, since both marine and non-marine deposits alternate, and the series may be separated on the basis of breaks in sedimentation which have been correlated with crustal movements. The four series thus are not the equivalents of the Appalachian divisions, though they do represent at least partially contemporaneous epochs.

The break between the Mississippian and Pennsylvanian sediments is a large one, for sufficient erosion had taken place during the interval to develop distinct hills and valleys on the old Mississippian surface before the initial Pennsylvania deposits were laid down over them. Other features of Pennsylvanian time are the concentration of very thick sediments in the Ouachita and Appalachian geosynclinal troughs and the seemingly cyclical sedimentation in the interior areas, such as in Kansas and Illinois. This cyclical arrangement of the strata is shown in the accompanying illustration.

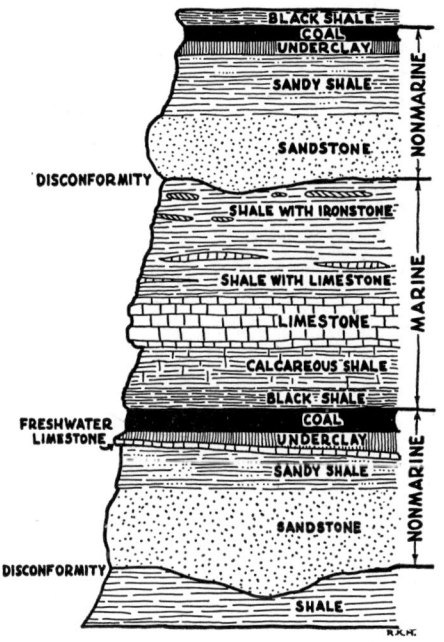

Simplified columnar section to show cyclical sedimentation in Pennsylvanian strata. Redrawn from R. C. Moore

NEAR the close of the period the seas withdrew from most of the interior of the continent, though deposition may have been continuous into Permian times in parts of the mid-continent area. There was mountain-building in western Texas, and crustal unrest literally around the world; but most of the areas affected were to suffer from still greater orogenic movements in the Permian.

THE Pennsylvanian formations are of great economic importance to man, since they are rich with many of the mineral resources on which his present civilization has been built. Chief among these resources are coal, oil and gas, building-stone, and the raw materials which form the basis of the manufacture of clay products and cement.

FIRST in importance in this group is coal (see Plates 45, 46, and 47), which is so typically Pennsylvanian that it is little wonder that these beds were early dubbed the *Carboniferous*, or carbon-bearing, rocks. Between 80 and 85 tons out of every 100 tons of coal now produced in the world have been mined from the Pennsylvanian strata. The chief foreign mining areas in which the coal is of this age are the Saar and Ruhr basins, the Donetz Basin of Russia, the British Isles deposits, and the fields of Silesia and Belgium. In the United States the coal fields of this age occupy close to 250,000 sq. mi. and are the largest in the world. As can be seen from the accompanying map, the fields all lie in the eastern part of the continent. Moreover, practically all of the production comes from states east of the Mississippi. This fact can be readily appreciated when we learn that, although the mid-continent field is the largest in areal extent, it yields only about $50,000,000 worth of coal a year, as compared to the Pennsylvanian total of over $1,000,000,000 annually.

Pennsylvanian coal fields of the United States. A small area of poor coal in Rhode Island is not shown

The most spectacular of these American coal areas is the small *Anthracite field* of northeastern Pennsylvania. Comprising a district of only approximately 500 sq. mi., it has nevertheless accounted for over 20 per cent of the coal produced in North America. In this basin the coal has been preserved in the down folds of the mountain structures which were produced near the close of the Paleozoic era. Some of the coal seams are remarkable for their great thicknesses, the Mammoth vein being over 100 ft. thick in some places and generally 40 ft. in thickness. Because these coal beds have been involved in orogenic movements, they have been subjected to great pressures. These have served to drive off some of the more volatile constituents of the coal, the result being *anthracite*, or hard coal. Here again we see how the organic and the physical commonly have united in determining the formation of products important to man. The physical conditions had to be just right for the development of the coal swamps, for the growth of the plants whose cellulose was to make up the coal, for the subsequent burial of the ancient bogs, for the growth of folded mountains out of the geosynclinal area, and for the preservation of the synclinal coalbearing folds from the 200,000,000 yr. of erosion since the mountains were formed. This is a lot to ask of Nature—and a great deal to receive!

We shall describe the coal-forming floras themselves in some detail in chapter 38, but it may be interesting to note at this point just how prevalent some of the ancient forests must have been. The *Appalachian coal field*, which is the greatest *bituminous*, or soft-coal, area of the world, has some half hundred of coal seams, of which about ten are extensively mined. Most extraordinary of these is the *Pittsburgh seam*, which underlies an area of well over 20,000

Mid-Pennsylvanian seas in black. Greatest coal swamp area indicated by white dots. Generalized after C. Shuchert

sq. mi. in western Pennsylvania, southeastern Ohio, and northern West Virginia. It is mined actively over an area of about 6,000 sq. mi., in which it is from 6 to 14 ft. thick. This means that the coal swamp in which grew the vegetation which now makes up the Pittsburgh seam was as large as the state of West Virginia, and its greatest development was in a district as large as Rhode Island and Connecticut combined. **From this Pittsburgh seam alone, man has mined a total of nearly 4,000,000,000 tons of high-grade coal,** the value of which is sufficiently great to stagger the imagination of Midas himself.

THE Pennsylvanian rocks of Kansas, Oklahoma, and Texas prior to 1925 were the country's great oil-producing beds, their total production being well over 2,000,000,000 barrels of the "liquid gold." In more recent years the Ordovician sands have accounted for the most spectacular oil and gas developments in this area, as we have earlier seen. Nevertheless, Pennsylvanian oil is still of prime importance, not only in the mid-continent area, but in Ohio, Pennsylvania, Illinois, Indiana, Kentucky, Wyoming, Utah, and New Mexico. Thus, more than 200,000,000 yr. ago the Earth lavishly stocked her natural coal bins and her oil reservoirs against the needs of our modern industrialism.

THE **Permian** system was established by Sir Roderick Murchison at the czar's command on the basis of the formations exposed in the Russian province of Perm. The period is one of continental elevation and reduced seas. **It is a time of change, of physical revolution and organic evolution, of aridity and of climatic refrigeration.**

There are relatively extensive North American deposits. The *Dunkard* formations of West Virginia, Ohio, and Pennsylvania represent deposits similar to the underlying Monongahela beds. The mid-continent area contains the *Big Blue* and *Cimarron* series, chiefly of marine Permian formations. But above the normal marine beds occur great deposits of salt, gypsum, and red beds which were formed in a desiccating arm of the retreating Permian sea. Western United States is also marked by scattered deposits of Permian red beds, some places as much as 2 mi. thick. Phosphate rock, however, occurs in Wyoming, Utah, and Idaho; and other marine deposits in Arizona indicate that Middle Permian seas were fairly extensive in the Cordilleran trough. But by late Permian times the Paleozoic geosynclines were dry land, and the only marine waters on the continent were those in the southwestern salt-forming embayment.

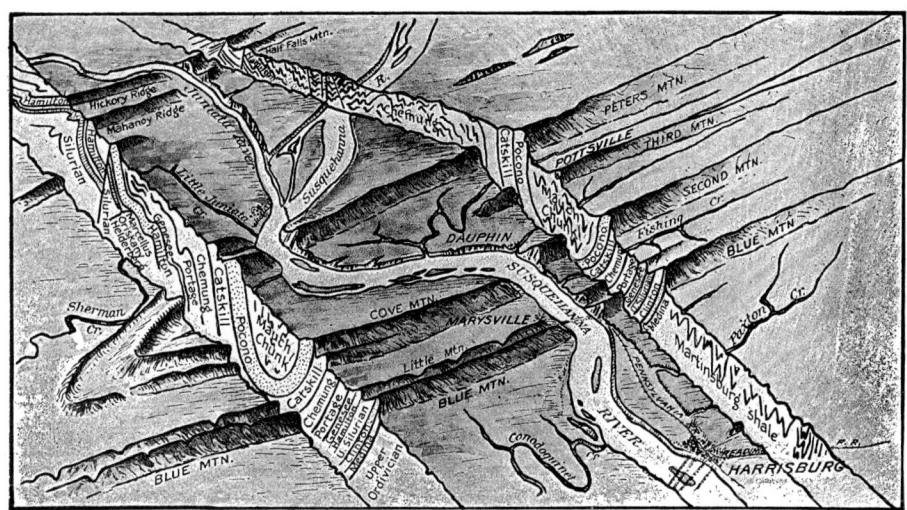

Section through the Appalachian Mountains between Harrisburg and Half Falls Mountain to show the structures formed in the Appalachian revolution and their influence on the topography. After Stose, Jonas, and Ashley

DURING the Permian the ancient land mass of Appalachia was thrust northwestward against the old geosynclinal area. The latter was thus folded into a great mountain chain that extended from the Gulf to the Canadian maritime provinces. The Ouachita Mountains of Oklahoma and Arkansas also may have been disturbed at this time, though their chief folding probably was Pennsylvanian. Crustal instability likewise was marked in Asia, Australia, and South America. The Urals were folded at this time, and the Variscan chain was disturbed again. Volcanic activity was marked at a number of places, notably along our own west coast from California north to Alaska. These mountain-making movements are generally grouped under the term **Appalachian revolution.**

PERMIAN economic products include the great oil and gas resources of western Texas and the salt deposits of Kansas, Oklahoma, Texas, New Mexico, and central Europe. These probably are greater than those of any other period, our own southwestern Permian deposits alone having been estimated to contain about 30,000,000,000,000 tons of salt! When one tries to picture this amount of salt in terms of the sodium chloride on the dinner table, he realizes the enormity of the quantity of sea water which must have been evaporated to supply this great total. Potash salts also occur in large amounts in Germany, where they have long been mined extensively, and in Texas and New Mexico, where their discovery is a fairly recent one. Calcium phosphate, copper, and coal are other Permian natural resources of some consequence.

Permian salt basin shown by white dots on the black of the Permian embayment. Note the great size of the continent. Generalized after C. Schuchert

WE HAVE now completed a brief survey of the eventful years of the late Paleozoic. The time was fraught with happenings of great importance to man in spite of the fact that during this period his remote ancestors had as yet scarcely gotten their feet well planted on the solid land. In the next two chapters we shall first inquire into the story of the Paleozoic terrestrial plants that made the first animal invasion of the land possible; and, secondly, we will consider the story of the animals of the late Paleozoic, some of which cast their lot with the life of the land while others chose to remain in their ancient home, the sea. In chapter 40 we shall investigate further the mountain-building and climatic changes in the late Paleozoic, and see if we can evaluate the impact of these physical factors upon the totality of the organic world.

CHAPTER 38

THE FOREST PRIMEVAL

> This is the forest primeval.
> —LONGFELLOW

"THE murmuring pines and the hemlocks" may well have seemed primeval to Longfellow, but they were by no means typical representatives of the earth's early forests. Most of the elements of that ancient floral assemblage are either extinct or found greatly restricted in the modern ecology of the plants.

IT IS difficult even for the scientist who is familiar with the record of the past to conjure up in his mind's eye an earth totally devoid of land plants. Yet, we know for certain that forests worthy of the present designation did not appear to clothe the Earth until as late as the Devonian period. But since Devonian coals are mined at a few places, there is good reason for believing that even some of the truly "primeval forests" represented lush growths of these early land plants. This naturally suggests a long and complicated history in the plant group prior to the Devonian. We have touched on a few of the high lights of this story in our discussion

.... to clothe the Earth ....

of the pre-Cambrian. In this chapter we propose to review briefly the entire history of the plants from the time of their origin to the end of the Paleozoic era. In order to do so, it becomes necessary also to describe briefly some of the principal features of the four main divisions of the plant kingdom. For convenience a simplified outline classification is given in the accompanying table, which also indicates the geological range of the main groups. The spermatophytes are included to make the table reasonably complete, although the higher members of the phylum do not appear until the early Cretaceous.

Phylum	Class or Order		Geological Range	Common Name or Living Example
IV. Spermatophyta (seed plants)	Angiospermae (covered seeds)	Dicotyledones	Cretaceous to present	Oaks, maples, elms
		Monocotyledones	Cretaceous to present	Grasses, grains
	Gymnospermae (naked seeds)	Cycadales	Triassic to present	Cycads
		Coniferales	Permian to present	Pines
		Ginkgoales	Permian to present	Ginkgos
		Bennettitales	Permian to Cretaceous	Cycadeoids
		Cordaitales	Devonian to Permian	Cordaites
		Cycadofilicales	Devonian to Jurassic	Seed ferns
III. Pteridophyta (fern plants)		Lycopodiales	Devonian to present	Club mosses
		Sphenophyllales	Devonian to Permian	Sphenophyllums
		Equisitales	Devonian to present	Horsetails
		Filicales	Devonian to present	Ferns
II. Bryophyta (moss plants)		Musci	Mississippian? to present	Mosses
		Hepaticae	Silurian? to present	Liverworts
I. Thallophyta (thallus plants)		Fungi	Silurian? to present	Fungi
		Diatomeae	Jurassic to present	Diatoms
		Brown algae	Pre-Cambrian to present	Seaweeds
		Blue-green algae	Pre-Cambrian to present	Blue-green algae
		Bacteria	Pre-Cambrian? to present	Bacteria

SYNOPSIS OF THE GEOLOGICAL HISTORY OF PLANTS

THE reader must here recall an important point which we discussed in chapter 36, namely, that the geological range of an organism is usually determined, and thus restricted, on the basis of direct fossil evidence. For instance, the range of the fungi as given in the table, is "Silurian? to present." This merely means that the first evidence of the presence of the fungi is found in the Silurian rocks; the question mark indicates that not all scientists agree that the evidence is conclusive. All would subscribe, however, to the statement that the fungi had a long *unrecorded* pre-Silurian history. In the following pages the discussion of the various plant groups will reveal a number of other cases in which the geological range of a plant group is obviously much less than the duration of time in which the plant actually existed.

THE FOREST PRIMEVAL

THE **thallophytes** are plants of soft, non-woody structure, which have no circulatory system or vascular tissue. Hence they are unable to leave the water. They range in size from single microscopic cells to seaweeds of enormous dimensions.

The *bacteria* are one-celled thallophytes of great economic importance in fermentation, disease, and decay. Today they are the chief agents of decomposition of organic matter, a rôle which they apparently have played for hundreds of millions of years. Plant tissue from Pennsylvania coal beds in some cases shows destruction of cell walls which is ascribed to bacterial action. And, had it not been for the general activity of the bacteria throughout the past, we certainly would have a much more complete record of the earth's former plant and animal assemblages. Bacteria doubtless originated early in the Archeozoic, but we do not have positive evidence for their presence at that time. Nevertheless, the deposition of the pre-Cambrian iron ores is quite generally ascribed in part to bacterial action. The ubiquitous character of these plants is well indicated by the fact that they occur in Paleozoic petroleum deposits. Feeding on the organic matter in the rocks and transforming it, in part, into petroleum and gas, these plants have existed for millions of years and through billions of generations deeply buried in the earth.

Blue-green algae are unicellular thallophytes which may in some cases be lime-secreting. Such forms are important today as rock-builders. They commonly grow in felt-like colonies and precipitate calcium carbonate in globular masses which show concentric laminations in cross section. Structures similar to the modern ones occur in some pre-Cambrian rocks; and they, of course, have been ascribed to the depositional activity of these algae.

The *seaweeds* are algae which undoubtedly contributed largely to the relatively high carbon content of some of the pre-Cambrian rocks. Carbonaceous and siliceous remains of these plants are common in rocks from the Cambrian on, and Silurian seaweeds have been reported more than 100 ft. in length and 2 ft. in diameter. Thus we may well assume that the modern sargassum type of seaweed is a very ancient plant development. It must be confessed, however, that the pre-Cambrian seaweeds are poorly represented by actual fossils, though they do occur.

The *diatoms* are microscopic thallophytes which are enclosed in two valves which are impregnated with silica. They occur in both fresh and salt water, and even in damp soils. Today they comprise a rather important part of the *plankton*, or the drifting mass of organisms at the surface of the ocean. They live in such countless numbers that their microscopic skeletons in the aggregate make up large deposits known as *diatomaceous ooze*. These deposits are known from the Jurassic to the present. That diatoms have not yet been found in any Paleozoic formation is a fairly certain indication that the group did not build siliceous skeletons at that time, for they doubtless were in existence then.

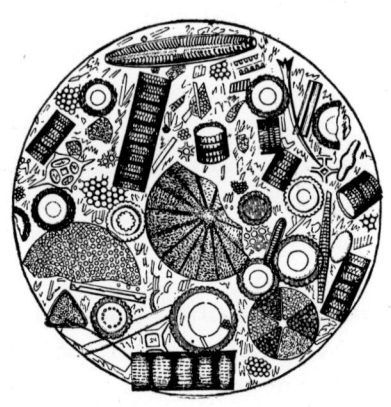

Tertiary diatomaceous earth highly magnified. After Dana

The *fungi* are thallophytes distinguished by the fact that they have no chlorophyll. Thus they must live on organic, rather than inorganic, matter. Since common fungi are yeasts, toadstools, and molds, one would not expect much of a geological record of this obviously ancient group. The branching threads of fungi, called mycelia, have, however, been reported as occurring under the bark of Pennsylvanian club mosses, and their work is possibly seen in some of the borings found in Silurian brachiopod shells.

THE **bryophytes,** which include the mosses and liverworts, lack vascular tissue, being like the thallophytes in this respect. As a consequence the representatives of this group are small and are confined to moist areas. Since in alternate generations they reproduce by means of gametes, which also require moisture, this group has been unable to alter its habitat markedly for hundreds of millions of years. The liverworts may have been the earliest of the

THE FOREST PRIMEVAL

land plants, though, of course, this is purely a philosophical assertion not backed by direct evidence. It has been considered that pre-Cambrian fresh-water lakes provided the site of the migration of primitive aquatic plants on to the land. These early ponds must have been restricted in size during dry weather, and thus the lake margin became first a swamp and finally merely marshy or damp ground. Here plastic types of algal cells may have developed cell extensions or rootlets downward from the under side of the thallus into the soil in the search first for moisture and then for food. Cells exposed then developed a cuticle against the desiccating effect of air; and a breathing apparatus, the stomata, was formed. Sexual generation, requiring water for the union of male and female cells, could only occur during the wet season; but the asexual spores were best developed and scattered during times of relative drought. Seasonal effects of this type probably were important physical urges in the early evolution of the bryophytes.

Almost certainly on the land in considerable development during the early Paleozoic periods, fossil representatives of the bryophytes have doubtful representatives in the Silurian and Mississippian; but no unquestioned fossils appear until as late as the Jurassic.

THE **pteridophytes** comprise the club mosses, the ground pines, the horsetails and scouring rushes, the extinct sphenophyllums, and the true ferns. In this phylum the roots, stems, leaves, and vascular tissue, through which the sap circulates, are all well differentiated. Reproduction is by means of spores borne on the under side of leaves. These spores must be surrounded by a film of water before they can germinate.

The early fossil record of the pteridophytes may begin with the story of the **Psilophyta,** a group which seems to be intermediate between the algae and the ferns. The representatives of this most primitive of the *definitely* known land floras were small, bushy, leafless plants growing to a height of several feet. Specimens are found well preserved in a silicified peat bog in the Devonian rocks of Scotland, and they occur in rocks of the same age on the Gaspe peninsula. Such fossils show that these early land plants had *stomata* or breathing pores such as occur in the leaves of higher plants, and that a vascular system was present. But the underground root system was not a true root at all, but rather a horizontal runner bearing small rootlets. Truly, this most primitive of "primeval forests" would have seemed a great disappointment to any poet!

One of the members of this primitive flora so closely resembles the later club mosses that it may well have been ancestral to that group. These club mosses attained heights of 30–40 ft., and horse-tails were 10–12 ft. high during the later Devonian. The true ferns also were already differentiated by this same time. Thus the "primeval forest" had a rapid development. The ferns, however, did not reach their climax until the Pennsylvanian, when fronds 12 ft. and more in length were common, and central trunks more than 2 ft. in diameter occurred. But some of these early so-called "ferns" have recently turned out to be seed ferns, or *cycadofilicales*, about which more later.

Psilophyton, of the primitive forests. After Dawson

By the close of the Paleozoic era the true ferns had suffered a marked decline, in part due to the Permian glaciation. Yet, some members of the group, such as *Glossopteris* and allies, characterized by smooth-edged, tongue-shaped leaves, seemed well adapted to the rigorous climate; and they had a great distribution, especially in the southern hemisphere.

DURING the Mississippian and Pennsylvanian periods the giants of the forests were two enormous club mosses called *Lepidodendron* and *Sigillaria*. The former is characterized by diagonal rows of leaf scars, the latter by vertical rows. The lepidodendrons had large needle-like leaves and branched and expanded tops, so that they looked something like a modern walnut or elm. The

THE FOREST PRIMEVAL

sigillarias, on the other hand, resembled some of the modern palms, having had a cluster of long, broad tongue-like leaves at their tops. Both of these trees were as large or larger than the common representatives of the modern forests. For example, an actual trunk of a lepidodendron 114 ft. in length, although minus the crown, was found in an English coal mine; and fossil stumps up to 4 ft. in diameter are not uncommon. These dominant plants of the coal-forming forests, as we have seen, had their origin in the Devonian; and a few sigillarias live on into the early Mesozoic era. But in the main, the group was unable to stand the unusual physical conditions of the Permian.

BY LATE Devonian times the horsetails also had differentiated from the early Devonian generalized flora. Today there are about twenty-five species of these plants, which are characterized by numerous unbranched, hollow, jointed stems. The modern representatives of the group are generally small, but the late Paleozoic floras literally the world around were marked by large bamboo-like horsetails. The trunks are called *Calamites;* and the circlets of leaves, which were borne on small branches originating at the nodes on the main stem, are designated *Annularia*. These ancient horsetails attained a height of at least 75 ft. and grew in a gigantic counterpart of our modern canebrakes or bamboo thickets.

The coal floras are also characterized by an extinct group of the pteridophytes known as the *sphenophyllales*. These plants had slender stems and beautiful wedge-shaped leaves set in whorls like the horsetail foliage. They have been regarded by some as aquatic plants; by others, as vines. First recognized in the Devonian, they are extinct by the end of the Paleozoic era, and apparently have left no descendants.

THE **Spermatophyta** include all the plants that reproduce by means of seeds. The flowers of this group contain the ovules and produce the pollen. The pollen fertilizes the ovules, and the latter then develop into seeds which contain the embryos. These grow to mature plants under favorable growth conditions. The *gymnosperms*, or naked-seed plants, are spermatophytes which have small flowers and bear seeds unprotected by membranes. The *angiosperms* have large and open flowers, and the embryos are protected by seed coats. Angiosperms are not represented in our "primeval forest"; but a number of gymnosperms were present, notably the seed ferns, the *Cordaites*, the ginkgos, and the primitive conifers.

A typical coal swamp flora. After Dana

The *cycadofilicales*, or seed ferns, or cycad ferns, are so called because in some respects they seem transitional between the ferns and the cycads and cycadeoids. As has been mentioned, these plants are with difficulty distinguished from true ferns, especially when only the leaves are found. The earliest of these are known from the Middle Devonian beds at Gilboa, New York, where the fossils occur in such profusion that eighteen stumps of large size were found in an area 50 ft. on a side. But, despite the abundance of these seed ferns in the Devonian, they were even more characteristic of the Pennsylvanian, which was apparently their time of climax. A few representatives of the group, however, lived on until the Jurassic.

Cordaites was a primitive gymnosperm confined to the late Paleozoic. Related to the ginkgos, which have similar reproductive organs, it was also possibly ancestral to the conifers. Originating in the Devonian, this ancient, large-leaved evergreen attained great size in the Pennsylvanian period, when specimens probably reached a height of over 100 ft. The stem of the tree was characterized by a wide pith center, and the foliage was made up of great grasslike leaves in some instances 6 in. across and as much as 6 ft. long. Reproductive organs, both male and female, were catkins intermediate between the primitive reproductive fronds of the seed ferns and the cones of the conifers.

THE FOREST PRIMEVAL

THE earliest member of the *coniferales* is the Permian genus *Walchia*, apparently an ancestral Araucarian conifer, which originated as a result of the changing physical conditions near the close of the Paleozoic era. Ancestral sequoias also were present during the Permian, and primitive members of the *ginkgoales*, or maidenhair trees, had put in their appearance. Similarly, the *cycadeoids*, which were ancestral to the *cycads*, the dominant early Mesozoic plants, had already representatives in the relatively restricted latest Paleozoic forests. Plate 45 illustrates some of the plants of the Paleozoic forests.

Foliage of *Walchia*, an ancestral conifer. After Emmons

Lepidodendron, Sigillaria, Cordaites, Calamites, Annularia—what terrible jawbreakers to inflict on the unsuspecting reader! But, after all, every time you burn Pennsylvanian coal you are releasing energy stored up by these and many other representatives of the "primeval forests." And their names, being what they are, we must describe them as such, rather than as Tom, Dick, or Harry, or by any other set of pleasant aliases.

In later chapters we shall have occasion to refer to the more recent developments in plant evolution on this terrestrial sphere.

CHAPTER 39

CROSSING THE STRAND

> The interest in a science such as geology must consist in the ability of making dead deposits represent living scenes.—HUGH MILLER

WHEN you come in from a brisk swim, you experience no great difficulty in crossing the narrow zone which serves as a boundary between sea and land. But when the *aquatic plants* pushed across this seemingly insignificant barrier, away back in the pre-Cambrian, it was a red-letter day indeed for the organic world. Similarly, when the first *animals* crossed the strand, in the late Silurian, there was at last something new under the sun. Much as we vertebrates may dislike to admit it, **the pioneer who led the first animal invasion of the land was an invertebrate, and a scorpion at that.** *Paleophonus*, meaning "ancient murderer," is the rather lurid name which has been applied to this early conquistador, who was surprisingly like his modern scorpion descendants. Critics have pointed out that these first land animals must have been lonesome souls, and they have mildly inquired as to what the scorpions might have found to eat. But then, as now, the strand must have been an ample storehouse of all sorts of food washed up from the sea. The real difficulty is that we cannot definitely prove that these earliest known scorpions and their compatriots, the *millipeds*, or "thousand-legged" arthropods, were actually air-breathers. They are, however, so nearly identical with the modern land-living types that the presumption that they were truly terrestrial is strong.

.... early conquistador ....

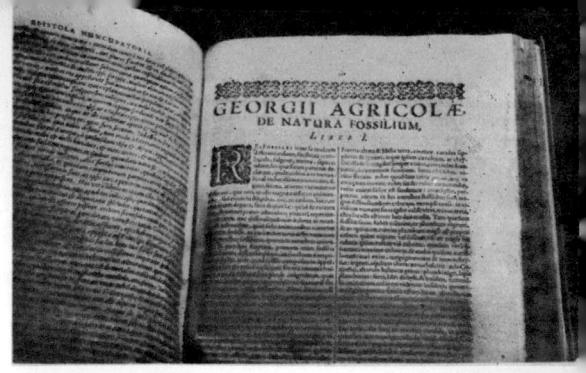

PLATE 33
DE NATURA FOSSILIUM

The true nature of fossils cannot be explained by a few illustrations, but this plate shows several common types. *Top, left:* Pelecypod interior, original shell preserved. *Above:* Cast of interior of the same shell, also a fossil. Since the original shell material is commonly dissolved by ground-water action, casts are more numerous than actual remains.

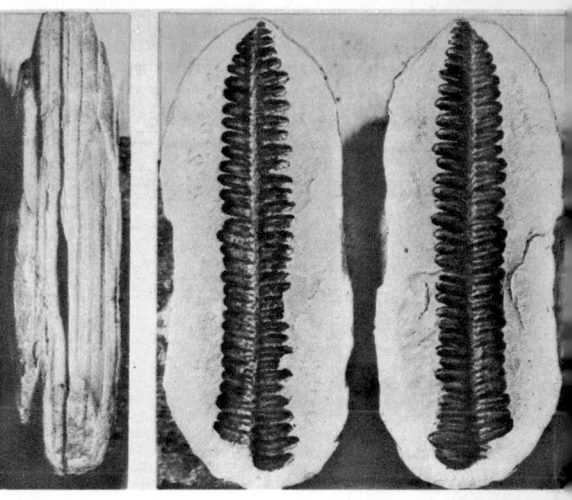

Left: Mold of exterior of same shell. *Above, left:* Concretion which, when broken, reveals the carbonized frond of one of the Pennsylvanian coal-forming plants. The film of carbon is all that remains of the original cellulose of the plant.

Left: Exterior of shell which made the mold seen above at left. *Above:* Another concretion with a carbonized plant fossil. Such concretions are commonly found associated with coal. The ones on this plate came from Mazon Creek, Illinois, a famous collecting locality.

PLATE 34
UNIVERSAL CEMETERY

Above: A slab of Ordovician limestone which served as the burial ground for countless brachiopods and gastropods.

All the world *is* a cemetery, for the rocks are commonly the last resting-place of all kinds of life. *Above:* A Devonian trilobite interment. *Below:* A Pennsylvanian plant remain. (Photos by courtesy of U.S. National Museum.) Such specimens show that the rocks *are* "tombstones on which the buried dead have written their epitaphs."

Above: A petrified section of a Douglas fir log. Grand Coulee district, Columbia River area. (Photo courtesy Northern Pacific Railroad.) *Below:* Slab of rock from Agate quarry, Nebraska, with abundant Tertiary rhinoceros remains. This type of preservation indicates clearly the abundance of some of the early mammal groups.

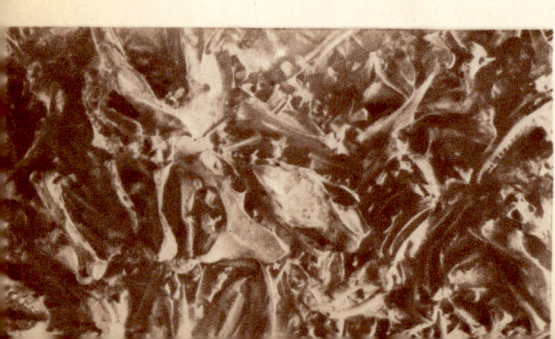

PLATE 35
OLDER FLOODS

Above: Mount Robson, British Columbia (photo courtesy of Canadian National Railways).

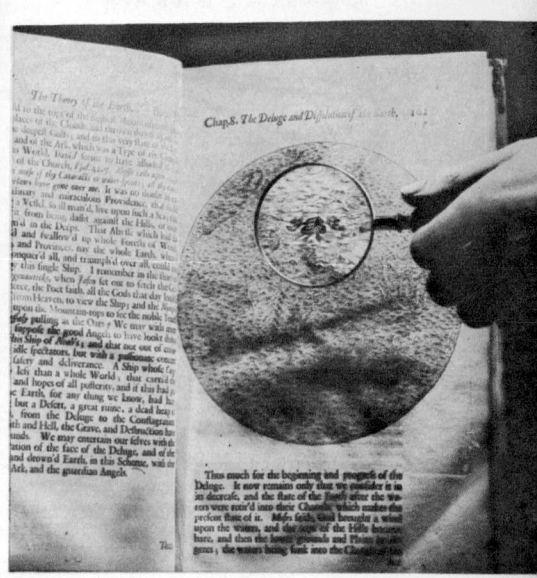

There have been almost countless older floods. Their records are written in the sedimentary rocks which may be seen in highest mountain, or deepest canyon. *Above:* Thomas Burnett's conception of an older flood. *Left:* Grand Canyon from Bright Angel Point. (Photo courtesy of Union Pacific Railroad.) This canyon reveals an unmatched story of many ancient marine inundations.

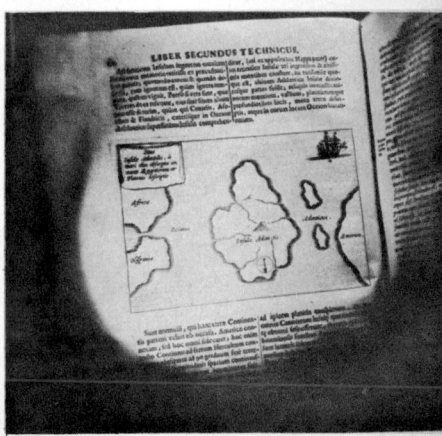

Above: Kircher's conception of Atlantis, before it was engulfed by the sea. Regardless of the validity of the Atlantis legend, we *know* that the rocks at the summit of Mount Robson were formed beneath an older flood, and the great pile of sediments shows that it was of considerable duration.

PLATE 36
MOUNTAINS

Above: Mount Rainier has been built up to a height of 14,408 ft. above the sea by volcanic action (photo courtesy National Park Service). But the complexly folded and faulted mountains near Exshaw, Alberta (*above, left:* photo courtesy Canadian Pacific Railroad) are the result of crustal movements, as are almost all linear mountain ranges such as the Appalachians and Rockies.

Above: The Wasatch Front of the Rockies (photo courtesy United Air Lines). *Below:* Mount Warren, Jasper National Park, a mountain elevated many thousand feet without much folding (photo courtesy Canadian Pacific Railroad). Extremely high mountains of little-folded or faulted sediments are somewhat uncommon.

Above: Mountains near Lake Winola, Pennsylvania, are largely erosional remnants (photo courtesy American Airlines). Such mountains may be composed of essentially flat-lying strata. Naturally all mountains are erosional so far as their topographic expression is concerned.

PLATE 37
GARGANTUAN CALENDAR

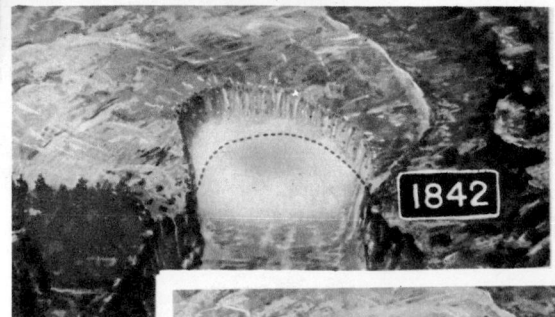

Geological time can be measured in a number of ways, as explained in chapter 32. One of the common methods is by observing the rate of recession of waterfalls such as Niagara, whose retreat may be calculated, on the basis of old maps, over a period of several centuries.

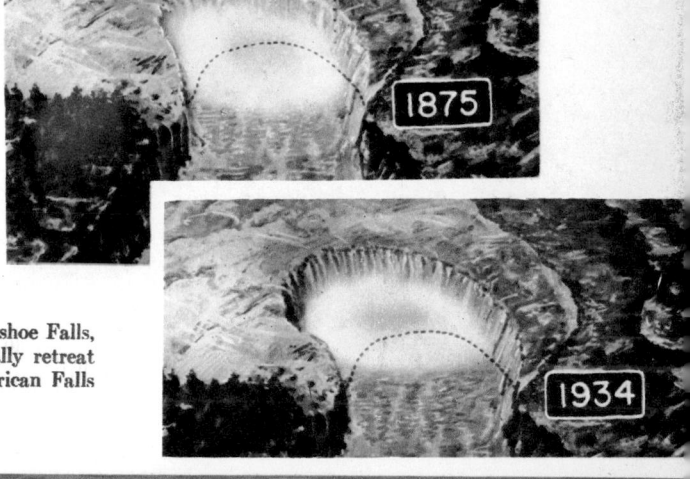

Niagara's rocky rim has been receding at a rapid rate, as can be seen from the position of the brink in 1764 as contrasted with the 1934 rim (photos from the moving picture "Work of Running Water"). *Below:* Niagara as it appears today from the air (photo courtesy American Airlines). Most of the present erosion is concentrated in the Horseshoe Falls, and consequently they will eventually retreat to a point at which the higher American Falls will be abandoned.

PLATE 38
THE PRE-CAMBRIAN

Above: Archean rocks exposed in inner gorge of Grand Canyon near mouth of Bright Angel Creek (photo courtesy National Park Service). Archean strata are also revealed in a number of other deep valleys.

Above: Proterozoic folded rocks seen from Bright Angel trail, Grand Canyon (photo courtesy National Park Service). *Right:* Varves or annual layers in pre-Cambrian rocks show that the seasons had their prototypes millions of years ago (photo from F. J. Pettijohn).

Below: Grain of country, or roots of pre-Cambrian mountain range, as seen from the air, Southern Canadian Shield (photo courtesy of Canadian Topographical Survey). These ancient folds mark the trend of a great pre-Cambrian mountain system.

Above: Fossil algal reef in Bass limestone of Proterozoic age, east side Garden Creek Canyon (photo courtesy National Park Service). *Below:* Barren lands of pre-Cambrian shield on Hudson Bay Railroad. This area was peneplained before the Cambrian, and it has not been involved in mountain building since that time (photo courtesy Canadian National Railways).

PLATE 39 EARLY PALEOZOIC FORMATIONS

Left: Gorge of Niagara River, cut in Silurian rocks. Lockport dolomite is seen at top, the Clinton group crops out in the middle, and Medina sandstone occurs near bottom of the gorge. (Photo courtesy Canadian National Railways.) The strata at the top of the gorge are widely distributed in the central states, Chicago, for instance, being built on Niagaran rocks. The Medina beds, however, crop out more characteristically in the east, and make many Appalachian ridges.

Above: Montmorency Falls, near Quebec, occur at a fault between pre-Cambrian and Ordovician rocks. The latter are seen at the right, dipping away from fault (photo courtesy Canadian National Railways). *Right:* The Cambrian Potsdam sandstone in Ausable Canyon, New York (photo courtesy New York Central Lines). Rocks of the same age, and of somewhat similar composition, crop out over large areas in the upper Mississippi River Valley, where they are known as the St. Croix sandstone.

PLATE 40
CAMBRIAN LIFE

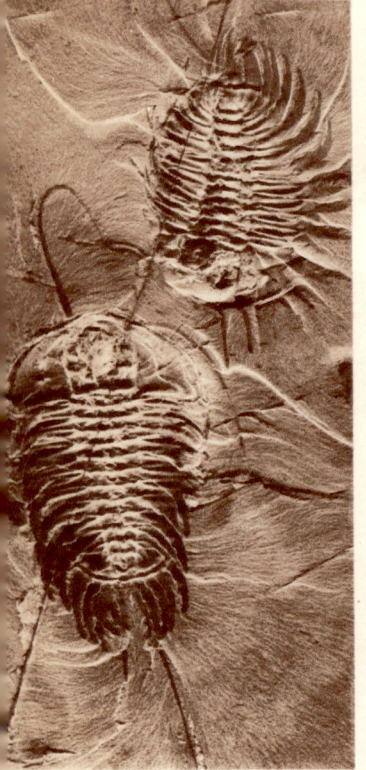

Above: Restoration of the fauna of the mid-Cambrian Burgess shale. *Left:* Two fine specimens of the trilobite *Neolenus* from the Burgess rocks, showing perfection of preservation. *Right:* A slab of fossiliferous Cambrian rock from China. *Below:* Another Burgess restoration. Inasmuch as Cambrian rocks are the oldest which contain an abundant fossil record, their faunas have been extensively studied. Since they yield specimens of the arthropods, the highest of the invertebrate phyla, it is assumed that this first well-recorded fauna had a long pre-Cambrian history.

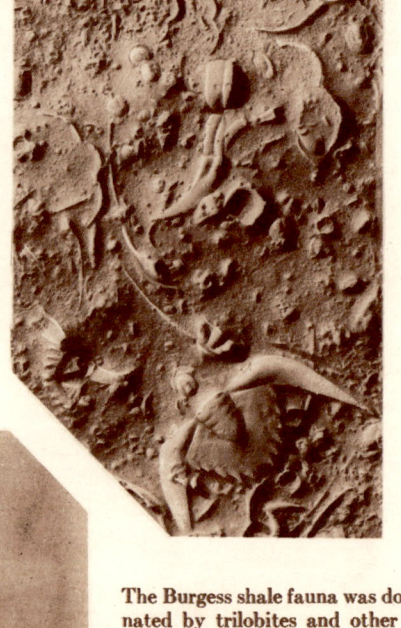

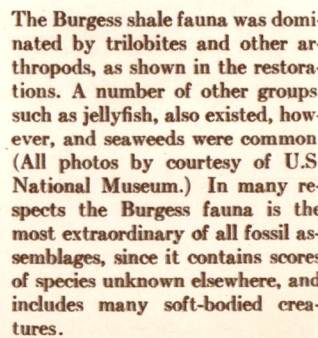

The Burgess shale fauna was dominated by trilobites and other arthropods, as shown in the restorations. A number of other groups, such as jellyfish, also existed, however, and seaweeds were common. (All photos by courtesy of U.S. National Museum.) In many respects the Burgess fauna is the most extraordinary of all fossil assemblages, since it contains scores of species unknown elsewhere, and includes many soft-bodied creatures.

PLATE 41
EARLY PALEOZOIC GIANTS

The giants of the early Paleozoic seas were the cephalopods and the eurypterids, which were among the largest invertebrates of all time. Some of the cephalopods probably were more than 20 feet in length, and a few eurypterids were almost half as large.

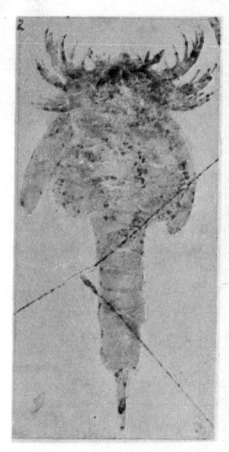
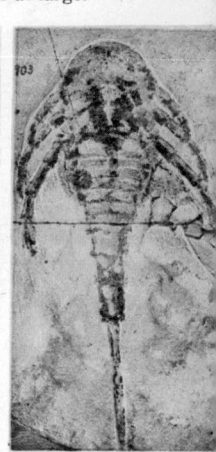

Above: Late Silurian eurypterids of the most common type. *Right:* Other Silurian representatives of this extinct group of animals related to modern scorpions. *Below:* Ordovician beach on which giant straight cephalopods (*Endoceras*) are stranded. (From a painting by C. R. Knight. Copyright, Field Museum of Natural History.) Although the time of eurypterid climax was late Silurian, they lived on almost to the end of the Paleozoic era; and the cephalopods too had a long subsequent career which we illustrate in part on Plate 55.

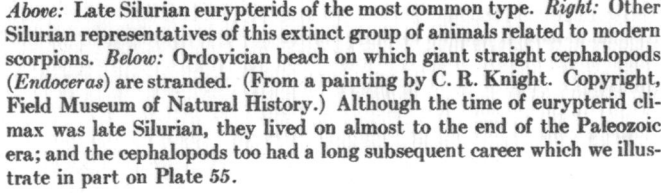

PLATE 42 EARLY PALEOZOIC ANIMALS

Above: Dorsal and ventral views of an Ordovician trilobite (*Triarthrus*) restored by C. E. Beecher to show the appendages. See Plate 40 for illustration of actual specimen with soft parts preserved. All trilobites had appendages of this type, but specimens which actually show the ventral anatomy are rare.

Above: Receptaculites, sometimes called the Galena sunflower, is an index fossil for the mid-Ordovician, despite the fact we do not know whether it is sponge or coral. It commonly occurs in thick layers in the Trenton limestones which crop out in the north-central states.

Above: Some Silurian trilobites were equipped with eyes set on the ends of stalks, but others, which assumed a burrowing habit, were nearly or entirely blind.

Above: Brachiospongia, an Ordovician sponge, sometimes used as an index fossil. Its finger-like processes are responsible for its appropriate name.

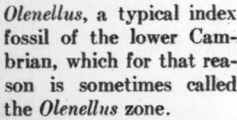

Olenellus, a typical index fossil of the lower Cambrian, which for that reason is sometimes called the *Olenellus* zone.

Lower left: A representative of the genus *Calymene*, a generalized, Ordovician and Silurian trilobite. *Right:* One of the relatively rare Ordovician cystoids with both arms and stem preserved

Many other invertebrate types swarmed in the early Paleozoic seas. Of the groups not shown here the most prominent were the brachiopods, the bryozoans, and the cephalopods. But of still greater interest were the first primitive vertebrates which made their appearance during the Ordovician period but were still insignificant at the end of the Silurian.

PLATE 43
LATE PALEOZOIC FORMATIONS

Above: Beartooth Butte, Wyoming, containing richly fossiliferous Devonian beds (photo courtesy Northern Pacific Railroad). *Below:* Pennsylvanian shaly sandstone resting disconformably on Devonian limestone near Rock Island, Illinois (photo courtesy Illinois Geological Survey). Most of the erosional breaks in the interior of the continent are of this type. That is, they record withdrawal of the sea and subsequent planation but there is no evidence of major earth movements. In the geosynclinal areas, however, angular unconformities are the rule.

Above: Madison (Mississippian) limestone, dipping to right, overlain with recent spring deposits, near Cody, Wyoming (photo courtesy Northern Pacific Railroad). *Below, left:* Rocks of four eras can be seen in the great gorge of Grand Canyon. *Below, right:* Late Paleozoic formations crop out on rim, Mesozoic rocks in desert areas above canyon brink (photos courtesy National Park Service). Although late Paleozoic formations such as these crop out widely in North America, the best-known rocks of this period of earth history are the Pennsylvanian coal-bearing sequences shown in Plate 47.

PLATE 44
LATE PALEOZOIC INVERTEBRATES

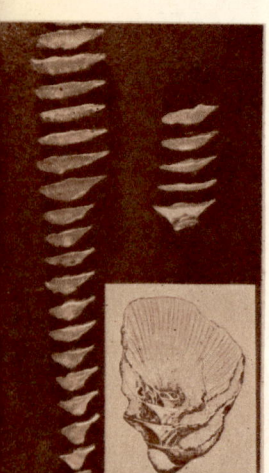
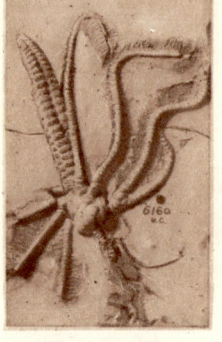
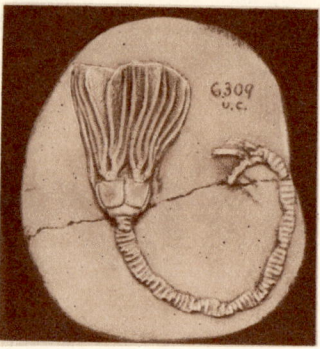

Left: Spiral support of a bryozoan index fossil of the Chester epoch. *Above:* Three crinoids showing arms and stems, and a primitive sea urchin. *Right:* Another crinoid showing a parasitic gastropod attached to the cup. Echinoderms were common rock builders in the late Paleozoic seas, and showed a tremendous amount of diversity at the time. But with the diminution of clear seas in the Pennsylvanian period, the stalked echinoderms were gradually reduced, the cystoids and blastoids becoming extinct, and crinoids were rare.

Above: Devonian cup coral of the type which was very common in the seas of the time. *Left, above:* Colonial, reef-forming coral of the same period. Corals were important rock builders in the clear seas of both the Silurian and the Devonian.

Below: Primitive insect wing, and (*right*) complete insect in Pennsylvanian concretion. These early land-living invertebrates were common in the ancient coal swamps.

Above: Shells of productid brachiopods which swarmed in late Paleozoic seas. *Below, right:* A Pennsylvanian ancestor of the horseshoe crab. *Below, left:* The simple type of trilobite which survived to the end of the era, the highly specialized species having become extinct in the Devonian.

PLATE 45
THE FOREST PRIMEVAL

Left, above: Composite Devonian flora, a restoration of the primeval forest made under the direction of Professor W. A. Parks and reproduced by courtesy of the Royal Ontario Museum. *Below:* Pennsylvanian, or coal-forming forest, built at Field Museum under the direction of Professor A. C. Noé. A giant dragon fly is seen in middle foreground (copyright, Field Museum of Natural History). As one would expect, the oldest workable coals are the same age as the first forests, that is, Devonian. But great coal deposits were not formed before the Pennsylvanian.

Above: Fossil bark of club moss, *Lepidodendron*, a large trunk of which appears at bottom of restoration at right. Each scar marks the point where once grew a needle.

Below: Fossil specimen of *Pecopteris*, one of the common coal-forming plants. *Lower left:* Fossil bark of the club moss, *Sigillaria*, a giant among the primeval plants. All of these primeval forests are so well preserved that we know a great deal about the plants which comprised them; and in some cases it is possible to study them more or less like modern botanical specimens (see Plate 46).

PLATE 46
COMPOSITION OF COAL

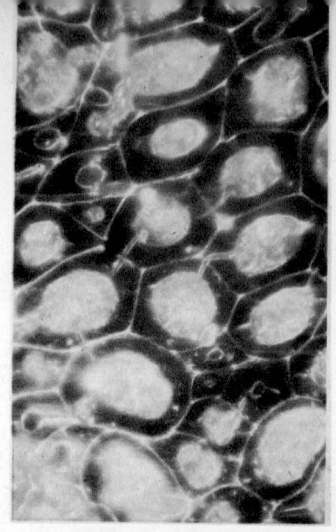

Today man is not content simply to mine coal. He must study its composition in order to throw light on its origin and determine the most satisfactory use for each type of the fuel. Many of the modern investigations involve detailed studies of coal structure and examinations of the plants whose remains have been incorporated in the coal itself.

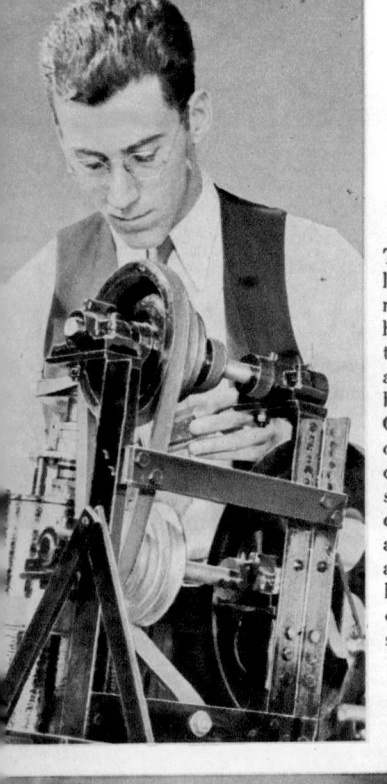

When coals are ground down into thin sections and studied under a high-powered microscope, as shown at the left, a new world of fascinating structures meets the eye; and we see that coals which superficially look alike may have very different compositions. *Above left:* Section showing a twig greatly enlarged. *Above right:* Section revealing cell structure. The two sections below show laminar structure and resin bodies in the coal. Some of the coals studied in thin section actually contain charcoal, which thus gives a certain record of Pennsylvanian fires in the early coal-forming forests.

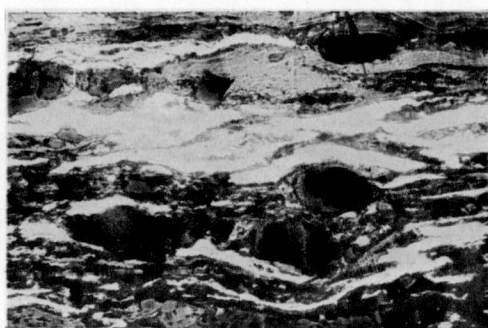

The thin section at the lower right reveals a minute plant sprig which has been flattened. Details of cell structure also are shown. (All photos by courtesy of Illinois Geological Survey, one of several organizations conducting detailed researches in coal.) All the coals shown in Plates 46 and 47 are Pennsylvanian in age, but coal also has been formed in other geological periods, notably in the Jurassic and Cretaceous.

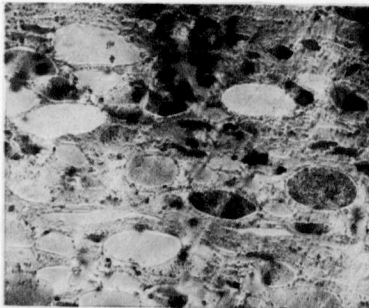

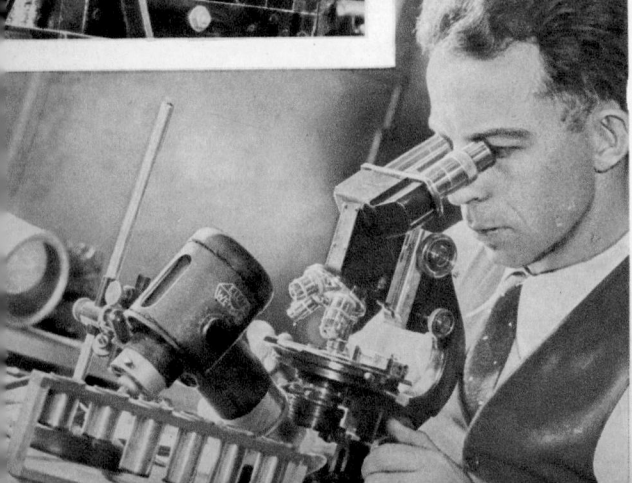

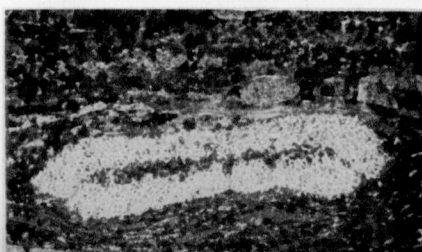

PLATE 47   CANNED SUNLIGHT

Coal, the "canned sunlight" of past ages, is still the most important of man's fuels. Here we show how he wins it from the earth and prepares it for market, by cleaning the coal and screening it into various sizes.

Preparing the coal for loading. A typical scene in a bituminous coal mine (photo by courtesy of the Norfolk and Western Railroad).

Above: A model of an anthracite mine showing the layout of the workings (Philadelphia and Reading Coal and Iron Company). *Below:* Three typical scenes in open-pit coal mining. *Left:* Removing overburden by shovel. *Center:* Many shovel buckets have a capacity of 15–20 cu. yd. *Right:* Coal, freed of overburden, is ready for loading. The three pictures at the right show (*above*) the screens used for grading coal, the loading mechanism, and No. 6, Illinois lump coal. (Photos by courtesy Illinois Geological Survey.) With the improvement in shovels and other mechanical equipment, the open-pit style of mining is increasing in popularity. Where the coal seam is not too deep, the method can be used to great advantage since a higher percentage of the coal is recovered than by any other style.

PLATE 48 LATE PALEOZOIC LAND VERTEBRATES

Copyright Field Museum of Natural History.

With continental elevation at the close of the Paleozoic land vertebrates greatly increased in number and diversity. Perhaps strongest of all were the sail-backed reptiles shown in the C. R. Knight mural above. A mounted skeleton of a carnivorous type, *Dimetrodon*, is pictured below at right. (Photo courtesy U.S. National Museum.) There were also herbivorous types which simulated the carnivores. One of these, with cross-bars on his dorsal spines, is seen in the middle foreground of the above restoration.

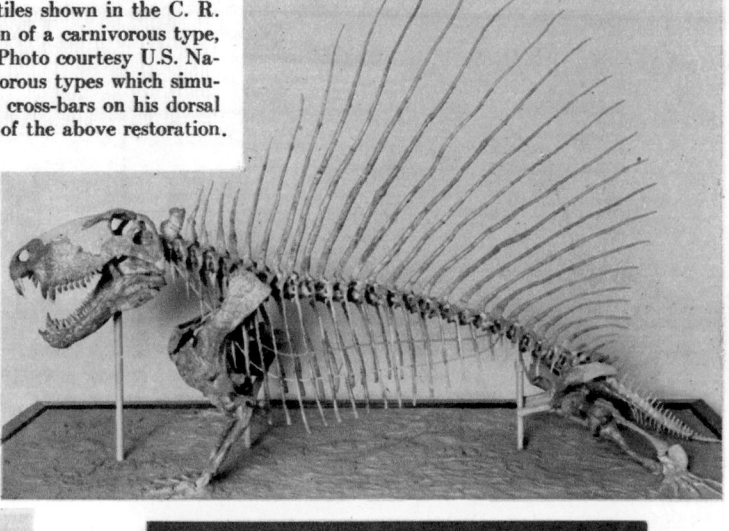

Above: Collecting vertebrate fossils in the Karoo "desert" area, South Africa.

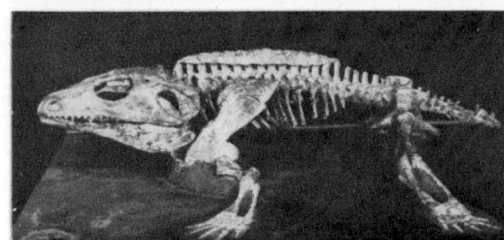

Above: Cacops, a Permian amphibian from Texas.

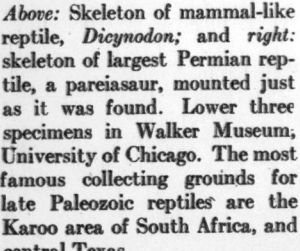

Above: Skeleton of mammal-like reptile, *Dicynodon;* and *right:* skeleton of largest Permian reptile, a pareiasaur, mounted just as it was found. Lower three specimens in Walker Museum, University of Chicago. The most famous collecting grounds for late Paleozoic reptiles are the Karoo area of South Africa, and central Texas.

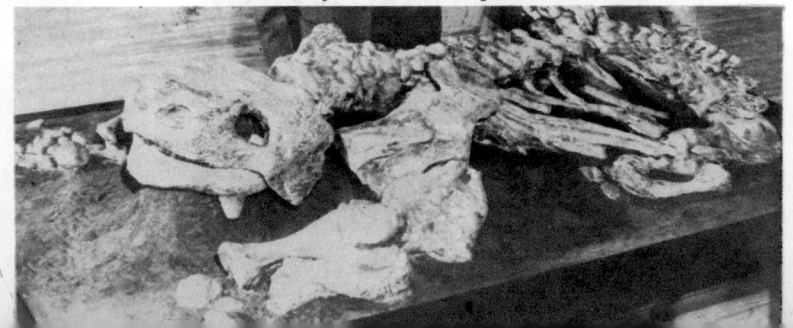

CROSSING THE STRAND

IN THE preceding chapter we have seen that the primeval forests were Devonian in age. In the earliest of them spiders were represented by nearly a score of known species. It is therefore apparent that, regardless of the validity of our belief in Silurian air-breathers, **the plants had paved the way for terrestrial arthropods at least as early as the early Devonian.**
From this time on to the end of the Paleozoic the invertebrate conquest of the land was a rapid and well-recorded one. The cockroaches were so numerous in the Pennsylvanian, for instance, that some eight hundred species have been described, and the older writers often called this period the "Age of Cockroaches." Since many of these ancestors of the modern pests were 3–5 in. in length, we may well be thankful that the group has gone in for size reduction throughout the ages. The primitive dragon flies, which are represented by about a dozen species, are the largest insects of all time, some having had a wing spread of approximately $2\frac{1}{2}$ ft.! The Permian rocks near Elmo, Kansas, have yielded the greatest amount of information concerning these ancient insects, about six thousand fine specimens having been collected there.

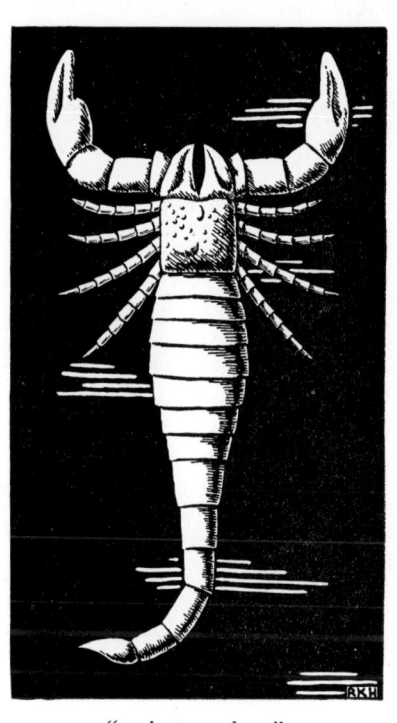

.... "ancient murderer" ....

The molluscs also had gone a long way toward their conquest of the land, *land snails* being recorded from Pennsylvanian rocks; and the clams had advanced as far as they ever were to get, that is, they had become fresh-water types.

But what about our own ancient vertebrate lineage? From whence did it spring? When and where did it first appear in the geological record? As might well be expected, the answers to these questions take us *back across the strand into the sea.* We frankly do not know precisely how the vertebrate stock originated, but all investigators agree that it sprung from some marine invertebrate strain near the beginning of the Paleozoic era. Credit for originating this prodigious offspring has been given to the coelenterates, to the annelid worms, and to the arachnids; but this may well be cheating the echinoderms, since the larva of the acron worm, which is one of the most primitive of modern chordates, is very similar to the larva of echinoderms.

Naturally, these early ancestors of the vertebrate group were probably without true bony skeletons. Possibly, therefore, we shall never discover more primitive members of the group than have already come to light. But at any rate, we can go back in the record reasonably far and still find in the fishes a fairly typical development of the essential structural features of all vertebrate skeletons, as shown in the accompanying diagram. Let us therefore briefly examine the fishes, the first-known vertebrates, which, as the direct ancestors of the amphibians, were the first to cross the strand, and thus become the distant progenitors of the reptiles, birds, and mammals who have inherited the terrestrial portions of this sphere.

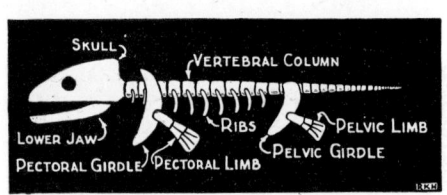

.... structural features of vertebrate skeleton. After Swinnerton

THE fishes are primitive, cold-blooded, aquatic vertebrates which breathe by means of gills. The body is typically streamlined and thus well adapted to an active life in the water. Many fish possess two sets of paired ventral fins and unpaired dorsal, ventral, and tail fins. The skeleton commonly is cartilaginous in the lower groups, but a well-developed bony framework is characteristic of the highest class. For the purpose of our brief discussion the fishes may be grouped as follows:

Pisces
{
Primitive jawless vertebrates (Agnatha)
Extinct armored fishes (Placodermi)
Sharks and their relatives (Chondrichthyes)
Bony fishes (Osteichthyes)
}

THE primitive jawless vertebrates are represented in modern seas by the lamprey eels and the hagfish. The **ostracoderms,** or extinct bony plated fish, **are the oldest representatives of the class and the first vertebrates known from the fossil record.** Their remains have been found in Ordovician rocks at Canyon City, Colorado, and elsewhere. The sediments in which their remains are entombed, as well as their fragmentary character, suggest that they lived in the waters of the land and had been washed out into an area of marine deposits by the very stream in which they had met their death. If the vertebrates were of fresh-water origin, it is, of course, futile to try to find their still earlier record in typically marine formations. Ostracoderms and their relatives reached their climax in the early Devonian, but became extinct at the close of that period.

The **placoderms,** like the ostracoderms, are primitive Middle Paleozoic armored fishes; but, unlike the latter, they have jaws and are related to the sharks. A notable member of the class is *Dinichthys*, which had a length of more than 20 ft. and was, indeed, the "terrible fish" of the Devonian and early Mississippian. *Pterichthys* was another member of the class unusual in that, although armored in front, it had scales in the rear, as can be seen in the accompanying restoration.

Armored fishes, *Pterichthys* in rear, *Cephalaspis* in front. After C. L. Fenton

THE **sharks and rays** have a cartilaginous skeleton and a body covered with granules or spiny plates. Unlike in other fish, however, there are no overlapping scales, the mouth is commonly on the underside of the head rather than at its front, and the gill slits are visible externally. In the main, these features are of little importance in dealing with extinct sharks, since only rarely is the body outline preserved. The spines, teeth, and parts of the dermal skeleton, are, however, commonly found. Such remains are rare in the Silurian, but they are abundant in the rocks of the Devonian and Mississippian systems.

THE **bony fish** have a bony skull structure, a body covered with overlapping scales, gill openings protected by an operculum, and an air bladder. The two more primitive groups, of the three making up this class, are the *lungfishes* and the *lobe-finned fishes*, both of which have only partially ossified skeletons, and both of which reach their climax in the late Paleozoic. The *ray-finned fishes* form the third subdivision of the class and comprise practically all of the modern species. Since the chief episodes in their story occur in Mesozoic to Recent times, we will only mention them briefly here. They are found first in the Devonian, and had achieved a position of dominance among the fish by the Mississippian, a dominance which they have never relinquished. They are characterized by completely ossified skeletons and, at least in most modern species, thin flexible scales. The gar pikes and sturgeons are modern representatives of the more primitive ray-finned fishes, and the bass and the trout are characteristic members of the more advanced group.

The lungfish were common in the Devonian terrestrial waters. No doubt they represent the end-product of a struggle to survive in areas of stagnant pools and sluggish streams with channels which changed with each succeeding rain. The three surviving genera live today under similar conditions. They manage to exist during periods of drought in a cocoon of mucus and mud, and thus they can really invalidate the old saying regarding "fish out of water." A cellular sac serves as a fairly well-developed lung, and thus this respiratory organ makes them essentially amphibious.

The lobe-finned fishes have pectoral and pelvic fins with a bone structure similar to that seen in the primitive amphibians. The teeth also are similar to those of some of the early land vertebrates in that they have a complexly folded internal structure called *labyrinthine*. They had scales of the "ganoid" type which are characterized by a shiny enameled surface and a bony underlayer. These fish, which first appear in the record at about the same time as the lung-

CROSSING THE STRAND

fish, are extinct by the end of the Mesozoic era. Consequently, we have no good method of examining their anatomy. It is regarded as probable, however, that they had some such lung development as that found in the modern lungfish, especially as there are internal openings for the nostril in the roof of the mouth of the fossil skulls of these ancient fish, and the supposed remain of a lung has been discovered in one fossil specimen.

THE *amphibians* are the descendants of fish and, remotely, the progenitors of all the later land vertebrates, including man. They lay eggs in water, whereas the reptiles characteristically lay them on land. The eggs, moreover, are without the shell or protective membrane found in the eggs of higher vertebrates. In the case of the frog, the embryo hatches at an early stage as a tadpole equipped with gills and a fishlike stream-lined body. When partially grown, a lung replaces the gills, limbs develop, and the creature leaves the water and becomes a land type. But it must return to the water to propagate its kind. The modern amphibians are an inconspicuous group characterized by such creatures as the frogs and the salamanders. But in the Late Paleozoic many *stegocephalian*, or "armor-headed," amphibians existed, nearly one hundred different species being known from the Pennsylvanian system of North America alone. Such types became extinct in the early Mesozoic, so that the amphibian reign was relatively a short one.

The oldest known amphibians are of this stegocephalian type. Their existence during the Devonian was first based on a single footprint, which has been called *Thinopus*. Not much to look at, it nevertheless was of great importance as being a reasonably authentic record of a vertebrate crossing of the strand at this early date. Geologists predicted that the actual skeletal remains would also turn up

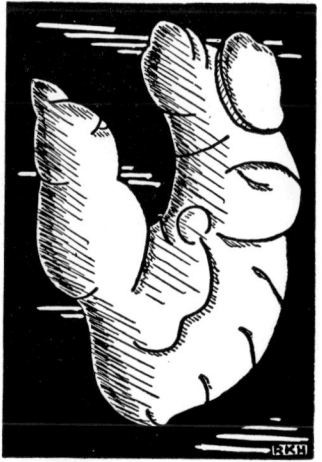

.... not much to look at ....

in Devonian rocks. But years elapsed, and *Thinopus* fell into disrepute along with the geologist's reputation for first-rate prophecy. And then in 1932 Upper Devonian stegocephalian skeletons were found in eastern Greenland by members of the Lauge Koch expedition. The amphibians obviously had come out of the sea during the Devonian period, but out of which marine vertebrate stock had they sprung?

THE Greenland stegocephalians are, as one would expect, more primitive than any other amphibian remains thus far discovered. Their skull is similar to that in some of the lobe-finned fishes. The structure of their teeth also is labyrinthine, and the arrangement of bones in the fins is similar to that in the primitive amphibian limb, as can be observed in the accompanying diagram. How could these ancient lobe-finned fish have given rise to the first land-living vertebrates? During the Devonian, when this important forward evolutionary step was made, there were apparently seasonal droughts, such as are experienced in the modern lungfish habitats. Hence, as the pools dwindled during the dry seasons, some of the more venturesome apparently crawled from one foul and nearly desiccated puddle to a nearby larger, fresher one. The only fish with a fin structure which even conceivably would have permitted them to have made such a journey were the lobe-finned forms. Once their lungs were operating efficiently, and their fins had altered even part way into clumsy limbs, the changes became more rapid. The time and the place for such an epoch-making metamorphosis were right. The change was made. We know at least the close relatives of the very ani-

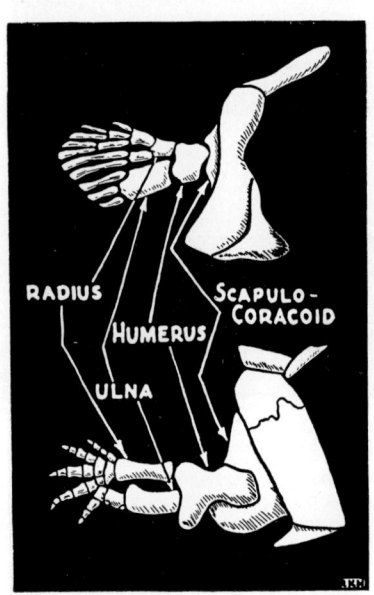

Comparison of bones in fin of a lobe-finned fish (*above*), and the bones in an amphibian limb

mals involved in this pioneering. But, of course, we did not see the changes take place, and it is difficult to evaluate all of the physical and biological factors involved in the urge to come out of the water.

THE **reptiles** were the lords of the late Paleozoic. The earliest members of the class are not found before the Pennsylvanian, when these primitive types are so like some of the stegocephalians that they are distinguished from them only with great difficulty. They had, however, made a great advance over the amphibians in that they had developed an egg which freed them from the necessity of returning to the water. This reptile egg possesses a firm protective shell which is porous, so that one of the two membranes within can act as a crude lung. The other membrane serves as a water reservoir to prevent the drying-up of the embryo. Food is supplied by the yolk substance within the egg. You may consider this egg as just another bothersome detail; but until its development, the vertebrate crossing of the strand was something of a moral victory. Without this egg the vertebrates were still chained to the water. With this new structure the reptiles set out to conquer the face of the earth. They soon had a better skeleton, a higher muscular system, and a more complicated circulation. Even their brains begin to show rudimentary cerebral hemispheres. And their success was almost instantaneous, for in a single geological period they crowded the amphibians out of the position of dominance among the animals of the earth. In this connection refer to Plate 48.

THE high land and dwindling streams toward the end of the late Paleozoic were bitter medicine for the stegocephalians, but the reptiles found their new freedom from the water to their liking and deployed rapidly. Most interesting among these reptilian developments were the appearance of the fin-backed *pelycosaurs*. Some of these forms probably were as much as 10 ft. in length and had great bony spines on their back. These highly specialized reptiles were extinct by the end of the Permian. Another reptilian group of much greater evolutionary significance were the *theriodonts*. These prob-

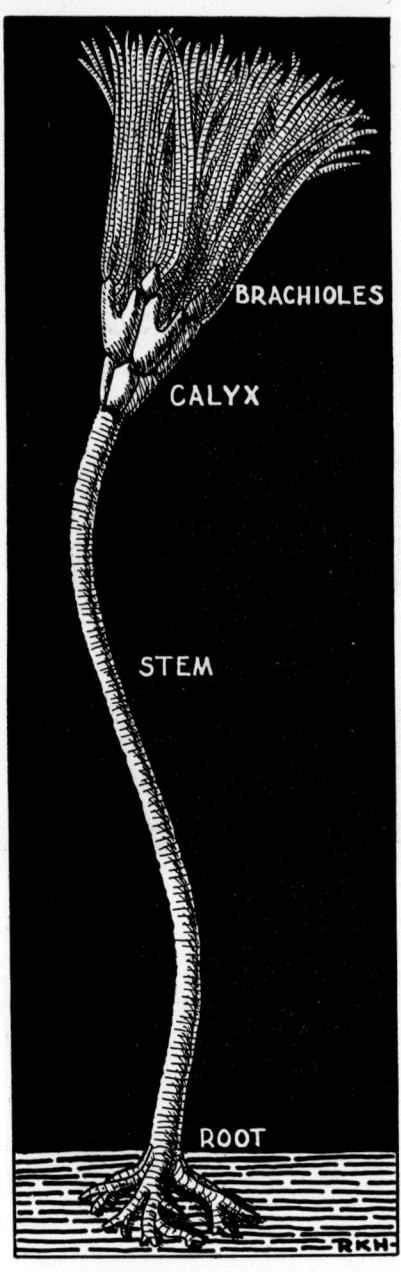

A typical Mississippian blastoid. Divested of brachioles, the calyx resembles a flower bud

ably were the animals which gave rise to the mammals in the next era.

It is obvious that not all the animals became terrestrial in the late Paleozoic. Many groups continued to work out notable careers in their ancient marine habitat. It is not possible, however, for us to consider more than a few of the high lights of their complicated stories here.

Corals, which had been important in the Silurian period, have an even greater flare in the clear waters of the Onondagan seas. These animals are coelenterates, which during life have radial partitions extending out into the body cavity, thus separating it into a number of divisions. Hard parts are secreted as a reflection of this soft structure. As a consequence the external skeleton of the individual coral, or polyp, is a cup with a number of radial partitions, called *septa*. There are many Devonian individual, or cup, corals; but the compound types were more important as reef-formers. After the Devonian, corals were no longer common in Paleozoic seas, the ancient, or tetraseptate, type becoming extinct at the close of the era.

The *echinoderms*, which, as we have seen, had a momentary flare in the Silurian period, go into a partial decline in the Devonian. But in the Mississippian they have their ancient climax. These echinoderms are all characterized by a fivefold symmetry so well displayed in the familiar starfish. Characteristically, they have a skeleton of calcareous plates which fit together like the pieces of a mosaic. The stalked echinoderms include the *cystoids*, *blastoids*, and *crinoids*. All of these classes are typically Paleozoic forms, although the crinoids have survived to the present. On the other hand, the free-moving echinoderms, such as the *starfish* and *sea urchins*, were unimportant in the Paleozoic but are of increasing consequence from that era down to the present. Of these echinoderm groups, the crinoids had their greatest development in the Valmeyer epoch of the Mississippian, at which time they lived in such numbers that they made up much of the limestone deposited in some areas. The blastoids had their great climactic in the Chester epoch of the same period, but declined into an unaccountably rapid extinction at its close. The starfish, which had

The tribolites continue to the end of the Paleozoic. In the Devonian period there are many highly spinose forms like the one above, but the last survivors are simple types such as is illustrated in Plate 44

been first introduced in fairly modern garb in the Ordovician, were relatively common in the Devonian; and at St. Louis a good many sea urchins are found in the Mississippian strata.

In the late Paleozoic the superficially coral-like *bryozoans* are noted for their lacy, or fenestelloid, structures and, particularly during the Chester epoch, for their unusual screw-shaped (*Archimedes*) or lyre-shaped (*Lyropora*) central supports. *Archimedes* and a number of other late Paleozoic vertebrates are shown on Plate 44.

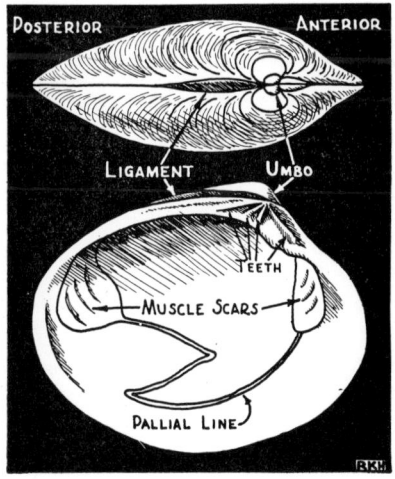

Parts of a typical pelecypod

Our old friends, the *brachiopods*, reach their climax in the Devonian; but in the later Paleozoic periods their *diversity* diminishes. Still numerous until the close of the era, they are, during the Carboniferous, characteristically of the produced, or *productid*, type. *Trilobites* have fallen to a relatively insignificant position by the Devonian, but some highly spinose types still persist. During the later Paleozoic periods only a few stragglers of the generalized type survive.

The molluscs are well represented in the late Paleozoic. The *clams*, which, like the starfish, appear in the Ordovician with startlingly modern appearance, are abundant in the near-shore deposits of the late Paleozoic periods; and the *gastropods* also make notable advances during this time. *Cephalopods* were likewise important; but, as we will discuss their history later in chapter 45, we need not elaborate here.

Finally, we must mention briefly the single-celled calcium-carbonate-shelled *foraminifera*. These seemingly insignificant animals became common as rock-builders in the mid-Mississippian, at which time their countless skeletons made up a part of the rock already referred to as the Indiana building-stone. By the Pennsylvanian they have evolved a spindle-shaped type called *Fusulina*, which looks like a grain of wheat. These are important rock-formers literally around the world.

THUS we see again that, whether crossing the strand or remaining in the sea, the plants and animals throughout most of the late Paleozoic were, as ever, marching ahead hand in hand with the physical world. But life advances by means of many retreats, and in the next chapter we must pause to investigate the influence of the unusual physical vicissitudes of the Permian on the organisms struggling for life at the very end of the Paleozoic era.

CHAPTER 40

AN EARLIER DEPRESSION

> There is some soul of goodness in things evil
> Would men observingly distill it out.
> —S<small>HAKESPEARE</small>

"LIFE allows you four depressions," is a sentence concocted by an astute advertising man. It compels attention. Suggesting uses of adversity, it is somehow vaguely reassuring. We read on: "The average investor's life spans eight to ten depressions. Three or four are gone before he knows what to do with them. Yet, if he is wise and able, he may profit substantially from the remaining ones. During every major decline thousands of investors have established the foundations for future independence."

But this merely has to do with depressions in the trifling affairs of humans. What about the more important "bad times" which the organic world as a whole has spanned? The late Paleozoic was such an early period of organic depression, when many an animal group found world-quotations on its life-expectancy embarrassingly low and, like so many humans under similar stress, gave up the ghost completely. But some animals and plants did profit substantially from this Permian depression. In fact, during all the "depressions" which the world has been "allowed" throughout the hundreds of millions of years since the first successful organic synthesis, there has always been an upward-trending, though ragged, curve in the graph of the life-complexity. In times of stress the weak and the

.... many an animal .... gave up the ghost ....

specialized organisms have died out, but the generalized and the strong have always emerged from the troughs of trouble more powerful than ever. Modified to fit the changing environment, they have been ready to take advantage of the return of "good times."

WE HAVE already seen that the continents, as well as nations and businesses, have had their ups and downs; and, of course, their areas have changed remarkably throughout the past. They have presented bold, swashbuckling outlines when they stood high; but they have made sorry, attenuated showings during their periods of depression. And the parable from the past is more clearly understood and more definitely encouraging when we remember that the Earth has not only risen above her earlier depressions but has generally risen higher, rejuvenated and youthful after each succeeding deluge. But the record of earlier depressions is, of course, only in part a physical story. Even more pertinent comparisons may be drawn with the panics which life itself has encountered and survived Anteus-like, with strength redoubled. Paradoxically enough, however, the times of continental depression are "good times," at least so far as marine life is concerned. The periods of mountain-building, on the other hand, are apparently "hard times" for the organic world, although new stocks are evolved during these times of stress.

LET us now inquire into some of the features of Permian mountain-building and glaciation which were responsible for the late Paleozoic organic depression. We have already learned in chapter 37 that the general crustal instability of the late Paleozoic culminated in a great Permian spasm of orogenic movements. In the area of the old Appalachian geosyncline, mountain-building reached its peak of complexity. The State Geological Survey of Pennsylvania has estimated that the rocks between Altoona and Philadelphia were so complexly folded and faulted that the section was shortened approximately 100 mi.! The size of some of the folds also gives an idea of the heights these Permian mountains may have attained. Reconstructions of these anticlines suggest heights of between

AN EARLIER DEPRESSION

20,000 and 30,000 ft. That any of the peaks achieved the latter height may well be doubted, since erosion actively tended to reduce the mountains as they were being folded up. But there can be little reason for thinking that the late Paleozoic Appalachians were any less majestic than the present Rockies.

The so-called Paleozoic Alps of Europe involve a large group of minor modern mountain areas, such as those of southern Wales, the Vosges Mountains of France, the Black Forest, the Harz, and Thuringian ranges of Germany, and the Erz, Riesen, and Sudetes ranges of northern Czechoslovakia. Although isolated today, they apparently were folded into one great system in the late Paleozoic. The eastern end of this system extended at least as far as Bohemia, and may have joined the late Paleozoic ranges north of the Himalayas. A great branch of this enormous system extended northward to the Arctic in the position of the present Ural Mountains. Westward the chain crossed northern France, southern England, and Ireland, and continued an undetermined distance out into the area of the present Atlantic Ocean.

The southern continents were emergent in the late Paleozoic, so that the record of their Permian orogeny is poorly written. Some evidence has been accumulating, however, which seems to show that eastern Australia was again affected and that there were Paleozoic ranges just east of the present Andes of South America. With all this mountain-building and continental elevation, climatic refrigeration was inescapable.

THAT the Permian was the time of the greatest period of glaciation that the world has known is a fact based on direct observation. Ice sheets were literally world-wide, though scattered in distribution, as can be judged from the positions of the probable ancient ice sheets shown on the following map. The Southern Hemisphere, however, was affected to a far greater extent than the Northern Hemisphere, a point of considerable interest since precisely the reverse situation obtained in the most recent glacial period. In fact, few of the evidences for Permian continental glaciation north of the Equator are absolutely certain. There is a small deposit of late Paleozoic glacial till with seasonally banded sediments at Squantum, near Boston. This may be Pennsylvanian rather than Permian in age, and there is a possibility that the deposits represent valley glaciers rather than continental ice sheets. Striated boulders are found in conglomerates of Permian age on Prince Edward Island, and similar occurrences are known at several places in Alaska and in Texas. Other possible Permian glacial areas are found in England, the Urals, the Alps, Germany, Afghanistan, and southern Russia. The chief ice sheets, however, were in India. Here the evidence is conclusive, the thick *Talchir* tillite underlying the marine Permian. Not only did this glaciation occur within a few degrees of the Equator, but the ice moved northward away from the present torrid zone.

Areas in which there is evidence for late Paleozoic glaciation

SOUTH of the Equator, South Australia and South Africa are the sites of the best-known areas of late Paleozoic glaciation. In Australia the ice moved north across Tasmania, Victoria, and New South Wales. Interestingly enough, the ancient tillites are here interbedded with sediments carrying coal of Permian age. In the provinces of the Union of South Africa, and in Southwest Africa and Madagascar, the late Paleozoic *Dwyka* tillites occur sporadically over an area in excess of 600,000 sq. mi.; and their areal extent must have been much greater prior to Mesozoic and Cenozoic erosion. Since the striated and grooved floor over which the ice moved is known in a number of places, it has been determined that the direction of glacial movement was toward the south and southwest. In other words, here, as in India, the ice moved out of the equatorial district.

In South America late Paleozoic tillites are found from the Falkland Islands on the south northward through Uruguay, western Argentina, southeastern Brazil, to southeastern Bolivia within a few degrees of the Equator. Not all these deposits are necessarily of pre-

AN EARLIER DEPRESSION

cisely the same age. Most of them, however, indicate ice movement in the same direction, that is, from east to west. Certain writers consider the South American glaciation to have been less important than the other Southern Hemisphere ice episodes of this time.

SOME scientists have explained late Paleozoic glaciation by taking liberties with the present position of the poles, but this solution of the problem introduces new ones by bringing most of the South American glaciated areas into the tropics. Some have suggested a mere reduction of temperature for the entire earth. The troubles with this solution are the same as those encountered in framing a similar explanation for recent glaciation. Permian ice sheets were widespread, but their rarity in the Northern Hemisphere is difficult to explain under the theory of general temperature reduction. Lowered temperatures do, however, accompany increased land elevations, as we have seen before; and this general cooling was undoubtedly a factor of some consequence in glaciation. **Permian glaciation, like that of the Quaternary, was characterized by interglacial periods**—another thorn in the side of almost all explanations of ancient and recent ice ages.

SOME geologists have conjured up an ancient land mass which has been dubbed **Gondwana.** This Permian continent, it is postulated, extended from Brazil on the west across the South Atlantic, Africa, India, and the Indian Ocean, into Australia, and thence trended southward into Antarctica. If such a land mass actually existed in the late Paleozoic, it would have greatly changed ocean currents, general climate, and precipitation. It would have been possible for snow and ice to have accumulated in the high interior of Gondwana, where glaciers would have formed. Further precipitation would have caused these ancient ice masses to move outward toward the sea. Not only the distribution of the glacial deposits themselves has been cited in support of the existence of Gondwana, but the *Glossopteris flora*, which we have earlier mentioned, has a distribution most easily explained by the supposed existence of this great east-west continental mass. Moreover, both land and marine faunas are so distributed that they could be explained readily if this land mass once actually existed.

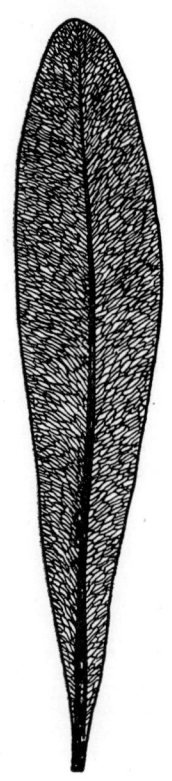

Leaf of the Permian "tounge fern," *Glossopteris*. From Steinmann

Disregarding the fact that the name itself has an unfortunately phoney ring, the existence of Gondwana perpetrates about as many geological conundrums as it solves. We must therefore regard its existence as still being in the realm of the problematical rather than in the domain of the proved.

Other climatic changes were almost inevitable in the late Paleozoic. Great orogenic movements, with their attendant continental uplift and consequent temperature reductions, played havoc with climatic stability. The shallow seas had been drained, and with them went their ameliorating influence on the weather. New land bridges had resulted from the uplift of the continents, as well as from the withdrawal of the seas because of the formation of glacial ice caps. As a consequence oceanic currents were markedly changed. Lofty mountain ranges cheated the great areas to their lee of moisture, and deserts of great areal extent were formed.

In North America the widespread Permian dune sands, red beds, and salt and gypsum deposits suggest a southwestern interior of great aridity. Central and western Europe suffered the same fate, the salt deposits of the Stassfurt district representing almost complete desiccation in that area; and sporadic red beds are found from England across Germany to the Urals. The North China and the Australian coal fields of Permian age indicate, however, that there was plenty of moisture in these areas. The red beds of the Karoo in South Africa, with their great assemblage of fossil reptiles, also indicate clearly that you cannot trust red beds implicitly as indicators of aridity.

AN EARLIER DEPRESSION

THIS, then, is the stage setting for an earlier depression, which was attended by organic disturbances of the most far-reaching sort. In fact, the destruction of life at this time was actually so great that the early geologists thought that all living things had been blotted from the face of the earth. We now know that, severe as the "depression" was, a few strong, generalized types survived it. Their descendants repopulated the earth with new and more vigorous inhabitants.

LET us first briefly examine the results of this depression in the marine invertebrate stocks. Naturally these were hardest hit of all. Their ancestral home, the shelving seas, had been almost completely drained. The few types that survived this radical change in habitat were crowded together under inhospitable conditions, since such few basins as remained were likely to be saline, and in any case all the waters were colder than usual. Nevertheless, there were some harbors of refuge in which Permian faunas were well developed. Evidence for this statement is preserved in the rocks of the Island of Timor, in the Permian beds of the Salt Range of India, and in our own Texas Permian area. But, although the change from the Paleozoic to the Mesozoic in the marine invertebrates was probably not as cataclysmic as once was considered, still it was a most notable one. Least affected by the change were such creatures as those which lived in the open surface waters and could thus migrate with ease. The *ammonoid cephalopods* are typical of this group—they probably weathered the depression more successfully than any other marine invertebrate stock. Many groups that had been outstanding were entirely gone; or, if they still remained, they had been reduced to the status of a private in the rear ranks of the general invertebrate army. By the opening of the Triassic period the trilobites, proud masters of the Cambrian seas, had passed out of the picture entirely; so had the Paleozoic type of coral, the eurypterids, and the cystoids and blastoids. Moreover, the once ubiquitous brachiopods were at long last greatly reduced, and never again were they to occupy any position of prominence.

ON THE extensive lands the invertebrates fared far better, most of the air-breathing arthropods, such as the insects and arachnids, getting a good start for the battle with the next era's vicissitudes. The land vertebrates also were able to muddle through the hard times fairly successfully. The stegocephalians, representing the undistinguished amphibian stock, clung tenaciously to the few stream areas of the arid regions. Reptiles, which had been freed of the curse of returning to water to propagate their kind, found the going to their liking. We have already seen that they deployed rapidly under the very conditions which proved so difficult for the marine animals. Nevertheless, their specialization operated against them, since many of the Permian types failed to carry over into the Mesozoic era. Chief interest of course attaches to the *therapsid* stock which did survive to give rise to the mammalian line in the next period.

THE dominant Pennsylvanian flora lived on into the Permian, at least in the Northern Hemisphere; but most of these ancient plants were little suited to aridity and cold. Hardy new plants with scanty foliage soon took the place of the older, luxuriant, swamp-living plants. The true conifers became important as the ancestral *Cordaites* gradually passed out of the picture. Ancestral sago palms also became more widespread. But in the Southern Hemisphere the *Glossopteris* flora were the outstanding plants. These relatives of the cycads may have lived in interglacial periods entirely, but there is a persistent belief that their thick skins represent a direct response to the unusual climatic conditions. After the Permian glaciation these plants migrated to Eurasia, and the club mosses also had a momentary flare-up before their reduction to their present position of plant ignominity. Thus the archaic plants of the true forests primeval scarcely survived the "depression." But a more modern plant stock had taken firm root along with the more advanced reptiles and the ever increasing horde of land-living invertebrates. They all moved triumphantly out of a period of depression into a time of organic expansion.

CHAPTER 41

EARLY MESOZOIC EVENTS

> Not to know the events which happened before one was born, that is to remain always a boy.—CICERO

MANY a well-trained American geologist has had little or no experience with the rocks of the early Mesozoic. But scarcely a European earth scientist of any consequence has failed to have a part of his training on the rocks of the **Triassic** and **Jurassic** systems. "Strata" Smith himself established the foundations of stratigraphical geology by working with Jurassic sediments; and, in fact, most of the "great" in European geology got their first lessons in the science on the Oölite series, as Smith informally designated the rocks we now call Jurassic.

William "Strata" Smith. From the History of the Geological Society of London

This discrepancy in the training-grounds of the European and American geologists is a simple reflection of the position of land and sea some three weeks before the end of that Gargantuan geological year to which we have so frequently referred. **There simply were no Triassic or Jurassic seas in eastern North America; consequently there are no marine deposits of these systems east of the Mississippi.** But whereas North America was standing high during this time, Europe was submerged. Deposition is the rule for the European early Mesozoic scene; erosion, the keynote of the North American history of the times.

IN VIEW of the points made in the last paragraph it is not surprising that we must look to Europe for the series names for the three epochs of the Triassic period. In the accompanying table these European series are shown in their general relationships to the few rocks of the same age in North America.

General Series Names	German Series Names	Western United States	Appalachian Fault Troughs
Upper Triassic	Keuper	Marine beds on Pacific Coast; terrestrial beds in Rocky Mountain states	Newark
Middle Triassic	Muschelkalk	Marine beds in California; chiefly absent elsewhere	Mostly absent
Lower Triassic	Bunter	Marine beds on Pacific Coast; terrestrial beds in Colorado Plateau	Absent

THE Carboniferous beds of Great Britain lie between red beds below and above. Those beneath the coal-bearing strata are, as we have seen, the Old Red Sandstone of Devonian age. What could be more natural than that the red beds *above* should be called the *New* Red Sandstone? These latter sands are chiefly Triassic (in part Permian). When traced eastward into Germany a limestone appears in the middle part of the sequence and separates the lower red beds, or **Bunter** series, from the upper one, called **Keuper**. The separating marine limestone is the **Muschelkalk**. The Swabian geologist von Alberti, recognizing the threefold development of these beds in 1834, gave them the name Triassic. Southward from the German exposures the Alpine Triassic sequence, in which there are six recognizable *marine* horizons, becomes prominent.

In North America the Triassic beds also have a "red" aspect throughout the general Rocky Mountain region. Thick cross-bedded red sandstones and red shales occur over large areas and are responsible for such scenic features as the "Painted Desert" of Arizona. Their greatest development is in southwestern Utah, where, in Zion Park, they are more than $\frac{1}{2}$ mi. thick. But from this general region the beds thin both to the east and west. In the latter direction they grade into the marine beds of the Pacific geosyncline. In early Triassic time this structural trough apparently was of no great consequence in California, although marine waters did spread eastward from that state into Nevada and Idaho. During the middle and late Triassic epochs the sea was more definitely confined to a true elongated geosyncline, but during this time the waters did not spread farther east than Nevada. Volcanic activity was of considerable consequence near the Canadian border during the late Triassic, thus paralleling similar and contemporaneous activity in the eastern Triassic area.

With the close of the Paleozoic era the Appalachian geosyncline passed out of existence. But as the Appalachian Mountains, which had been folded up out of the trough, were being reduced by erosion, a new basin of deposition formed along the eastern flank of the range, but actually on the old rock platform of the land mass of Appalachia. During the late Triassic this elongate trough must have been somewhat similar to our own present valley of California. The late Triassic rivers deposited their burden of sand, gravel, and mud in this basin, which extended from the Acadian area on the north to South Carolina on the south, thus building up the red beds which now largely comprise the **Newark series.**

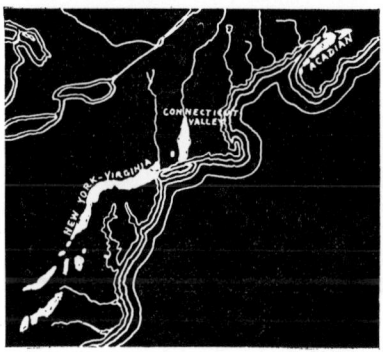

The Newark areas from the Acadian district on the north to South Carolina

Many kinds of dinosaurs have left their footprints on these sands of time, especially in the Connecticut Valley area. And in the Richmond basin small coal swamps with ferns, cycads, and early pines made perfect habitats for amphibians and small mammal-like reptiles. Throughout the great length of this valley of deposition there were numerous lakes in which ganoid fish abounded, but **marine waters never invaded the trough.**

During the time these clastic sediments were being laid down, igneous activity also was relatively important. Great flows of

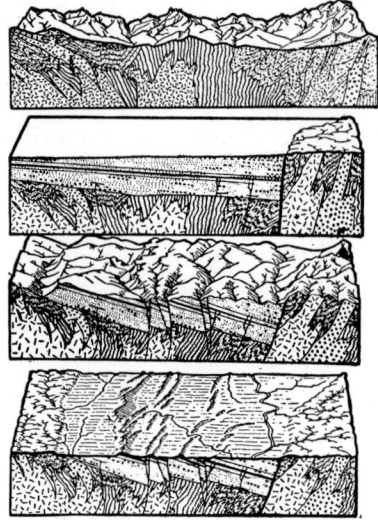

Block diagrams representing a few stages in the history of the Connecticut Valley. From top to bottom: (1) Early Appalachians made up of complexly folded Paleozoic and pre-Cambrian rocks. (2) Triassic sediments have been deposited in a downfaulted trough. (3) The area after the Palisades disturbance. (4) Present situation with Triassic beds preserved in fault blocks in spite of great amount of erosion since their deposition. From R. C. Moore after J. Barrell.

basalt were poured out on the surface or were intruded as sheets between the sedimentary layers. The Palisades of the Hudson, with their columnar jointing, were formed of such late Triassic igneous material. At the very close of the period the seas were withdrawing from the western depositional troughs, and block faulting was common in the eastern Triassic lowlands. Undoubtedly some of these latter movements were contemporaneous with the sedimentation, but the faulting was sufficiently marked at the close of the period to merit a distinct name. Appropriately, the term which has been chosen for this period of earth movements is the **Palisades disturbance.** The movements involved, interestingly enough, were the result of *tensional*, rather than *compressional*, stresses, as was the case at the close of the Paleozoic era. Some of the highlights in this early Mesozoic history of the Connecticut Valley are shown in the accompanying diagrams.

THE economic resources of the Triassic period are of no great importance in North America. There are extensive gypsum beds of this age, however, in Nevada, Wyoming, Montana, and South Dakota; and they have been exploited to a considerable extent. The first coals actively mined on this continent were those of the Newark series in the Richmond, Virginia, basin. Still mined for local consumption, they are of little consequence today in the total production of the country. When brownstone houses were the rage, the eastern Triassic sandstones also were extensively quarried; and salt deposits of Triassic age are still exploited on a fairly large scale in Germany.

As was the case for the Triassic, so for the Jurassic period also, there have not yet been designated generally accepted series names of North American origin. The general European terms are indicated in the accompanying table.

General Series	European Series
Upper Jurassic	Malm *or* White Jura
Middle Jurassic	Dogger *or* Brown Jura
Lower Jurassic	Lias *or* Black Jura

The exact origin of most of these miners' or quarrymen's names, such as **Dogger** and **Malm,** is lost; but Leopold von Buch gave the designations "Black," "Brown," and "White" on the basis of the predominant color of the beds. "Strata" Smith, as we have seen, called the same rocks the "Oölitic," since many of the English Jurassic limestones are oölitic in composition. But Brongniart, the French geologist, called the rocks of the same age the "Jurassic," from their fine exposure in the Jura Mountains, and this name has stuck in the literature.

The Jurassic rocks, so widespread and so richly fossiliferous in Europe, are represented only poorly in North America. *East of the Colorado-Kansas line there are no deposits of this age at all.* Throughout most of the period all of the continent was undergoing erosion. Moreover, at this time the eastern shore of North America lay out on the present continental shelf.

North America in the early Jurassic. Generalized after C. Schuchert

Therefore all of the sediments removed from the eastern part of the continent during the Jurassic were deposited in an area which is now beneath the sea. But the west coast of Jurassic times was also considerably east of its present position. Thus, in reconstructing the geography of the period in the mind's eye, one must imagine the entire continent lying somewhat east of its present position. As a result, there are west-coast marine Jurassic beds representing a fairly complete sedimentary sequence for the period.

AT THE beginning of late Jurassic time, however, a new geosyncline started to form in about the position of the Paleozoic Cordilleran trough. This basin of deposition was to widen and deepen until in the late Mesozoic it was to be the site of a strait extending from the Gulf to the Arctic. But during the Jurassic this sea had spread southward only as far as to Utah and Colorado. In this sea, whose outline is indicated on the accompanying map, there were deposited the shales, sandstones, and impure fossiliferous limestones of the *Sundance* formation. The marine embayment consequently is sometimes called the *Sundance sea*. With its retreat at the close of the Jurassic, the streams flowing out of the rising lands to the west filled the basin with the continental deposits known as the *Morrison formation*. These beds have turned out to be the most famous dinosaur tomb known in the world.

Late Jurassic Sundance sea. Generalized after C. Schuchert

On most continents Jurassic deposition was common and crustal stability the rule. In North America, on the contrary, not only did erosion predominate throughout the period, but it was terminated by an orogeny of considerable violence. The early Mesozoic geosyncline along the Pacific Coast was severely folded in the first important mountain-building episode to disturb this area since pre-Cambrian times. Volcanic activity continued throughout the period; but at the close of the Jurassic, extrusive volcanism gave way to the intrusion of granite batholiths, particularly in California and along the west coast of Canada. To this episode of mountain-building and volcanism the term *Nevadian disturbance* has been applied. Some of the physical features of the early Mesozoic are depicted on Plate 49.

IN NORTH AMERICA the Jurassic is anything but a coal-producing system, although coals of this age are common in northern Alaska, where there are many seams of a rather low-grade bituminous coal. But when the other continents are examined, we find that this system is of considerable consequence in the production of the fuel. The most important coal beds of Australia are Jurassic; and coal of the system occurs extensively in China, India, Siberia, Hungary, Spitzbergen, and eastern Greenland.

As we have seen, the Nevadian disturbance, at the close of the period, introduced granite batholiths of great size. This igneous activity not only metamorphosed the enclosing shales to slates, but the fluids and gases ascending from the igneous bodies were responsible for the introduction of the gold-quartz veins of the Sierra Nevadas. This was the very gold which lured the "Forty-niners" to California after Sutters' epoch-making initial discovery. Much of the early recovery of the gold was from placer, or river-gravel, deposits which were formed during the next era by the erosion of the gold-quartz veins and their enclosing rocks. The so-called "Mother Lode" is a series of these veins cropping out in the western Sierra Nevadas. The Mother Lode veins, and the placer gold derived from them, form the basis for California's importance as a gold-producing state.

WE HAVE now briefly examined the physical features of the early Mesozoic and must next turn our attention to the closing events of this great era. In some respects it would be logical to discuss at this point the life-history of the two periods whose chief physical events we have just catalogued. Since there is a certain very real unity in the life of the entire Mesozoic era, however, we are postponing this organic story of the times to chapters 43–45.

CHAPTER 42

END OF AN ERA

> Here's a world of pomp and state
> Buried in dust. . . . .
> —F. Beaumont

WE LAUNCHED a seagoing anachronism, the liner "Queen Mary," into the Devonian seas in order to bring home the importance of ancient inundations. During the early Mesozoic this vessel might well have been out of commission so far as North America was concerned, since there were only a few small western continental seas in which to operate her. But now, during the Cretaceous, once again the "Queen Mary" could have steamed majestically from the modern mouth of the Mackenzie on the north to the present delta of the Mississippi on the south; or, if the demand were great, she could have anchored off Duluth, Minnesota, or Cairo, Illinois. But as the period wanes, and we come again to the end of an era, we find the Earth is once more racked by paroxysms of mountain-building. Father Time, it seems, is putting the screws on this unimportant sphere again. With the birth of great mountain chains, and the elevation of the continents, the last great epicontinental seas are drained and our continent-going liner is laid up for good. Or at least until the next great cycle of inundation.

Father Time puts the screws on this unimportant sphere

BUT, wait. Let us use our vessel for a short modern voyage and steam up the English Channel past the great white cliffs of Boulogne, Brighton, and Dover. The chalk beds making up these bold and colorful prominences comprise a part of a great system of

END OF AN ERA

rocks which was named the **Cretaceous** (Latin, *creta* = chalk) by D'Halloy in 1822. Although more chalk was formed during the Cretaceous than in any other geological period, it is by no means the most important type of sediment in all or even most Cretaceous areas. To this extent, then, the name is a poor one—and, like the Triassic, it fails to have a geographic significance.

AS WAS the case for the early Mesozoic periods, the Cretaceous standard sections are also European in origin; but for this latter period there are generally two epochs recognized, rather than the more characteristic three to which we have become accustomed. In the accompanying table a few of the important American Cretaceous sequences are listed, and their general equivalency shown:

	Central Plains and Texas Area	Pacific Coast Area	East Gulf and Atlantic Coast Area
Upper Cretaceous	Laramie	Absent	Absent
			Ripley
	Montana	Chico series	Selma
	Colorado		Eutaw
	Dakota	Absent	Tuscaloosa
Lower Cretaceous	Washita	Shastan series	Potomac series
	Fredericksburg		
	Trinity		
	Absent		

It must be remembered that this table, like all the similar ones in this volume, has been prepared merely to simplify the following discussion and to help the reader visualize the general age relationships of the various series or formations. Exact correlations are not implied.

BY THE beginning of the Cretaceous period the Appalachian Mountains had been eroded down nearly to baselevel. Warpings of the mountain belt at this time caused the streams of the area to quicken their erosional pace. They consequently carried clastic materials eastward and dumped them on the coastal plains. Thus was formed the terrestrial **Potomac** series which rests unconformably on the Triassic or older rocks and is overlain disconformably by the late Cretaceous sediments.

Farther to the southwest the Cretaceous sea had spread widely over Mexico, while Texas was still dry land. Finally the marine waters, moving steadily northward, engulfed the Texas area. Here the **Comanchean** series, comprising the *Trinity, Fredericksburg,* and *Washita* groups, was laid down. These formations, which are about 1,500 ft. thick, are chiefly marine. Forming farther north, at about the same time as the Washita beds, were the Lower Cretaceous terrestrial coal-bearing rocks (Kootenai) of Alberta, which contain more than a score of workable coal seams.

AFTER the Nevadian disturbance the resulting highlands supplied an abundance of detrital material to the long geosyncline which lay to the west of the folded Sierras and extended from California to Alaska. In this basin of deposition the thousands of feet chiefly of sandstone and shale of the **Shastan** series accumulated during the early Cretaceous. These beds commonly are 2 mi. thick, and at some places thicknesses of 5–7 mi. have been reported. The amount of sinking in the depositional trough, therefore, was prodigious. In spite of the generally shallow-water character of the beds, well-developed faunal zones show that the oceanic waters had ready access to the basin.

LATE Cretaceous marine sediments occur throughout the Great Plains and a large part of the Rocky Mountain states. In general these formations become thicker toward the west as they merge into terrestrial deposits formed at the base of the western highlands. In some such general setting as this were formed the *Dakota, Colorado, Montana,* and *Laramie* beds. The Dakota sandstone represents the sands and shales which were spread over the geosyncline by aggrading streams just prior to its inundations by the great Cretaceous flood. The Colorado group is one of the most widely distributed of the late Cretaceous marine sequences, and it represents the initial sediments of widespread seas. The formation known as the *Niobrara* chalk is perhaps the best known of these beds, since it is world-famous for its vertebrate remains. The Montana group consists largely of marine shales and sandstones; but toward the old shore line on the west, non-marine beds become common and coal seams and land-vertebrate remains occur. The Laramie group includes most of the great coal deposits of our Rocky Mountain area. The beds also contain plant remains in profusion and many dinosaur skeletons, but only a few marine sediments are known. Thus, the Dakota group represents the initial sedimentation in the formation of a great depositional area; the Colorado beds are deposits made during the maximum extent of marine inundation; the Montana group, although largely marine, indicates by the character and distribution of its sediments that the seas are withdrawing during its deposition; and the Laramie sequence comprises deposits made at a time when the withdrawal of marine waters was essentially complete. **Since this sea was drained, marine embayments have never again spread over notable portions of North America.**

.... since this sea was drained—the Cretaceous (Colorado) seas in black. After C. Schuchert

IN THE eastern Gulf states the *Tuscaloosa, Eutaw, Selma,* and *Ripley* groups were being deposited at essentially the same time that the central western beds were laid down. The Tuscaloosa beds are largely terrestrial, but the entire sequence is dominantly marine, the Selma chalk being the most widespread formation. When these beds were forming, the mouth of the ancestral Mississippi was somewhere in the vicinity of Cairo, Illinois. Florida, as well as the entire Gulf and South Atlantic coasts, as we know them today, were submerged.

During the late Cretaceous there was some deposition of coarse clastics in the Pacific geosynclines, where the **Chico** series was laid down. The formations also contain volcanic materials in the Queen Charlotte Islands.

DIASTROPHIC movements at the close of the Cretaceous were those typical of the end of any era. They were (*a*) **general uplift and gentle warpings** and (*b*) **mountain-building and volcanic activity.** The Appalachian Mountains by the close of the Cretaceous had been reduced nearly to baselevel—their relatively smooth surface has been designated the *Cretaceous peneplain.* But at this time the entire area is bodily uplifted, with little or no folding of the beds, to a height of nearly $\frac{1}{2}$ mi. The present ruggedness of the Appalachian area results from the subsequent erosion of the softer layers, but evidence of the old peneplain is still preserved in the even crests of the old ridges of harder rock. The area has also experienced post-Cretaceous uplifts.

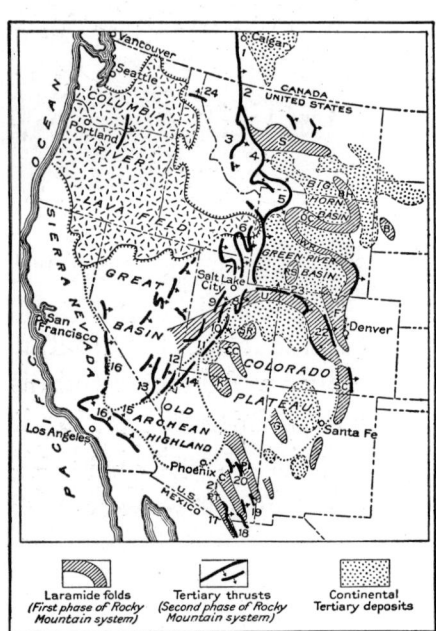

Orogenic movements at or slightly after the close of the Cretaceous. As has been pointed out before, all mountain-building involves a considerable lapse of time. Most of the thrust faults originating in the late Cretaceous were active in the early Tertiary. The Lewis overthrust, shown on Plate 50, is number (2) on the map above. After Paul Billingsley

As the marine waters withdrew into the abyssal depths of the oceans, the continents emerged. The seas were withdrawn from the coastal areas, and the continental outline grew until it was larger than today. But, while this gentle rejuvenescence of the continents was taking place, there were more heroic earth processes in operation in many parts of the world. In North America, the Cretaceous seas, which had an areal extent to be measured in millions of square miles, were not only drained, but the great Cordilleran geosyncline, in which their thickest deposits had formed, became the site of mountain-building on a grand scale. The orogenic movements extended from Alaska to Mexico and involved an area of at least 1,500,000 sq. mi. In its intensity it most closely approximated

END OF AN ERA

some of the great pre-Cambrian orogenies which we earlier described. This great mountain-building has been designated the **Laramide revolution.**

One notable feature of this revolution was the overthrusting which accompanied it. Along the eastern edge of the Rockies the rocks of the mountains are commonly separated from those of the plains by great thrust faults. In these overthrusts the older strata have been shoved eastward over younger sediments. Thus in such areas Proterozoic rocks may rest on Cretaceous beds, or early Paleozoic strata on early Mesozoic rocks. Faults of this type occur from Utah on the south to well into the Canadian Rockies on the north. Some of the faults have been traced a distance of more than 200 mi., and they commonly show an eastward shove of from 10 to 35 mi. The *Lewis overthrust* is perhaps the best known of these faults, because it can be observed in the vicinity of Glacier National Park, where it has been pointed out to many a tourist; but there are several other thrusts of equal or greater magnitude.

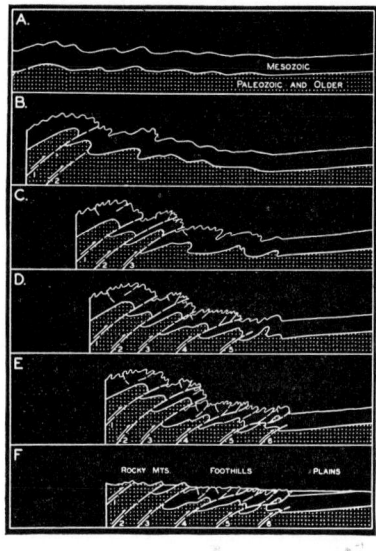

Series of sections illustrating the orogenic history of the eastern Rockies and their foothills, in the area numbered (1) on the preceding map. *A*, gentle early fold; *B*, first overthrusts from west; *C*, further overthrusting; *D*, disturbed belt moves eastward with fifth fault forming; *E*, final movement; and *F*, the area after erosion. After T. A. Link

The Rockies of today owe a part of their grandeur to general Cenozoic uplift, but the birth of their structure is late Mesozoic. Igneous activity was marked at the time of their origin, there being volcanic activity of the extrusive type in most of the western states. Great granite batholiths also were intruded at a number of places. The *Boulder batholith* which underlies the copper-producing area of Butte, Montana, and the *Idaho batholith*, which crops out over a large area in central Idaho, are the most notable of these intrusions. *The granite core of the Rockies*, however, at most places *is pre-Cambrian in age;* and the uplifts which the mountains have experienced, at the close of the Mesozoic and during the Cenozoic, have merely permitted erosion to cut down to these ancient granites.

WESTERN India was the site of a great series of fissure eruptions of basalts during the late Cretaceous. These lavas spread out over an area of more than 200,000 sq. mi. in the Dekkan district of southwestern India. Near Bombay the lavas are nearly 2 mi. thick, but individual flows are usually less than a score of feet in thickness. Originally these great lava beds may have extended into Africa across the area of the present Arabian Sea and Arabia, although some consider these latter lavas to be somewhat older than the Indian ones. In any case, many geologists are of the opinion that this late Mesozoic igneous activity in the area bordering the Indian Ocean is directly connected with the sinking of some of the segments of the hypothetical continent of Gondwana, and which, they believe, now form the Indian Ocean floor.

In addition to these igneous activities outside of North America, other continents also suffered from the Laramide mountain-building. This is chiefly true, however, for those countries which border the Pacific Ocean. The eastern Andes were folded at this time, though, like the Rockies, they owe their modern ruggedness to later uplifts. The origin of the Greater Antilles mountain arc is also contemporaneous with the formation of the Rockies, and northeastern Asia and Antarctica also were folded at the same time.

AGAIN we find in the Cretaceous the warm equitable climate characteristic of periods of submergence. Palms and the breadfruit tree lived in western Greenland, and thus climatic zones obviously were not nearly so well developed as at present. With the building of late Cretaceous mountain ranges and the general continental uplift, however, the modern type of climate, with its marked differentiation into zones, had its inception, although notable climatic fluctuations also occur in the Cenozoic era.

END OF AN ERA

CRETACEOUS economic products are many and diversified; but coal, oil, and water are of the greatest importance. The Cretaceous coals have a tremendous development not only in United States, where the reserves are to be measured in billions of tons, but in Canada, Alaska, Australia, New Zealand, and Germany. Unfortunately, not much of this coal is of sufficiently high grade to compete successfully with the product from Pennsylvanian strata. The dry parts of Australia and our own Great Plains get much of their water from wells driven into the Cretaceous sandstones, which, at many places, form suitable artesian aquifers. Approximately 16 per cent of the world's enormous total of oil produced to date has been taken from Cretaceous reservoirs. The early Cretaceous limestones of Texas and Mexico are especially prolific sources of the fuel, one of the Mexican wells actually having produced oil from these beds at the rate of a quarter million barrels a day.

WE HAVE now come to the end of another era, and we have finished another major cycle in earth history. At the beginning of the Mesozoic we found the continents high—we leave them in the same condition at the end of the Cretaceous. But in the meantime many of our old Paleozoic physical features have vanished. The Appalachian geosyncline was gone at the opening of the era; the Cordilleran trough was finally abandoned at its close. And during the cycle the old familiar land masses of Appalachia, Cascadia, and Llanoria foundered beneath the sea. They had played an important part in supplying sediment to the continent for many millions of years. With their removal North America rounds into a fairly modern appearance, despite the fact that 60,000,000 yr. have elapsed between the end of the Mesozoic era and the time you read these lines.

CHAPTER 43

MEDIEVAL LIFE

> Death and life, in ceaseless strife,
> Beat wild on this world's shore.
> —CAROLINE NORTON

WE ARE likely to think of the World War as the time of unparalleled carnage, strife, and bloodshed. But millions of years earlier the medieval animals, without benefit of tanks or tear gas, converted the Mesozoic scene into a pretty good imitation of a modern battlefield; and, what is more, they put on a reasonably continuous performance. For this was the time when might meant right, and there were many of the mighty. Certainly it was a poor time for the small fry, although, as usual, the amazing fecundity of the oppressed prevented their extinction.

.... a pretty good imitation of a modern battlefield .... from an old engraving

In the main this story of medieval life is a reptilian saga, for the reptiles were powerful both on the land and in the sea. Moreover, they demonstrated their versatility by developing a flying legion and by giving rise to the birds. Thus the Mesozoic has been well designated as the "Age of Reptiles." We must therefore devote most of this and the succeeding chapter to these outstanding creatures of the times. The reader should remember, however, that the invertebrates, too, have quite a story to tell of their increasing complexity, and that the evolution of the plants also goes on apace during this era. Unfortunately, there is not space to recount the triumphs of all the animals and plants in the style they deserve; and so we must largely confine our present account to the annals of the mighty.

The first reptiles were sluggish creatures with many of the lowly water-loving habits of their amphibian ancestors, but by the early Mesozoic they had already started to explore the possibility of radiation into more exciting habitats. Those that stayed on the land were chiefly the "ruling reptiles" of the dinosaur type, which are so important that we are devoting the entire following chapter to their discussion. But these powerful land animals were really the conservatives among the reptiles. Once having become creatures of the continents, they forever remain land-livers or, at worst, marsh-dwellers. Their more venturesome cousins, however, moved back into the sea or invaded the air. It is their stories that concern us here.

THE **plesiosaurs** are seagoing monsters of extraordinary mien, for they resemble a turtle through which has been strung the body of a python. Furthermore, the plesiosaur's neck got a considerably larger portion of the snake's body than the tail. Appearing in the Mesozoic seas at about the end of the Triassic period, the plesiosaurs apparently reached their climax, so far as numbers are concerned, in the Jurassic; but their greatest size was attained in the Cretaceous, when individuals 40–50 ft. long existed. *Elasmosaurus*, one of these fine fellows that lived in the seas that flooded the present Kansas area, had a length of about 45 ft., of which approximately half was head and neck. The latter structure was an extraordinary one with great flexibility. And, no wonder, for it was made up of seventy-six vertebrae, the all-time record for necks! Gastroliths, or stomach stones, are commonly found with the skeletons of the plesiosaurs. These counterparts of the modern fowl's "gizzard stones" commonly include peculiar types of rocks that the ancient marine monster could only have picked up in the limited areas in which the types or rock are known to occur. Thus these stones in some cases can be made to serve as a "log" of the ancient sojourns of *Elasmosaurus* in the Cretaceous seas.

PLESIOSAURS used long oarlike flippers in their locomotion. These doubtless were effective; yet not nearly so efficient as a well-developed fishlike tail which was the chief organ of propulsion

Ichthyosaur .... externally like the fishes or dolphins ....

of the **ichthyosaurs.** These latter reptiles, which were far better adapted to their marine habitat than the plesiosaurs, had a great external resemblance to the fishes, or to those modern marine mammals, the dolphins. In these three vertebrate strains—the fishes, the ichthyosaurs, and the dolphins—which are so dissimilar as to stock, so like as to outline, we see the perfect example of convergence of form in different groups as they become modified to fit the same habitat.

Ichthyosaurs, unlike most reptiles, probably gave birth to their young alive. Not only does their skeletal structure indicate that they could not have gone on shore to lay eggs, but immature ichthyosaurs have been found within the body of the fossilized female which did not live to bear them.

MESOZOIC marine reptilian types also include the **mosasaurs** and the **turtles.** Mosasaurs were long and slender marine lizards looking much like the prototype of the accepted standard for the sea serpent of modern stories. Many specimens indicate that these monsters were commonly more than 35 ft. in length. Contemporaries of these reptiles in the Cretaceous seas were the marine turtles which, as in the case of *Archelon*, were large enough to cover the average-sized living-room rug. Imagine the turtle soup one of these old mossbacks would have made.

We do not know precisely how all of these marine forms originated, but we have evidence of some of the evolutionary stages through which the groups passed. For in-

AWFUL CHANGES
MAN FOUND ONLY IN A FOSSIL STATE——REAPPEARANCE OF ICHTHYOSAURI

A Lecture.—"You will at once perceive," continued PROFESSOR ICHTHYOSAURUS, "that the skull before us belonged to some of the lower order of animals; the teeth are very insignificant, the power of the jaws trifling, and altogether it seems wonderful how the creature could have procured food."

As early as 1830 the ichthyosaurs were so well known in England that they were the subject of caricature

stance, by the Middle-Triassic, one of the reptilian lines, which was discontented with its new-found freedom of the land, was the group called the **nothosaurs.** Not yet completely adapted to an aquatic existence, they nevertheless had bodies and girdles very much like those of the plesiosaurs. The feet were webbed, but otherwise much less specialized than in the later marine reptiles. Already somewhat more advanced than the plesiosaurs in a few respects, they probably represent a group not very far off of the ancestral plesiosaur line.

Nothosaurus .... discontented with the land

WE HAVE observed that, when the lands become crowded, animals may take refuge in the seas. Apparently they have also looked fondly toward the air as a haven from oppression on *terra firma*. At any rate, more than a score of attempts to fly have been made, but only a few creatures have developed true flight, as distinguished from mere gliding ability. The arthropods, as we have seen, were the first aerial excursionists, so that the origin of flight is a Paleozoic accomplishment. But not until the early Mesozoic were the *vertebrates* to enter the air as a habitat. Since that time, the reptiles, birds, and mammals have all turned the trick. But since the birds have been described as mere "feathered reptiles," we might say that only two main groups have become bona fide flyers. The birds and bats are the two newcomers to the ranks of the successful flying vertebrates. We can still observe their habits. The pioneers of the early aviators were the flying reptiles. We can study these creatures only through their fossils. We know them so well because they, like all their comrades of the air, had a universal failing. They never were able completely to divorce themselves from the land. These flying reptiles, overestimating their powers of flight, commonly found themselves far out at sea. As they sank exhausted to its waters, and thence to the muds at its bottom, they started to write the story of their existence, a story so spectacular that Conan Doyle used it in his famous tale of the "Lost World."

LATE in the eighteenth century the almost incredibly fossiliferous white Jurassic limestones of Solnhofen began to yield a small animal which quite obviously had wings. These creatures began to be made known about 1784 by the naturalist Collini, who, through some perversity, regarded the ancient animals as marine types. Finally in 1801, Citizen Cuvier, the great and good friend of Napoleon, was bringing order out of the chaos he found in the classification of the fossil vertebrates, even as the Corsican corporal was codifying the then nebulous French laws. Both men were successful in their tasks. Cuvier recognized the winged fossils for what they were —flying reptiles. And, considering the extreme elongation of the fourth digit characteristic, he gave the name **pterodactyl.** Cuvier also pointed out that the power of flight was due to a wing membrane which was supported by this greatly elongated digit, whereas the other "fingers" remained short and were terminated with claws.

A pterosaur preserved in the Solnhofen limestone. After Mantell, 1839

Some of these pterodactyls, or **pterosaurs** as they are more commonly called, were as small as a canary; others were as large as a small airplane, and thus were the largest of all flying creatures. *Pteranodon* was the name of these ancient giants of the air. Sailing out over the ancient Cretaceous seas in the district which is now western Kansas, this remarkable reptile must have cast a fearful shadow over the ancient waters. For here was a sharp-billed fisheater with a wing spread of close to 30 ft.! When we remember, however, that the pterosaurs were equipped with pneumatic bones, and recall further that even the modern albatross, with a wing spread of close to 12 ft., weighs well under 20 lb., we see that *Pteranodon* was no flying behemoth. Even the fossilized bones of this great pterosaur weigh less than 5 lb., and the entire creature probably would have tipped the scales at little more than a prize turkey gobbler. Truly, the body of *Pteranodon* was nothing but a muscular appendage to the wings of this prehistoric rival of the roc of Arabian figment.

THE fabulous Solnhofen beds have yielded remains of other flying creatures. We are now speaking of the first birds of which we have any record. And they are so nearly like the reptiles out of which stock they sprung that, were a still more primitive bird to be discovered, it is doubtful whether true powers of flight could be ascribed to it. Fortunately, these remarkable ancient creatures are known from two relatively perfect fossil specimens. These demonstrate conclusively that **these early birds possessed feathers, the true avian prerogative, but in all other respects they were reptilian.** They have sharp teeth in the jaw like the lizards. The bones of the wing are reptilian, not avian. The structure of the entire skeleton is, in fact, reptilian, and the bones are not pneumatic. One fine Jurassic day, when one of the members of this "ancient wing," or *Archaeopteryx* clan, was winging his awkward way toward an attractive atoll in the late Jurassic sea, he found these solid bones too heavy and fell exhausted to his watery grave. As his body sank to the Solnhofen muds, it marked the first step toward the immortality that was to be his when at last his bones were finally discovered.

One of the two nearly perfect specimens of the first bird. After von Meyer

.... winging his awkward way toward an attractive atoll ....

THERE were also several toothed birds living in the Cretaceous period. *Ichthyornis*, about the size of a pigeon, was one of these. With a well-formed breastbone and minus the tail which tied *Archaeopteryx* to the reptiles, this bird was a fish-eating, gull-like, and surprisingly modern-appearing bird. But with a set of well-developed teeth, *Ichthyornis* made foolish the modern saying: "scarce as bird's teeth." *Hesperornis*, a contemporary, was remarkable in that already it had lost the very powers of flight that its ancestors had so laboriously acquired. Like the modern loon, this "bird of the west" was a diver, with strong webbed feet for swimming.

Although these typically American Cretaceous birds are the last of our Mesozoic avian aviators, their stock *sans* teeth was to flower remarkably in the next era. The flying dragons or pterosaurs, however, are forever, lost to our rapidly changing life-picture along with the rest of the truly medieval life of the era. Plates 51 and 52 illustrate some of the more important members of the organic world mentioned in this chapter.

A Solnhofen butterfly, only partially restored, demonstrates the perfection of preservation in this late Jurassic deposit. After Handlirsch

As we have hinted before, the reptiles were by no means the only creatures of the Mesozoic mélange. The mammals, like the birds, originated out of the reptilian stock at the very time that the plasticity of the group was at its height, but theirs is a story we must reserve for chapters 48 and 49. The only other Mesozoic vertebrates, the fish, became rather modern in appearance during the era; and many achieved a large size, specimens from 10 to 15 ft. in length being known.

The invertebrates also achieved a degree of modernity during the Mesozoic. The cephalopods have perhaps the most interesting record; and, indeed, this is sufficiently important that we have devoted most of chapter 45 to its elucidation. Other invertebrates of outstanding importance were the lobster and crablike crustaceans, the pelecypods, echinoderms, and protozoans. Lobsters and crayfish were not only numerous but modern in appearance. The oysters, representing the pelecypod class, were of extraordinary abundance, particularly in the Cretaceous, as were the sea-urchin types of echinoderms. Protozoans, of the foraminiferal group, despite their small individual size, actually make up a surprisingly large amount of the widespread Cretaceous chalks. Most famous of all the Mesozoic fossil-collecting formations is the Solnhofen limestone of Jurassic age. Long quarried as a lithographic stone near Solnhofen, Bavaria, these beds have now yielded about five hundred kinds of fossils. Like the Middle-Cambrian Burgess shale, this limestone also is capable of preserving soft parts, as well as skeletons.

Among the plants notable changes also took place during the Mesozoic. The cycads were so important during the Triassic and Jurassic periods that the early Mesozoic has often been called the "Age of Cycads." But during this time many of the early conifers also reached very large size. Those whose logs are today the feature of the Triassic petrified forest of Arizona may have stood nearly 200 ft. high, since some of the incomplete fossil logs are over 100 ft. long. The *deciduous* trees become important in the Cretaceous, and with them the other newly introduced angiosperms, like the grasses and cereals, made the world a reasonable place for the subsequent rapidly expanding evolution of the mammals. Here, again, we see the dependence of the animal world on plant progress.

AND now we ring down the curtain on a brief Mesozoic scene. The landscape has been a strange one, but in it we have recognized some familiar plants, and along the coast we have seen a number of invertebrates not so very different from those we know today. But we have really been amazed at the medieval vertebrates such as the great sea serpents in the waters, the flying dragons and strange toothed birds in the air, and those ruling reptiles, the dinosaurs, on the land. We have thus far paid scant attention to this most characteristic of all Mesozoic animal groups. We have been saving this interesting story until the end. We will discuss it in the following chapter.

CHAPTER 44

MEGALOMANIA

> How are the mighty fallen!
> —II Sam. 1:25

THE dinosaurs are as passé as the bustle, the hoop skirt, or the prairie schooner; yet they stare at us from many a billboard or magazine advertisement. They are as familiar as taxes and politicians, as much talked about, and no doubt as little understood. But, like the stories of knights in armor, their legends also remain as symbols of a glorious past. Even the records of the crusades and the stories of King Arthur's Round Table are not nearly so well written or so reliable as the tales the extinct dinosaurs have to tell. For there still is legitimate doubt as to the authenticity of the Arthurian legends, and there is many a dispute of the type of whether the bones of King Richard the Lion-hearted lie moldering in French or in British soil. But there is no longer much argument about the dinosaurs. Their skeletons were entombed in the Mesozoic rocks of many lands, and after more than a century of study of these indelible records we now know at least the main features of their story.

.... like .... knights in armor ....

The age of Dinosaurs lasted little more than a fortnight on the basis of the Gargantuan calendar we have so often consulted. But the more than 100,000,000 yr. involved in their history is nonetheless 100 per cent greater than the length of time since their extinction. Despite their seeming preoccupation with megalomania, or their tendency to do things in a big way, **they dominated the animal world for a span of time which was twice the length of the present mammalian dynasty.**

LATE in the Permian there originated a stock of primitive reptiles known as the *thecodonts*. They were small carnivorous forms having about the same range in size as the modern lizards. But they were, by the Triassic period, already developing a bipedal gait and displaying the internal features which demonstrate their significant position in the reptilian world. For from this virile and plastic stock there evolved, not only the crocodiles and the flying reptiles, but those ruling reptiles, the dinosaurs themselves.

The dinosaurs are the subject of much popular misconception. They are generally considered a single group of gigantic animals. Many dinosaurs did indeed achieve great size; but small members of the group were not uncommon, and a few bulked not much larger than a house cat. Moreover, even when they first appear in the geological record, they have already deployed into two distinct types, the **saurischians** and the **ornithiscians,** and members of these two groups show nearly as much differentiation as is seen among the modern mammals, a point demonstrated by the illustrations on Plates 53 and 54.

THE **saurischians include all carnivorous dinosaurs,** and primitively the group was doubtless *exclusively* flesh-eating. Some of the saurischians, however, became four-footed herbivores. But regardless of their habits, they *all had a triradiate pelvic structure.* The **ornithiscians were all herbivorous,** and many remained bipedal to the very extinction of the group. The armored and horned types, however, reverted to quadrupedal locomotion. *All had a tetraradiate pelvic structure*, since their pubis was double pronged. We may express the development of these two groups, and their important subdivisions, in a simple table.

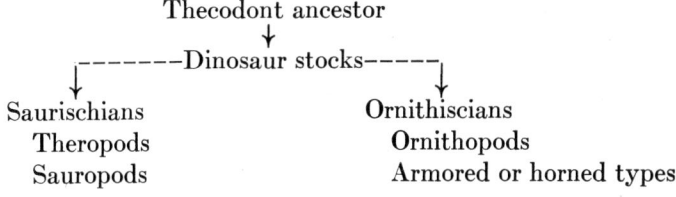

THE **theropod** dinosaurs, in the broadest sense, include all the flesh-eating terrestrial reptiles of the Mesozoic era. They were all bipedal with birdlike rear limbs and reduced fore limbs. Thus they closely resembled their thecodont ancestors. The most primitive among these theropods were the late Triassic and Jurassic forms such as *Ornitholestes* and *Compsognathus*. The former was about 6 ft. long but had no more bulk than a large dog, and the latter had a head less than 5 in. in length. Certainly, one could not accuse *these* dinosaurs of being megalomaniacs! Instead of being sluggish creatures, they were active, swift-running bipeds. They had light, hollow bones, like the birds; and the feet of these theropods were so similar to those of their avian cousins that their tracks could easily be confused. In fact, the tracks made by the horde of late Triassic theropods on the sands which were deposited in the Connecticut Valley *were* so confused. Collected and studied more than a century ago, these tracks were first pointed to as the imprints made by Noah's raven as he walked over the muds left by the receding waters of the Deluge!

Practically all of these small primitive theropod dinosaurs were extinct at the close of the early Mesozoic; but their descendants were represented in the Cretaceous by the queer ostrich-like dinosaur, *Struthiomimus*, and its relatives. In this group the teeth were lost, and a horny bill took their place. It has been suggested that this specialized dinosaur was an egg-eater that robbed the nests of other reptiles, a guess which has received some confirmation at least in the discovery in Mongolia of a crushed skull of one of these forms in a nest of fossil eggs belonging to a horned dinosaur.

Carey Orr in the *Chicago Tribune*

MEGALOMANIA

ALTHOUGH the ancestral theropods were small, there had appeared as early as the late Triassic a number of more powerful carnivores. In the Jurassic there were at least two types which are well known, *Allosaurus* and *Ceratosaurus*. *Allosaurus* was a powerful bipedal carnivorous dinosaur some 34 ft. in length and with a powerful skull well over 2 ft. long. Like his smaller ancestral types, he had a long balancing tail and fore limbs too much reduced to be of service in locomotion. *Ceratosaurus* was much like his contemporary, though somewhat smaller. But, unlike any other discovered carnivorous dinosaur, he was, as his name implies, a horned creature.

The Cretaceous witnessed the climax of this group of flesh-eaters in *Tyrannosaurus*. Fifty feet in length and standing 20 ft. high, *Tyrannosaurus* had a head considerably larger than the bucket on a cubic-yard steam shovel, and it was armed with powerful scimitar-like teeth as much as 6 in. in length. With such a bulk *Tyrannosaurus* must have been a sluggish creature, indeed, as compared with the early theropods. But once he sank his fangs into any of the herbivorous dinosaurs, the battle doubtless was of short duration. In an era of bloodshed *Tyrannosaurus* was king over all. He was, without question, the most powerful predaceous creature ever to walk the face of the earth—the epitome of Nature's sanguinary efforts in a sanguinary age. But, powerful as he was, *Tyrannosaurus* and his relatives had a relatively short reign. With the extinction of the herbivorous dinosaurs on which they preyed, and thus helped to annihilate, their food was gone; and they, too, went the way of all megalomaniacs.

THE **sauropods** are Jurassic and Cretaceous saurischians which became quadripedal and herbivorous, and achieved the largest size of any of the terrestrial animals. The earliest known representatives of the strain lived in the late Triassic; but, although they show some sauropod features, they are still regarded as theropods. But in the Jurassic, and particularly in the Morrison beds earlier referred to, are found the skeletons of monsters such as *Brontosaurus* and *Diplodocus*. *Brontosaurus*, like the rest of his sauropod group, had an elephantine body with massive limbs and a long slim neck terminating in a small inoffensive-looking head. His tail, like the neck, was also of great length. Some individuals of this genus probably weighed as much as a freight-car load of coal, though in not any of them was the cranial capacity equivalent to that of a small modern dog.

Diplodocus was a near relative of *Brontosaurus*, but it had more of a society figure. Slimmer by far than its cousin, *Diplodocus* was the *longest* creature that has walked on the face of the earth. The specimen in the Carnegie Museum of Pittsburgh is 87 ft. in length. And, going again on the well-founded principle that the biggest fish are never caught, there certainly must have been some granddaddy *Diplodocus* somewhere in the late Jurassic lowlands which was at least 100 ft. long. But a third member of the sauropod group was the bulkiest of them all. *Brachiosaurus*, found in East Africa, as well as in the United States, was probably 80 ft. long despite the fact that his tail was short. All of these creatures had a pelvic ganglion far larger than their brain, and they must have been animals largely motivated by reflexes rather than reason.

IN GENERAL the weight of an animal varies in proportion to the cube of a linear dimension. The strength of its leg, however, is proportionate to its cross section, which therefore increases by squares. Thus, when a late Jurassic sauropod became twice as long as he had been during his youth, he found that he weighed about eight times as much. But, unfortunately, his legs were only four times as strong. To be more specific, *Brontosaurus*, in growing from a mere youngster of 35 ft. to an old fellow 70 ft. long, increased his weight from 5 tons to nearly 40 tons. Thus, as an adult each of his limbs would have to carry a weight of 10 tons, rather than a mere ton and a quarter. This strongly suggests that, large as were the sauropod limbs, they were not strong enough for their great load. Hence it has generally been supposed that the sauropods were amphibious creatures living in marshes and lagoons where the water buoyed them up sufficiently to solve their great problems of locomotion.

It is not surprising, therefore, if their habitat has been correctly interpreted, to find the sauropods rapidly diminishing during Cretaceous times. Particularly is this true toward the close of the period, when continental elevation drained their swamps and reduced their habitats. Had they only known it, the sauropods would have viewed with sadness the growth of the Rockies. As a matter of fact, few of these master megalomaniacs of North America were able to survive the physical disturbances at the close of the Jurassic period, let alone those which terminated the Cretaceous.

.... viewed with alarm ....

THE ornithischian dinosaurs include the essentially bipedal **ornithopods** and the various quadrupedal armored or horned types. The ornithopods come on the scene somewhat later than the saurischians, but by the middle Jurassic their record is a fairly good one. *Iguanodon* is a characteristic European form notable for the spikelike extension of the bone in the "thumb." More than a score of these dinosaurs were found in a crevasse in a Belgian coal mine. Apparently, an entire herd of the animals running across a Mesozoic landscape fell into the fissure and were unable to extricate themselves. Their misfortune became good luck for the scientists, since their discovery represents one of the most remarkable of all dinosaurian finds. But the North American mummified remains of *Trachodon*, the duck-billed relative of *Iguanodon*, are almost as spectacular. A web of skin found between the digits of the foot indicates that *Trachodon* was an amphibious type. Although this general group was numerous in the late Mesozoic, they all became extinct with the birth of the Rocky Mountains. Mountain-building, draining of Cretaceous swamp areas, and changes in the plant life, were at least some of the factors which caused the extinction of the ornithopod dinosaurs.

THE armored ornithischians comprise such forms as *Stegosaurus*, *Palaeoscincus*, and the *ceratopsian* dinosaurs. *Stegosaurus* and its relatives were moderately large dinosaurs, 20 ft. or more in length. They had general sauropod proportions except that their necks and tails were considerably shorter. A double row of alternating plates ran the entire length of the neck and trunk. The tip of the tail, however, bore several pairs of spikes. Although these prehistoric bits of armament were doubtless effective defense weapons, the creature was nevertheless vulnerable to a flank attack. The members of this group have been called the "Dumb Doras of the Dinosaurs." Even among a superlatively ignorant race, they must have been noticeable for their lack of intelligence.

HEAVILY armored, squat, almost turtle-like quadrupeds, such as *Palaeoscincus*, took the place of the stegosaurs in the Cretaceous. With somewhat the proportions of a modern horned toad, they have been appropriately called "reptilian tanks." With practically no offensive weapons, they must nevertheless have been difficultly handled prey for the carnivores.

. . . . which laid "The" eggs . . . .

The horned or ceratopsian dinosaurs were confined to the latest Mesozoic. Their heads were exceedingly large, making up as much as a third of the length of the body. Much of this head size is achieved by extending a great frill of bone backward over the neck like a great flared and highly starched collar. This must have been good protection against the attack of the contemporary *Tyrannosaurus*, but the ceratopsian forms in general were elephantine in proportions, and probably too clumsy to maneuver cleverly. The most primitive of the ceratopsian forms is *Protoceratops* from the Cretaceous of Mongolia. This dinosaur is known in all stages of development, from the egg through the unhatched embryos to the

MEGALOMANIA

adults. It is the dinosaur which laid "the" dinosaur eggs which have been made the topic of common table talk by newspaper accounts.

IN SPITE of their complete dominance of the Mesozoic scene, the reptiles apparently were never able to resist the urge to do things in a big way. The reward for their megalomania was extinction of all but the most generalized types at the close of the Cretaceous, though undoubtedly other factors besides mere size and specialization played a rôle in their extermination. Here, again, we must warn the reader that these statements merely represent a convenient manner of speaking. No animal group, not even man, has yet been able sentiently to direct or misdirect its evolutionary course.

BY THIS stage in our story it has become apparent that the Mesozoic reptiles were masters of all the important habitats. The dinosaurs ruled the land, marine reptiles invaded and conquered the sea, and the "flying dragons" or pterosaurs were lords of the air. But scurrying underfoot of the giant dinosaurs were a few mouselike primitive mammals. They were subservient, indeed, to the gigantic masters of the moment, who, as is characteristic of the great (and especially the near-great), probably were totally unaware of the mammal's presence. But these small creatures, like some apparently insignificant individuals and many unpromising infant industries, had great potentialities. They proved their mettle at the close of the Mesozoic, when, as we have seen, the earth went through another one of her really great "depressions." We will take up *their* story in chapters 48 and 49, after we have prepared the physical background for their organic development.

CHAPTER 45

CEPHALOPODS

> Thy low-vaulted past.
> —Holmes

THE cephalopods have been building themselves "more stately mansions" for many a long year. And although some enterprising members of the group have indeed left their "low-vaulted past," architectural plans for the new mansions of the conservative members of the family have remained pretty much the same for nearly a half-billion summers.

.... architectural plans pretty much the same ....

Away back in chapter 36 we saw that the "strong men" of the early Paleozoic were cephalopods. And those who have any familiarity with the life of the sea today know that the modern cephalopods, like the squid and the octopus, are still important members of the marine assemblage. But, obviously, they are very different from their remote ancestors in the Ordovician seas. However, since the group does have continuity, since it is so abundantly represented in Mesozoic formations, and since it illustrates a number of paleontological principles, we are describing it here as a sort of good-will gesture toward the many important invertebrate classes whose geological history we have had to neglect.

Not only do the cephalopods comprise the most advanced molluscan class, but they are the largest and strongest of all the invertebrates. The group may be simply subdivided in the following manner:

Class Cephalopoda
Subclass **Tetrabranchiata**   Subclass **Dibranchiata**
  Order **Nautiloidea**   Order **Decapoda**
  Order **Ammonoidea**   Order **Octopoda**

CEPHALOPODS

Very briefly described, **the tetrabranchs have four gills and possess external skeletons.** Important in the Paleozoic and Mesozoic eras, this subclass is represented today only by the "pearly nautilus," *Nautilus pompilius*. **The dibranchs,** on the other hand, **have two gills and either internal skeletons or none at all.** Originating during the Mesozoic era, this subclass has expanded as the other one has dwindled in importance, until today the dibranchiate cephalopods are one of the most successful of all the invertebrate stocks.

SINCE the four-gilled cephalopods are the first to appear in the geological record, let us begin by briefly examining some of their structural features. We can best do this by describing the pearly nautilus, the sole survivor of a group which has, throughout the past, been represented by about twenty-five hundred species. As one might well expect, on the basis of our earlier experiences, this modern nautiloid is a relatively simple cephalopod confined to a limited habitat in the East Indian seas. The shell of this form is an expanding tube which is coiled in a plane and is bilaterally symmetrical. If the shell is sawed in two along the line of bilaterality, it is at once observed that the skeleton is a chambered one. The chambers are made by a series of calcareous partitions, called *septa*, which extend completely across the tube. These diaphragm-like structures are concave forward toward an unchambered portion of the shell which is occupied by the fleshy part of the creature. The line made by the juncture of each septum with the outer wall is designated the *suture*. Each septum is perforated near its middle by an opening which serves as the passage for a fleshy tube called the *siphon*. As the cephalopod grew, the body periodically

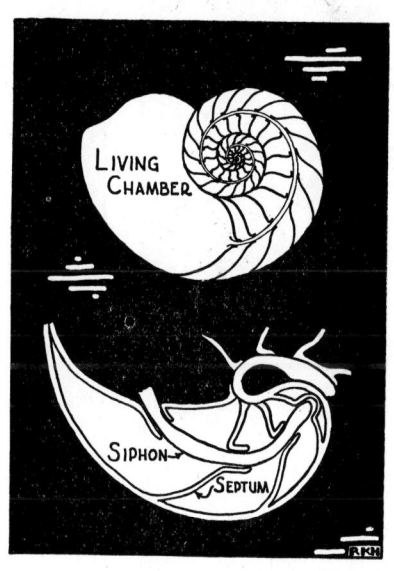

Nautilus pompilius in section. *Below:* The initial portion of the shell enlarged.

was drawn forward, and a new septum or base was secreted behind it. The animal, however, maintained a connection with its abandoned chambers by means of the siphon.

By examining the shell closely, we see that the suture lines, which are rather undulating in the later stages of growth, are relatively straight in the older parts of the shell. Similarly we notice that the first few chambers occur in a part of the shell which is nearly straight. These are interesting observations, since they are features seen in the adult stages of the *ancestral* nautiloids. In other words, *Nautilus pompilius*, like most cephalopods, in its individual growth crudely **recapitulates or summarizes some of the stages through which its ancestral stock has passed.** For, as we trace the nautiloids back through the long geological ages to the first of the group, we find less and less complex sutures and more and more simple shell shapes. These shell shapes and other cephalopod features are referred to again in Plate 55.

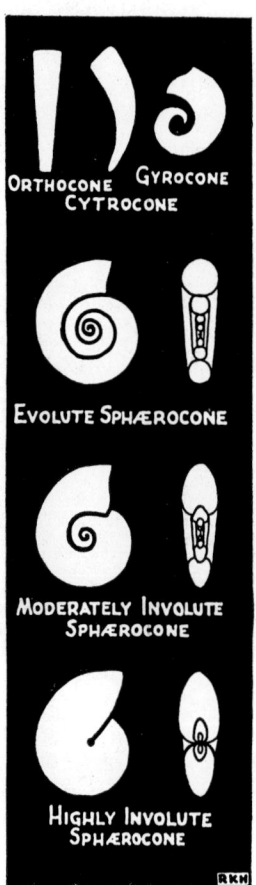

Typical tetrabranch cephalopod shell shapes

The early Paleozoic cephalopods were practically all nautiloids, that is, they were simple tetrabranchs with relatively simple suture lines. Many regard the elongated ice-cream-cone-like forms (*Orthoceras*) as the most primitive of these nautiloids. And although this generalized type is to persist for ages, the newly introduced forms show slightly curved shells. As time goes on, scimiter-shaped skeletons, and then loosely coiled ones, come into the scene. Finally, before the end of the early Paleozoic, some of the nautiloids were completely coiled forms, in which the whorls just touched each other. A little later still more complicated shell structure became common, as we shall soon see.

CEPHALOPODS

DURING the late Paleozoic there were two distinct groups of cephalopods. These were the nautiloids, whose outstanding position of supremacy in the Ordovician and Silurian periods has been lost (although they remain fairly common until the end of the era), and the newly introduced ammonoids, which are not to come to the pinnacle of their career until the Mesozoic era is well along.

THE ammonoids are regarded as tetrabranchs, though, since they became extinct by the end of the Mesozoic, no one has been able to prove definitely that they, like the pearly nautilus, possessed four gills. They exhibit most of the shell shapes that characterize the nautiloids, but they are different from the latter in internal structure. The only point which needs concern us, however, is that **the ammonoids go in for complicated suture lines.** Since their complexity increases as geological time goes on, the sutural expression becomes of extreme importance. It can be used as a sort of prehistoric fingerprint, not only to identify the individual that possessed it, but also to indicate the period in which the cephalopod which bore it lived.

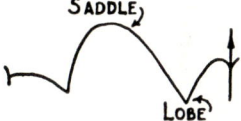
SIMPLE GONIATITE

ADVANCED GONIATITE

CERATITE

AMMONITE

Typical ammonoid sutures

In the late Paleozoic these ammonoid sutures are fairly simple, moderately curved and angulated lines being the rule, though there were exceptions, as we shall see. The swings of the suture line toward the aperture of the shell are called *saddles*, those that are directed backward are designated *lobes*. The Devonian and Mississippian ammonoids, with simple rounded saddles and pointed lobes, are generally called *goniatites*. The later Paleozoic types may show secondary crenulations on the lobes of the suture line. The cephalopods with such a sutural expression are known as *ceratites*. When such secondary complications of the suture are superimposed on both lobes and saddles, the shells are described as *ammonites*. Some of the late Paleozoic cephalopods have a ceratitic suture, but this type of shell does not reach its period of greatest abundance until the Triassic. Similarly, a very few of the late Paleozoic ammonoids have ammonitic sutures, but again this is not a dominant sutural expression until the Jurassic period.

BY THE late Paleozoic both nautiloids and ammonoids have developed shell shapes which are highly *involute*, that is, the last whorl in the coiled shell almost completely overlaps and conceals the inner ones. Throughout the Paleozoic, also, the nautiloids have toyed with queer apertural modifications. In the Mesozoic the ammonoids copied this style and also produced a number of unusual structures. But by the beginning of the Mesozoic the nautiloids are very much restricted. A few simple types, however, live on to bear witness again to the fact that the generalized survive. But as early as the Triassic period there were already about twenty-five hundred species of ammonoids swarming in the seas. Evolving rapidly, and dying out suddenly, these cephalopods make excellent index fossils, as we have previously noted. Not only are they numerous, varied, and short lived, but, being swimmers, they are widespread during life, and, since their shells may float, they may be carried far after death to further increase their geographic distribution. Consequently, **the European Mesozoic formations at most places are zoned, or correlated, on the basis of these invertebrates.**

DURING the Jurassic period these complex cephalopods were probably at the height of their career, but so rapidly had the ammonoid stock been changing that not any of the species were the same as those found in the Triassic system. In the Cretaceous they were still important, and many reached gigantic size. In fact, if some of the coiled forms, which are larger than an automobile wheel, were to be uncoiled, they would doubtless be as long or longer than their gigantic nautiloidean cousins who ruled in the early Paleozoic seas.

But the story of the "Age of Ammonites," as some have called it, comes to an unhappy ending with the close of the Mesozoic. Swift and ignoble extinction follows hard on a glorious climax which has been achieved only after a long, slow rise. And this rise, moreover, had begun away back in the middle Paleozoic, nearly 300,000,000 yr. before ammonoid extinction. The first hints of hard times for the ammonoid group are seen in the specializations some of them achieve during the Cretaceous. For instance, many species become partially uncoiled; some masquerade under the typical gastropod style of coiling into a spire; and some, such as *Baculites*, uncoil almost completely. In all such types, however aberrant, the ancestry of the group can be read in the sutures. As we have seen, these become more and more simple toward the early portion of each individual shell, and **in succession backward they represent the adult sutural expression of progressively older genera on the same cephalopod evolutionary line.**

.... some, like *Turrilites*, masquerade under typical gastropod coiling ....

Even before the ammonoids were at their climax, the dibranchiate cephalopods came into the Mesozoic picture. The two-gilled cephalopods with ten arms, the decapods, have the earliest geological record which dates back to the Triassic. Represented in modern seas by a host of squids and cuttlefish, the Mesozoic forebears of this group became numerous in the Jurassic. Although the internal shell of the modern forms has been reduced until only the "cuttlefish bone" remains, the older types had an internal shell of three parts. One of these was an anterior shoehorn-like shell; the second was a chambered cone, like a nautiloid, which filled into a notch in the front end of the third part of the skeleton. This third portion is known as

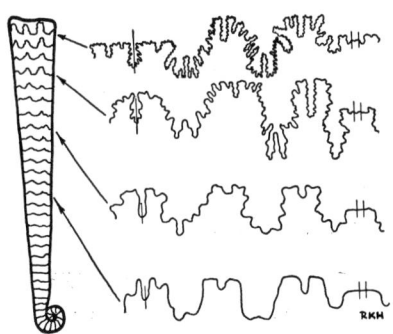

Baculites, with outer shell removed to show the sutures, which become progressively more simple toward the older part of the shell. After Brown

the *guard*. It is a cigar-shaped calcareous structure, which is largest of the three parts and is the most commonly preserved. Agricola gave them their name, *belemnites*, or "darts of war," nearly four centuries ago (1546), apparently with reference to their supposed resemblance to the thunderbolts which Jove was supposed to hurl to the earth.

An ink sac like that found in modern squids is in some cases discovered as a part of the best-preserved fossil specimens. The pigment has been put back into solution at least in one case, and a picture made of the fossil with the very ink the living form secreted about 175,000,000 yr. ago. These ancient belemnites were so common that the Jurassic rocks in some places are actually made up of their skeletons, and more than two hundred of the guards have been found within the body of a fossil ichthyosaur. Many of the larger creatures of the Jurassic seas must have found the ancient squids tasty morsels long before the modern ones became highly sought after by the fishes.

The octopods, or eight-armed dibranchs, have a poor geological record; but from their rare fossils we know that they were in existence in the Cretaceous. During the Cenozoic era, judging from their modern abundance, these forms, together with the decapods, doubtless continued with undiminished vigor; but owing to their decreasing skeletal parts, their actual fossil record is not a very good one. Meanwhile, after the extinction of the ammonoids at the end of the Cretaceous, the four-gilled cephalopods were represented only by a few stragglers in the nautiloid group. But if they are unimportant today, they nonetheless can show their upstart dibranchite cousins a family tree of great antiquity.

A restoration of *Belemnites*, to show position of the internal skeleton

THE complicated life-history of the cephalopods, to which we could do little justice in this account, is better understood than that of most animals; but so far as we now know, all the organic groups tell a somewhat similar story. It is a tale of youth, adolescence, maturity, senescence, and death. It is an account of life delicately adjusted to its environment, responsive to every minor episode in the physical world, mirroring every little change in enemies and each and every fluctuation in food supply. Changing rapidly in youth and adolescence, and giving rise to new stocks in robust maturity, a strain suddenly finds itself past the prime of life. It then seems to experiment wildly, but to no avail—the overspecialized only die out the more rapidly, and the generalized live on, though often to an inglorious and unproductive old age. Thus the family history essentially ends, or, if it is carried on, a junior branch of the house, once despised, is the new standard-bearer. But groups that, like the cephalopods, can persist not only through a geological period, or even through an era, but through a number of the major earth cycles should not be too ashamed of their record. Not any of the vertebrates have yet done as well.

CHAPTER 46

THE TERTIARY

> To one who thoughtfully ponders the centuries and
> Surveys the whole in the clear light of the spirit; ....
> Oceans and continents alone are of account.
>
> —GOETHE

WE HAVE now thoughtfully pondered many a century; and although our survey has been brief, we hope it has been conducted in the clear light of the true scientific spirit. And even if we do not agree with Goethe entirely, we see that with the beginning of the Cenozoic era the continents at least are of considerable account, for they stand high in rugged relief, so that the ever changing face of the Earth is wrinkled like that of an old crone.

....wrinkled like an old crone....

If we turn back to chapter 32 and consult the Gargantuan calendar once again, we find that the Tertiary began on the day after Christmas in that gigantic geological year of ours. Thus we can allot but seven short days to both the Tertiary and Quaternary periods. But it is a week of tremendous activity and momentous happenings. We propose therefore, to devote chapters 46 and 47 to the examination of its physical events and to tell its organic story in chapters 48 and 49. Before we commence with this physical account, however, it may be well to explain the origin of the nomenclature of the Cenozoic periods and epochs.

Giovanni Arduino, of the University of Padua, as early as the middle of the eighteenth century, proposed a fourfold division of the rocks of the southern Alps. These divisions he called **Primary, Secondary, Tertiary,** and **Volcanic.** For some time this classification was used for the rocks found in many different parts of the

THE TERTIARY

earth, the North American usage of the term "Primary" having previously been commented on in chapter 33. Later, when it was discovered that this classification had no real significance, except in the area to which it had first been applied, the name "Tertiary" was already too firmly imbedded in the literature to be dropped. Nearly a century after Arduino's day the name **Quaternary** was applied to the unconsolidated drift and terrestrial deposits that veneer much of Europe. And this term also has persisted, so that the Cenozoic era today is generally regarded as being made up of the Tertiary and Quaternary periods. There is no very real reason however for separating the last of our eras into anything more than a single period with a number of epochs.

A LITTLE more than a century ago (1833) Sir Charles Lyell, the great English geologist, studied the Cenozoic rocks of Europe, with especial reference to the Paris Basin. Here the various Tertiary marine embayments had left good rock records of these inundations, and the beds representing the various floods were well separated by disconformities. Lyell noticed that each set of beds contained a marine molluscan fauna in which certain of the species were like those inhabiting the modern seas. Probing the matter further, he found that **this relationship could be expressed as percentages in which the older beds had few species identical with the modern ones, and the younger rocks had increasingly more like the living types.** Accordingly he erected a classification for the Cenozoic era whose main features, brought up to date, can best be expressed in tabular form.

Sir Charles Lyell .... erected a classification ....

A little examination of this table will show that before it was completed its authors and revisers became somewhat embarrassed for comparatives and superlatives. Moreover, the percentages of

Periods and Epochs	Meaning of Terms	Percentage of Modern Species
Quaternary period:		
Holocene epoch...............	Entirely Recent	100
Pleistocene epoch.............	Most Recent	90–100
Tertiary period:		
Pliocene epoch................	More Recent	50– 90
Miocene epoch................	Less Recent	20– 40
Oligocene epoch..............	Little Recent	10– 15
Eocene epoch.................	Dawn of Recent	0– 5
Paleocene epoch..............	Ancient Recent	0

modern species are not as Lyell gave them, for they have been corrected to coincide with modern findings. So here again we have a departure from the standard usage of type sections and geographic names in our nomenclature, a departure which has sanction only because of its antiquity and universal usage. **But the very basis for this table speaks eloquently of the interrelationship of the organic and the physical worlds.**

.... never inundated more than a scant 5 per cent. .... North America in the Oligocene. Generalized after C. Schuchert

After the Laramide revolution at the end of the Mesozoic era, North America was emergent; and it has remained so, for **never during the Cenozoic was it to be inundated more than a scant 5 per cent.** Throughout the era, however, seas did occasionally spread over the borders of the Atlantic and Gulf states, as can be seen in the accompanying map. Marine waters also inundated relatively small areas along the Pacific margin of the continent in the western part of the area that is now Washington, Oregon, and Cali-

THE TERTIARY

fornia. Terrestial deposition was common in the western states; and volcanic activity, which has been responsible for many of our western ore deposits, also was notable in the same general area. Since the Tertiary story is so different for the various areas mentioned above, we will discuss the major features of the period by geographic divisions. These are: (1) the Atlantic and Gulf Coasts (including the eastern interior), (2) the Western Interior, and (3) the Pacific Coast.

THE Atlantic and Gulf Coasts have a very simple structure in that the formations all dip gently seaward. Thus the oldest rocks are exposed in bands on the inward side of the Cenozoic coastal belt, and the younger ones crop out on the seaward side of this belt, though here, as elsewhere, they may be covered with a veneer of Quaternary material. The area of Cenozoic rock exposure thus extends along the coast from New Jersey on the north, south to Yucatan. Here, then, is an Atlantic and Gulf lowland strip more than 3,000 mi. in length, and from a few miles in width to close to 500 mi. wide. The belt is narrow in New Jersey, but increases in width southward until it covers much of South Carolina, Georgia, and Alabama, most of Mississippi, and all of Florida and Louisiana; and it extends northward to the present mouth of the Ohio. It is also a broad belt in Texas and northern Mexico. Farther south in Mexico it becomes thin again, but it makes up the entire peninsula of Yucatan. Throughout the most of this coastal area the Tertiary beds are generally 2,000–4,000 ft. thick, but their maximum thickness in Louisiana and Texas is estimated to be nearly 6 mi.! The common materials of these Tertiary formations are sands, clays, and some limestones; but the sediments are in the main poorly cemented.

In Texas, Louisiana, and Mexico these Tertiary formations are commonly pierced by great plugs of salt which have intruded them from below. The resulting structures are known as "salt domes"; and they are of great economic importance, since, besides the salt, oil, gas, and sulphur may be associated with them. In fact, **about half the oil which has thus far been produced has been from Tertiary rocks,** and of this vast total much has come from salt-dome structures. Most of the world's sulphur is also obtained from the Gulf Coast salt plugs.

The origin of these great plugs of salt is not yet completely understood. We can discover through drilling and mining operations, however, that the main mass is relatively pure salt, and that it may be a mile or more in diameter, as well as considerably over a mile deep. The top of the salt mass is generally capped with gypsum and limestone, and less commonly with sulphur. The enclosing Tertiary beds are folded upward and faulted adjacent to the salt. The plugs apparently therefore represent upward flow of salt from some pre-Tertiary source. The salt has been set in motion by pressures whose origin is not definitely known, and it tends to move in the direction of least resistance with a flowage something like the flow of glacial ice. Some scientists have suggested that the pressures required for the movement are no more extraordinary than those which result from the great weight of the overlying Tertiary sediments themselves.

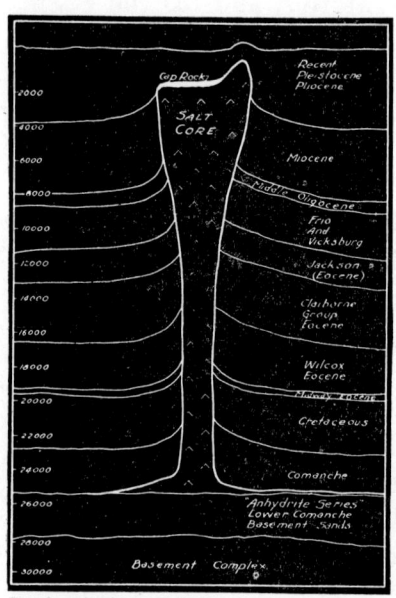

Idealized section through a Louisiana salt dome. After Howe and Moresi

At the end of the Cretaceous period the peneplain which had long been developing in the Appalachian area, was, as we have seen in chapter 42, destroyed by uplift and subsequent erosion. This subsequent erosion, of course, took place during the Tertiary period. Further uplifts during this period naturally have complicated the story, but probably nowhere in the world has the record of ancient erosion, as opposed to ancient deposition, been studied so thoroughly as in the Appalachians. The physiographer's knowing eye, to which we have earlier referred, is always trained on the Appalachian's even-crested ridges and benches, which tell him so much regarding ancient erosional surfaces. It is well to point out again in this connection, therefore, that **there is no deposition without attendant erosion.** We have been preoccupied with the depositional story merely because most of the old erosion surfaces are unconformities of some sort or another which lie buried beneath younger sediments. We can study these as they are revealed in canyon walls,

THE TERTIARY

but we cannot reconstruct them over large areas except in those districts in which there has been a great deal of drilling activity.

THE Western Interior is characterized by a complex mixture of continental water-lain sediments, aeolian deposits, and lava flows and ash falls. If we remember that the Laramide revolution left this area standing high, perhaps we can picture the physiographic setting in which these Western-Interior Tertiary beds were deposited. The mountains were rugged, and erosion thus was rapid. The material carried by the streams in many cases was deposited not far from its source in the form of fans and alluvial plains. Such deposits, as well as lava flows, commonly blocked valleys, just as they do today. Isolated basins resulted, and some of these were accentuated by continuing structural adjustments. Formed under such conditions, it is little wonder that these Tertiary beds are difficult to correlate from one basin of deposition to another. Many of them, however, are rich in mammalian or plant remains which permit some reasonably accurate age determinations of the isolated basin sediments. Some of these Tertiary volcanic and sedimentary deposits are illustrated in Plates 56 and 57.

The Eocene sequence contains the famous *Green River shales* whose seasonal banding we commented on in chapter 32. These Green River beds are the "oil shales" which the papers often tell us are to save the petroleum industry when the liquid petroleum has all been taken from the earth. Another important Tertiary sequence is the Oligocene *White River beds*, which infrequent, but torrential, rains have carved into the fantastic erosional forms which have long been known as the "Bad Lands" of South Dakota. Miocene beds also are involved in this unusual area; but they are perhaps even better known from the *Florissant*, Colorado district, where volcanic dust of this age, falling into a small lake, has been the medium of preservation for more than a thousand species of insects, plants, and other organisms.

Igneous rocks of Tertiary age are estimated to cover some 300,-000 sq. mi. in the western part of the continent. Lava flows are especially prominent in the Columbia and Snake river regions of

Washington, Oregon, and Idaho. In this area there are dozens of individual flows with an average thickness of something under 100 ft., and the flows have a total thickness of more than a mile. **Sandwiched in between the layers of lava are soils in which stumps of large trees are still standing.** Without straining our powers of deduction we conclude that relatively long erosion intervals elapsed between successive flows.

On the Pacific Coast both marine and non-marine deposition continued throughout the Tertiary epochs, especially in the area west of the present Sierra Nevada and Cascade ranges. The Eocene and Miocene beds, however, are much better developed than the Oligocene. The Miocene *Monterey* formation of California is notable for its great amount of diatom material. The diatoms are responsible for the great quantities of siliceous material in the formation, and some geologists even ascribe the origin of the great resources of California oil to these tiny plants. Beginning with the Miocene, igneous activity was prevalent in this area, just as it was farther east; and such prominent volcanoes as Mounts Shasta and Rainier came into existence sometime in the later Tertiary period.

Generalized section through the fossil forests of Amethyst Mountain, Yellowstone National Park. At least fifteen successive forests are shown. After Holmes

DURING the Miocene, crustal unrest was prevalent, especially along the Pacific coast. Thus there began the series of still rather poorly dated orogenic movements which we call the **Cascadian disturbance.** Probably culminating in the late Pliocene, or, as some think, even as late as the Pleistocene, these disturbances had a far-reaching effect not only in North America but in most of the rest of the world. We do not even know for certain that this great

THE TERTIARY

period of mountain-building is over. And, as we remember the modern "circle of volcanic fire" around the Pacific, and read of earthquake tremors of destructive violence occurring literally around the world, we are inclined to doubt that it is.

On other continents than our own we can see many prominent effects of the Tertiary orogenic movements. The modern Alps were folded and thrust-faulted during this period, and the Carpathian and Himalayan ranges also were formed. Marine Eocene formations still occur at heights of about 4 mi. above their depositional elevation in the latter mountains, giving a minimum measure of the great elevation that has taken place in that area since the early Tertiary. Nor was volcanic activity confined to North America during the Tertiary period. Great eruptions took place at many localities in the general North Atlantic area, in East Africa, and in the Mediterranean district.

ALL these volcanic disturbances, in addition to the great continental uplifts and attendant crustal unrest, may well have been partly, or even largely, responsible for the period of glaciation which beset the earth in Quaternary time. In the next chapter we will examine some of the important physical features of this most recent of the earth's many ice ages.

CHAPTER 47

THE QUATERNARY

*A frozen continent
Lies dark and wild, beat by perpetual storm.*
—MILTON

AN AVALANCHE speaks persuasively of the power of gravity. In fact, even the untutored knows that, when a stone along any valley wall becomes loosened, it promptly rolls downhill. Thus fragments of bedrock are characteristically found scattered along valley slopes *below* the zone where the rock itself outcrops. And yet, more than a hundred years ago it was noticed that in many areas the loose rock fragments occur not only below the outcropping beds but *above* them, and commonly far from their source. Here is a paradoxical situation which challenged early natural historians. An explanation obviously was demanded. So they called the displaced boulders and rock fragments *drift boulders*, and attributed their unnatural position to the Noachian deluge, which they thought had scattered the material far and wide and had swept some of it from low to higher elevations.

.... left in their unnatural position by the Noachian deluge ....

Contributory evidence which fitted in well with this concept was the fact that these boulders were commonly mixed with a heterogeneous deposit that veneered hills and valleys alike, with little regard for relative elevations. Any ordinary deposit laid by water, it was early realized, should be arranged in regular layers and should be deposited with some respect for differences in level. Thus the unusual drift obviously demanded an unusual origin. Only the great biblical flood could have furnished the necessary conditions, it was held; and there the matter rested.

BUT about 1840, when similar observations on the *drift* had been made in England, northern Europe, and North America, Louis Agassiz was critically examining the glaciers of the Alps. He found clear-cut evidence that in the not too remote past these glaciers had extended far out from their valleys onto the plains below. Moreover, the lines of evidence marshaled in support of this contention had a bearing on the interpretation of the drift-covered hills elsewhere, far from the Alps themselves. That is, the drift was quite similar in composition to the deposits left by the earlier Alpine glaciers, and Agassiz noticed that commonly the drift itself was sharply separated from the underlying bedrock just as it was in the case of the Alpine glacial deposits. In addition, the bedrock surface in both cases bore evidences of being smoothed, planed, or scratched, as though a giant hand had rubbed some monstrous abrasive over it. Agassiz's obvious conclusion, therefore, was that great sheets of ice also had overlain much of the British Isles, northern Europe, and northern North America. At first this explanation of the drift was too farfetched for most minds. Nevertheless, evidence piled on evidence, and eventually all had to concede that the arguments in favor of the fact that tremendous continental glaciers once covered many millions of square miles of Europe and North America were indeed irrefutable.

TODAY many hear about glaciation from childhood. In fact, it is almost an annual event for the Sunday supplements to warn us of the day, not far distant, when the glaciers will again push down from the north and the last shivering representative of *Homo sapiens* will succumb to its icy blasts. Most of these popular ideas about glaciation are either erroneous or at best vague, but apparently few find it difficult to believe that the earth has indeed suffered periods of glaciation in its ancient history. For this important contribution to our understanding of the geologically recent past we are indebted

.... indebted to Louis Agassiz ....

mainly to Louis Agassiz, whose painstaking observations, beginning about a century ago, finally changed what was at first regarded as an outlandish theory to a generally accepted fact.

But more was to follow. Investigators more than fifty years ago found that, by tracing the long ridges of débris left by the ice, it was possible to outline the actual size and extent of an ancient glacier and to map its relative positions as it advanced and retreated over the land. These ridges are the *moraines* we mentioned away back in chapter 10. Associated with them are outwash plains, kames, and all the other features that attest to deposition by glacial ice and its meltwater. Furthermore, by observing the directions in which the glacial scratches are oriented on the bedrock geologists found it possible to determine the direction from which the ice came. Thus were outlined the great centers of ice radiation, about which more anon. Meanwhile, however, T. C. Chamberlin, working in Wisconsin, observed another peculiar feature of the ice deposits. He found places where, between beds of glacial till, there were layers of plant remains, as if to indicate that during the very glaciation itself there had been vegetation present. But this, of course, was a rank impossibility in such frozen wastes. Further search disclosed that there must have been times of glacial advance, followed by ice retreat and the growth of vegetation, only to be succeeded once again by another ice advance. **Thus it was at last demonstrated not only that there had been a great glacial period but that four separate and distinct times ice sheets moved outward from their centers of radiation, and that between these advances there were periods of comparative warmth,** during one of which the climate was at least as mild as it is today.

THE QUATERNARY

IT IS remarkable in many respects that although there is not a trace of an icecap on either northern Europe or North America today, the evidences of their former occupation by ice are so well preserved that the glacial geologist needs only to visualize an icecap at a given locality to decipher completely the origin of a landscape whose every feature is the result of ice deposition or ice erosion. Studies by various geological surveys are available, in which detailed maps indicate many successive stages of advance and retreat of the several ice sheets. Indeed, more recently studies of seasonally banded sediments have enabled us actually to date the successive stages in terms of years, so that there now exists a year-by-year record of at least the waning stages of the last great ice sheet. We shall consider the evidence of these seasonal layers in greater detail shortly.

BY STUDYING the striae of the glacial pavement, by noting the alinement of the different types of moraines, and by plotting the distribution of certain peculiar rocks in the drift in order to determine their source, we have been able to locate on maps the very centers from which the great icecaps radiated. In America three main centers are recognized, the **Labradorian,** the **Keewatin,** and the **Cordilleran.** Each of these served as a center of ice radiation during more than a single advance; indeed, they were active during each of the glacial epochs; but, as conditions varied, first one and then another ice source dominated, so that the various glacial deposits from the three centers overlap to some extent.

As the accompanying map shows, the Labradorian center lies

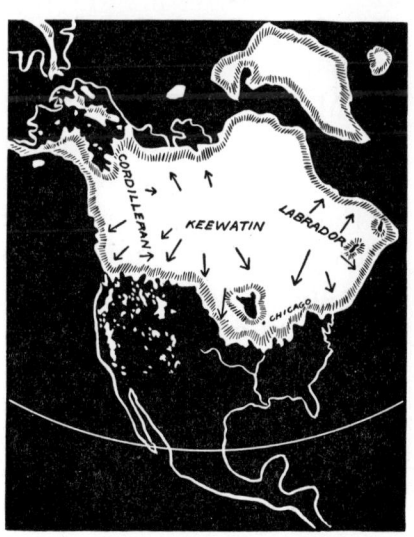

. . . . centers from which the icecaps radiated . . . .

east of Hudson Bay, and from the position of the arrows we may judge that much of the eastern part of the continent received its ice from that source. The Keewatin center, on the other hand, lay just west of Hudson Bay, and the ice from that center moved west, south, and southeast, overlapping the deposits of the Labradorian center in the general vicinity of the Great Lakes. Thus Wisconsin and Illinois at various times received glacial ice from both centers. The Cordilleran center stretched along the Cordilleran belt of mountains, and for the most part its ice was confined to the immediate vicinity of that mountainous area. Thus, in some respects this district was another Alps; and indeed, the number of glaciers still present in the region, combined with its variety of glacial erosive and depositional features, lend it the name "the Alps of America." Glacier National Park lies within this region, and the many readers who have been there will certainly recall the rugged Alpine grandeur of these ice-sculptured Rockies.

In Europe the great center of continental ice was in the Scandinavian peninsula, from whence ice radiated over much of northern Europe. The Alps, and the Pyrenees too, were centers of glacial ice which swept down from the catchment areas and spread far out over the plains at their base.

IT MAY be interesting to learn a little more about how these centers of ice radiation are recognized. We already know that glacial ice is a geological agent active in both erosion and deposition. As the evidences of glaciation are followed north from the outermost extension of the moraines, it is found that the region of dominantly depositional activity gives way to one in which glacial scour dominates. But as the ice centers are approached, undisturbed mantle rock becomes more common. Thus there is a central region of little glacial activity surrounded by a zone of intensive erosion, followed outward, in turn, by a region of deposition. In the centers from which the ice radiated we may therefore picture a vast cap of nearly stagnant ice, deploying outward under the influence of the surface gradient, actively eroding the land just beyond the ice center, and finally wasting away and depositing its great load of débris.

ONE may obtain a better idea of the enormous extent of recent glaciation by noting that in the case of the Keewatin center, for example, the ice moved southward for some 1,500 mi., and that during its maximum extent some 4,000,000 sq. mi. of our continent were covered with glaciers. The ice near the center of the modern Greenland icecap is about 8,000 ft. thick, and there is every reason to believe that similar, or even greater, thicknesses were common in continental glaciers which covered North America during the Pleistocene. Naturally enough, such a vast cake of ice, plus the ice in contemporaneous European glaciers, represented enormous quantities of water. Consequently the oceans, which supplied this water, were lowered considerably; and the continental platforms were weighted down under their great ice load to such an extent that elastic stresses, which are not even yet completely adjusted, were set up within the lithosphere. The actual results of both these phenomena have been clearly recorded, and we shall touch upon them later.

Icecaps of the Northern Hemisphere during the Pleistocene. After Antevs

WE HAVE mentioned that, in all, there were four great ice invasions during the glacial epoch that we are discussing. It may be well, therefore, to set up a chronological scheme of classifying this important part of geological history. As we have earlier seen, the ice age is a part of the Quaternary period, which also includes the present, so that our table, which, as usual, places the earliest events at the bottom, is as follows:

QUATERNARY PERIOD
Recent epoch
Pleistocene epoch

		North America	Europe
4.	Glaciation	Wisconsin	Würm
	Interglacial	Sangamon	Riss-Würm
3.	Glaciation	Illinoian	Riss
	Interglacial	Yarmouth	Mindel-Riss
2.	Glaciation	Kansan	Mindel
	Interglacial	Aftonian	Günz-Mindel
1.	Glaciation	Nebraskan	Günz

Following this table, which shows that **the European ice advances and interglacial episodes can be synchronized with our own,** we may outline a few typical events of the Pleistocene epoch. With the opening of the Quaternary period, we visualize a climate gradually increasing in rigor, with finally the accumulation and growth of vast icecaps in the great centers of glaciation. (Some of the possible causes of glacial climate were outlined in chapter 29.) With additional precipitation the ice became thicker; and as it deployed from the centers it engulfed large parts of the continent, moving as far south as Nebraska, from which the first stage received its name. When the ice had extended southward as far as conditions permitted, wastage at last essentially balanced forward ice movement, so that the front edge of the cap fluctuated to and fro, and thus deposited a series of festooned moraines. Then, as the climate became still milder, wastage predominated over ice movement, and thus the ice fronts retreated. Torrents of meltwater poured from the waning icecap and strewed outwash deposits along the valleys that bore the waters away. With the retreat of the ice edge, vegetation gradually moved back northward; and during the Aftonian interglacial interval, a temperate climate, perhaps not unlike today's, came into being. Following the Aftonian, conditions again swung toward frigidity, the Kansan glaciers swept over the land, and ice and snow once again held sway. Thus was the oscillation repeated four times, Yarmouth warmth following Kansan ice retreat, and Sangamon mild climate coming after the Illinoian gla-

THE QUATERNARY

cial episode. One of the interesting features of the early stages of Wisconsin glaciation was the accumulation of great deposits of wind-blown dust, called **loess,** along the Mississippi River Valley and its tributaries. And today, following the Wisconsin glaciation, we are perhaps in some stage of another interglacial period.

Many and diverse were the events that accompanied the great glaciations, and so strongly have they left their impress of the lands that much of northern North America and Europe owe the present detailed configuration of their landscape to the deposits left by the ice or to the powerful erosive effects of that agent. Among the more outstanding features due primarily to the effects of the glaciers are the Great Lakes. Theirs has been an interesting history, and we shall very briefly touch upon it here. As the ice retreated during the last, or Wisconsin stage, two broad lobes of ice, perhaps retreating down broad preglacial valleys, ponded waters between the limits of the ice and the more southerly moraines that marked the maximum extent of the Wisconsin glaciation. Thus arose the beginnings of Lake Michigan and Lake Erie. These lakes at first were drained to the Mississippi; in the case of the early stages of Lake Michigan, the waters poured out along the Illinois Valley. As the ice edge retreated still

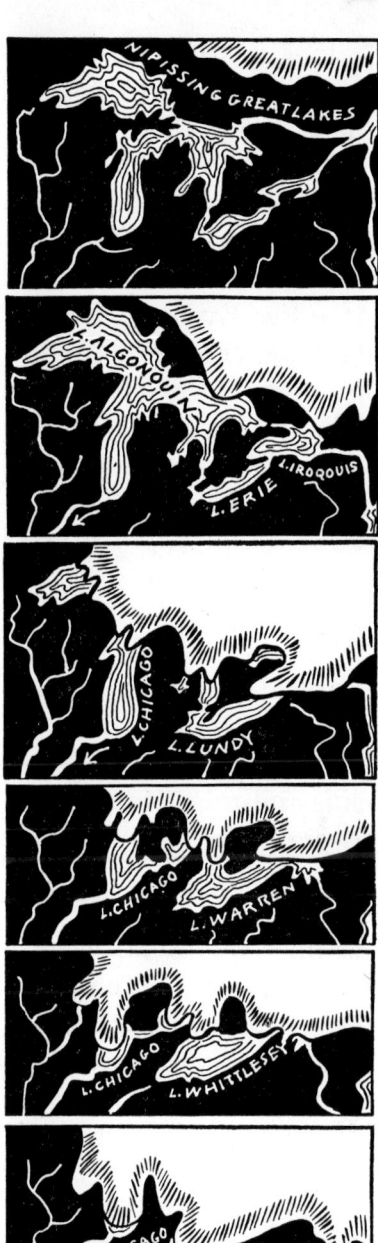

Outline history of the Great Lakes showing ice retreat and different outlets. Read from bottom to top. Based on Leverett and Taylor

farther north, the two lakes were united by a valley which extended across the state of Michigan, so that both lakes were drained through the Illinois Valley.

As it retreated still farther north, the ice edge exposed parts of the present-day Lake Superior and Lake Huron basins; and as, finally, the Lake Ontario area was freed from the ice, the ponded waters found a new and lower outlet along the Hudson Valley to the Atlantic. Eventually came the time when the St. Lawrence Valley itself was freed of ice, and the present drainage of the Great Lakes into the Gulf of St. Lawrence was established. Glacial Lake Agassiz, of which Lake Winnepeg is a remnant, had a somewhat similar history. Its waters at first drained into the Mississippi system; but after the ice retreated far enough to the north, its modern outlet into Hudson Bay was established.

IT MUST be kept in mind that during glaciation and subsequent ice retreat the land passed over by the glaciers may be greatly modified in its surface expression, so that lake basins, formed by the damming of meltwater by moraines, are usual features. Further, where more resistant layers of bedrock are exposed by glacial scour, cliffs and ridges may be formed, which later give rise to conditions favorable for waterfalls. Thus, in fact, were the falls at Niagara formed; the waters, draining from the Great Lakes above the falls, encountered the Niagara escarpment, which probably had been modified by glacial scour, and they poured over its edge into Lake Ontario. During the course of time the water eroded the escarpment back, forming the Niagara gorge. This retreat is still going on, as current newsreports indicate at intervals when large blocks of rock tumble from the edge of the falls. In fact, by observing the rate of retreat of this edge, it is possible, as indicated in chapter 32, to date approximately the time since the Niagara escarpment was first exposed by the retreating ice. As we have previously pointed out, studies of this nature indicate that 18,000–25,000 yr. is about the order of magnitude of time since that event—a time, great enough from a human standpoint, but only 3 or 4 min. on that Gargantuan calendar of ours.

THE QUATERNARY

AS WE have earlier stated, the attempt of geologists to fit an actual time scale in years to the geological calendar can only have approximate results for most of the ages long since past. The reader will recall some of the devices used (chap. 32) and will remember that many of these indicated only the order of magnitude of time rather than even a moderately accurate number of years. Yet, as we approach closer and closer to the present, we find that the much more extensive deposits of recent ages, combined with their greater freshness (lack of weathering and the like), permit us to set up chronologies that actually do permit of a year-by-year recording of events. Here, then, geology reaches part of its ultimate end: not only to record and explain the impressive events of the past, but to date them as well on some absolute time scale.

CURIOUSLY enough, the calendar of years since the Pleistocene is recorded in seemingly insignificant clay deposits that are found associated with glacial lakes. We saw that moraines could easily pond the meltwater, so that shallow lakes, generally tempo-

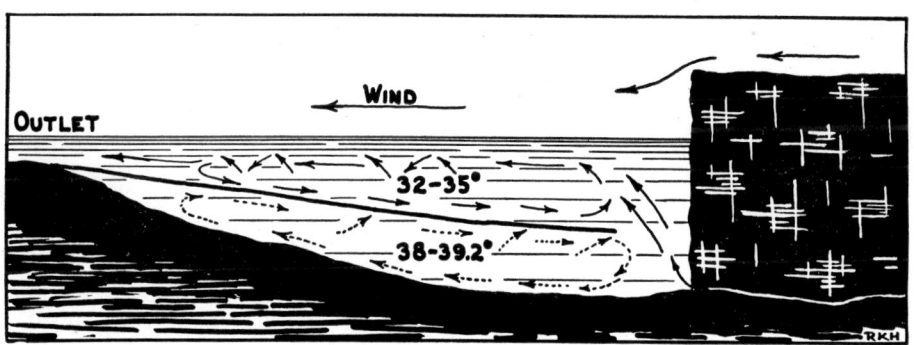

Circulation in water ponded at melting front of a glacier shown on right. In such positions varves are deposited. After Antevs

rary in character (they disappeared as soon as the retreating ice dam exposed an outlet), served as receiving-basins for great volumes of glacial débris, borne from the ice edge by the waters that circulated through the lakes. If we stop to examine what happens to these deposits in detail, we shall see what a fascinating field of possibilities is opened by a study of these old lake sediments. In the summer, when melting is active, the débris from the ice charges the water of the lake; and the coarser particles, such as sand and silt, are deposited over the lake bottom. As long as new material is dumped into the lake, the layer of sediment continues to accumulate; but in the course of events, the short summer runs its span and the lake freezes over again. Likewise melting of the glacier virtually ceases, and apparently all is quiet in the lake under its shell of ice. But is it? Recall from our earlier chapters on sedimentary processes that the very fine clay particles settle very slowly in water. Hence, when winter rolls around, the water of the lake is still turbid, owing to these clay particles, even though all the sand and silt has already reached the bottom. Agitation of the water (summer breezes from waves; also, additional meltwater sets up currents) has kept the clays in suspension, but during the quietness of winter this clay has a chance to settle out. As it does so, the layer of coarse sand and silt is buried under a layer of very fine clays. Again spring approaches, however, and the ice thaws, fresh meltwater with its débris enters the lake, and quite abruptly the coarse summer material is laid on the fine winter clays. Thus the round goes on season after season, so that for each year there is deposited one of these double beds which is made up of a coarse and light-colored layer below and a fine-grained and dark one above.

THE deposits we have been describing are the **varves,** which we have already casually met in chapter 32. These varves form the yardstick by which geologists date recent events. In actual practice it is found that the varves for a certain group of years remain strikingly similar over wide areas, so that it is possible to correlate them from one lake basin to another. Naturally enough, even slight fluctuations in the lengths of summers or their warmth will be reflected in the thicknesses of the annual layers, and thus these fluctuations will affect many lake basins simultaneously. Therefore, a painstaking measurement of varve series not only makes it possible to count the actual years, but, by tracing a particular varve to its point of origin, it is obvious that the very location of the morainal

THE QUATERNARY

front at that year can be determined. Indeed, this very thing has been done in great detail in parts of North America and Europe, especially in the Connecticut River valley, and in Sweden and Fin-

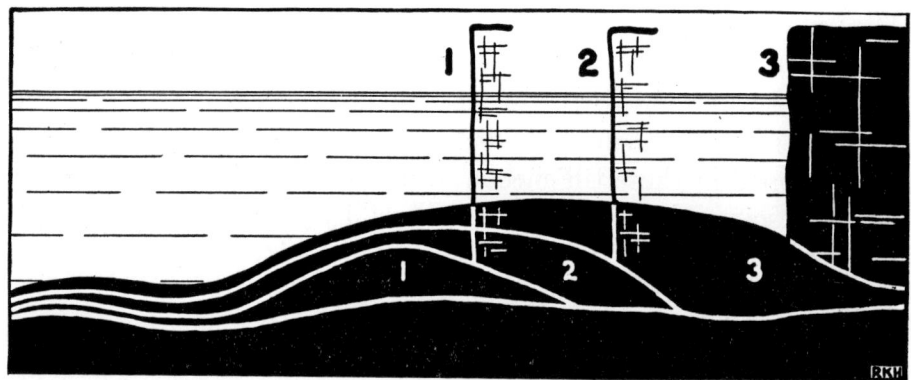

Showing ice retreat from position *1* to *3* and how the varves overlap one another in the direction of retreat. After Antevs

land. Gerard De Geer, noted Swedish geologist, was the first to explore this fruitful field. From his results and those of co-workers, it is now possible to draw lines across maps showing the precise location of the retreating ice edge through the years. In this manner more accurate dating devices are available than that furnished by Niagara; and yet the two lines of evidence point to results of a similar order of magnitude, so that the waterfall method is at least not entirely unreliable.

IT MAY be interesting to consider how far back these glacial varves may be carried. Obviously, they are best developed in connection with the last glaciation, because the deposits of that glacial retreat lie at or near the surface and continuous exposures may be found. Not so with the earlier stages, however, whose varves are buried beneath later deposits or were eroded by the later ice sheets, so that only fragmentary data are available. For the time since the last glaciation, however, including the retreat of the ice, a figure something like 20,000–29,000 yr. is reached for the eastern United States and Canada. **For older glaciations the determination of the time**

involved depends upon the extent to which the glacial till and outwash materials were weathered between successive ice invasions. But in all the 25,000 odd years that have elapsed since the close of the Wisconsin stage, till sheets deposited by that glacier have been weathered and altered to a depth of only 2 to 3 ft. from the surface, another indication of the slowness with which weathering changes take place. In the process of weathering, certain recognizable features, such as the leaching-out of calcareous material, are developed in these till sheets. **Gumbotil,** a dark sticky clay, is one of these weathering products. Thus, by observing the relative degree of weathering, it is possible to translate, at least roughly, the previous interglacial intervals into lapses of time which may be compared with the calculated time since the last glaciation. When studies of this type are made, it is estimated that, reckoning the time since the Wisconsin glaciation as unity, the time since the Illinoian is about 20 units, since the Kansan perhaps 50, and since the Nebraskan as much as 100 or more. Thus we have a time lapse since the beginning of the first glaciation which is something of the order of 2,000,000 yr., admittedly a rough approximation at best. But few, if any, geologists now set the figure at much under 1,000,000 yr.

WE ARE prone to forget that during this great expanse of time many interesting events, of course, were also transpiring outside the glaciated regions. To them unfortunately we can devote only the closing paragraphs of our chapter. Beyond the edge of the ice many rivers were loaded with outwash material; but other streams, flowing through non-glaciated regions, were forming normal river-laid sediments and terraces. Marine deposits were being laid down at the margin of the continent as always, so that one might be tempted to say that outside the areas of glaciation the ordinary sequence of events that transpire on continents and their borders were proceeding as usual. There were, however, some rather extraordinary features; and among these is great Lake Bonneville, which occupied a large area in the general region of Salt Lake City, Utah. Salt Lake itself is but the emaciated remnant of this great body of water, which reached depths of 1,000 ft. and covered an area of some 19,000 sq. mi. The formation of the lake depended upon several circumstances. First, the area lies in the so-called Basin and Range province, in which abound large areas that have no direct outlet to the sea. During the glacial episodes increased precipitation and reduced evaporation caused the water to accumulate in the basins, and lakes formed. Thus Lake Bonneville arose and remained long enough to form distinct wave-cut terraces, bars, and splits, and all the multitudinous phenomena associated with shore agents. These features are still conspicuous elements in the landscape in the vicinity of Salt Lake City and Ogden. As the climate again became more arid, and the Snake River outlet was abandoned, the waters evaporated, concentrating the salts they bore in solution, until finally Great Salt Lake, with its surrounding deserts of salt, remained.

Outline map of Lake Bonneville and associated Pleistocene lake basins, together with their drainage areas. After W. M. Davis

IN ADDITION to normal sedimentary deposits, there are evidences of considerable igneous activity during the Quaternary, especially throughout our Cordilleran area, from Mexico on the south well into Canada on the north. Numerous small cinder cones and lava flows abound locally. Their youth is clearly attested by their lack of erosion, by lava dammed streams, and by the absence of weathered material among the fresh igneous products.

WE MENTIONED, near the beginning of our chapter, that the seas were lowered during the maximum stages of glaciation. Evidences of this lowering are found in great submarine channels, as at the mouth of the Hudson River, suggesting a lowering of the sea level, so that streams could cut down much farther than at present. Recently some investigators are even having the temerity to assert that the great submarine canyons demand a lowering of oceanic level not of hundreds, but of *thousands*, of feet! Likewise evidences of the stresses set up by the great loads of ice reflect themselves in recent warpings of parts of the North American continent. Raised and tilted beach lines along the Great Lakes point to a rise of land to the north since the time of the lake stages earlier described, showing that, as the burden of ice was removed, the continent adjusted itself to the relieved situation by a sort of slow elastic rebound. It is not impossible that the minor earthquakes that occur along St. Lawrence Valley from time to time represent recent adjustments of this still-rising terrain.

WE HAVE now presented a mere thumbnail sketch of the physical events of the Quaternary. As may be expected, the glaciation of large areas of land must also have had profound effects on land life. And in the very seas themselves, cooled by the general lowering of temperatures, and certainly considerably restricted by withdrawals of water, there should have been reflections of this refrigeration on marine forms of life. In our next chapter we shall examine a few of the organic responses to these unusual physical conditions.

CHAPTER 48

THE WARM-BLOODED

> The choice and master spirits of this age.
> —SHAKESPEARE

THE mammals are indeed the choice and master spirits of this age, but by this time we are all aware that their reign has geologically been a very abbreviated one. According to our Gargantuan calendar, which we have perhaps too often consulted, they ascended to the seat of the mighty on December 26. Their time of dominance, therefore, has been those seven short days which make up the week of Cenozoic time. They made their first inconspicuous appearance on the terrestrial scene, however, some 16 days earlier, in the Triassic period. During the entire

.... playing David to Goliath ....

Mesozoic era they remained completely subservient to the reptiles; but by the end of the Cretaceous, the mammals were at last in a position to play David to Goliath Dinosaur.

It must be obvious that we cannot properly discuss the geological history of a group as diversified as the mammals in a few short pages. We can, however, point out several of the salient trends in the evolution of the class, and discuss in some small detail the development of several of its better-known subdivisions.

MANY scientists are of the opinion that the mammals were derived from the late Paleozoic reptilian stock known as the **Therapsida,** whose representatives were roughly intermediate in structure between some of the modern **monotremes,** or primitive egg-laying mammals, and the early armor-headed amphibians. The **cynodonts,** or dog-toothed reptiles, were members of this group which lived in South Africa during the Triassic period. These creatures had their teeth divided into incisors, canines, and molars unlike the typical reptilian even-tooth series; and they had other skeletal features in common with the mammals. If the mammals were not actually derived from any of the known cynodonts, it is at least likely that they may have sprung from some very similar reptilian stock.

THE physical urge which played a rôle in the change from reptiles to mammals is one at which we can only shrewdly guess. But during the refrigeration and aridity of the Permian, a premium doubtless was placed on the mammalian features of warm blood, a body-covering of hair, agility, and intelligence. Certainly sluggishness, scales, and cold blood fitted in rather poorly with the times.

As the mammals developed, they gradually deployed into all sorts of habitats, seeking food and protection. As a result, there finally originated such diverse forms as the carnivores and the herbivores, and such opposites as the bats and the whales. Naturally, such changes involved great modifications in tooth structure and in body outline. In these changes we may see several main trends. In so far as the animals have progressed along these various trends, just by so much are they different from their small **generalized mammalian ancestor which had a long face, a small brain, short legs, flat feet with five digits each, and 44 short-crowned teeth.** The trends followed are: (1) **specialization in, and reduction of, teeth;** (2) **specialization in limb structure and reduction of digits;** (3) **increase in size and complexity of the brain;** and (4) **general increase in size.** Naturally, some mammalian groups have progressed far along one of these trends but have shown little progress in the others. Herein lies the basis for the great diversity in the mammalian class.

BUT, however far the mammals followed these trends in the Cenozoic era, they made little progress in any of them during the Mesozoic. During the latter era there were two general types of mammals. One group, which appeared in the late Triassic period, was distinguished by the several cusps, or tubercles, on their teeth; hence they are called the **multituberculates.** These were

THE WARM-BLOODED

probably herbivorous forms, and possibly they were egg-laying. Although they persisted until the early Cenozoic, so far as we know they did not give rise to any modern descendants. The other group, originating in the Jurassic, was characterized by shrewlike **insectivores,** in which probably are to be found the remote roots of all the modern mammalian types, including yourself. Not any of these Mesozoic mammals were as large as a robust tomcat; but the brain, although primitive for a mammal, already was markedly in advance of the contemporaneous reptilian type. Moreover, then as now, the mammals probably took some care of their young in the trying days just after birth. This is a race-saving habit which few of the reptiles have taken the pains to adopt.

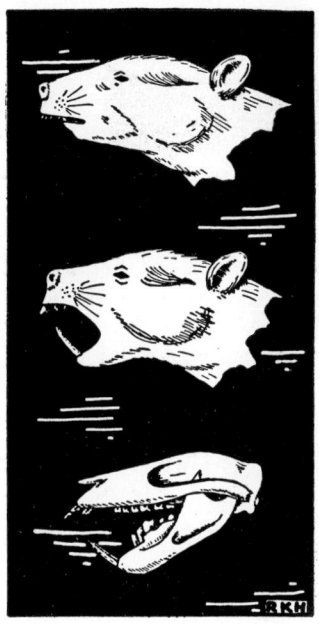

Three views of a primitive Jurassic mammal from the Morrison beds. The animal is shown about natural size. After G. G. Simpson

AS SOON as the ruling reptiles had passed forever out of the geological picture at the close of the Mesozoic, the mammals expanded with extreme rapidity. But at first, either the bad example which the dinosaurs had set or their new-found freedom was too much for them. For many of the early Cenozoic mammals also went in for megalomania in a whole-hearted way. One of these groups with a misplaced reliance in size was that one known as the **titanotheres.** Members of the odd-toed hoofed mammalian group, they appeared in the Eocene as small forms about the size of a house dog. But by the

Mammals take some care of their young .... a habit few reptiles adopt

First and last of the titanotheres. After H. F. Osborn

Oligocene they had become well worthy of their name, and were gigantic, usually spade-horned rhinocerine forms. Much studied because of the light they throw on evolutionary processes, they suddenly became extinct by the end of the very epoch in which they reached their climax. The **amblypods,** or blunt-footed, Eocene megalomaniacs, had an even shorter life-span, for they are confined to the one epoch. *Uintatherium* is probably the least prepossessing, and consequently the most interesting, member of the group. As large as a small elephant, and built like one, it had a head with three pairs of horns, and, in the males at least, scimiter-like tusks something like those of a saber-toothed tiger.

The American Museum's exploratory parties in Mongolia have recently found in Oligocene and Miocene rocks fossil rhinoceros-like forms to which the name *Baluchitherium* has been applied. These are the largest of all the land mammals, with a length of some 26 ft., standing some 18 ft. high, and having a head 4–5 ft. long!

Andrewsarchus and *Baluchitherium*

In addition to such mammalian monsters there were present at the beginning of the Cenozoic the *creodonts*, or prototypes of the later carnivores, and the *condylarths*, or light-bodied, hoofed herbivores. Although neither of these groups is known definitely to have given rise to the modern hoofed and clawed mammals, they were not

Creodont above; condylarth below. After Lull

far from the progenitor strains. It is therefore interesting to note how similar the two groups are at this early pre-specialization stage in mammalian history. Here, again, we see clear evidence of the fact that, **no matter how far apart the ends of the branches of life may be today, they all converge backward in geological time into the main trunk of the tree of life.**

THE ancient carnivorous line reached its climax in size in *Andrewsarchus*, the last of the creodonts. This was a Mongolian wolflike form with a skull 3 ft. long—and thus it was the largest of all mammalian flesh-eaters. The modern carnivores, such as the dogs and cats, are fairly well known since the Oligocene epoch; but, as would be expected, they, too, became less and less differentiated as the stocks are traced backward toward the beginning of the era. The true cats were established as a line separate from the saber-toothed cats early in the Cenozoic era. The lion, tiger, and common house cat are only a few among the modern representatives of the first group; but not a single saber-toothed cat lived on past the Pleistocene.

AMONG the many other mammalian groups which we have not mentioned is the strain of odd-toed hoofed creatures which has culminated in the modern horse. The oldest horse known is *Eohippus*, the "dawn horse," from the Eocene. A contemporary of some of the early amblypods, the little foxlike first representative of the equine line must have appeared insignificant, indeed, to the large archaic mammals. But although they did not know it, the horses left the amblypods literally stuck in the mud as they forged upward and ahead in the most completely known of the mammalian evolutionary sequences. *Eohippus* to *Equus*, the modern horse, is a far cry indeed; but the long journey was made by many small well-recorded intermediate steps. In this evolutionary advance we note that (1) **the limbs elongate;** (2) **the digits become reduced in number until only the middle toe carries the weight;** (3) **the muscles are bunched higher in the body, so that a sort of body streamlining results;** (4) **the teeth become higher-crowned, and are spe-**

cialized for cropping and grinding; (5) the head and neck elongate, keeping pace with lengthening limbs; and (6) the individuals become larger. Thus the horse group followed most of the mammalian trends which we earlier outlined, as can be seen in Plate 58. Although originating in North America, the horse line became extinct on this continent during the Pleistocene. Every horse you have ever seen is a member of an Old World strain which has been reintroduced by man into its ancestral home.

.... the horse left the amblypods stuck in the mud ....

THE even-toed hoofed mammals also have written a wonderful record of their development in the terrestrial deposits of the isolated Tertiary basins we discussed in chapter 46. Among these creatures were such familiar types as the camels, pigs, and deers. But the extinct piglike *oreodonts* were perhaps the most numerous of them all, their skeletons being common at many places, notable in the "Bad Land" Oligocene beds.

THE *proboscideans*, or proboscis-bearing mammals like the elephant, are on the wane today; but they were among the most widespread of the Cenozoic mammals. In fact, they are only recently extinct in North America, where it has been expansively said that their skeletons can be found in every bog. The accompanying diagram illustrates some of the high lights in the evolution of the group.

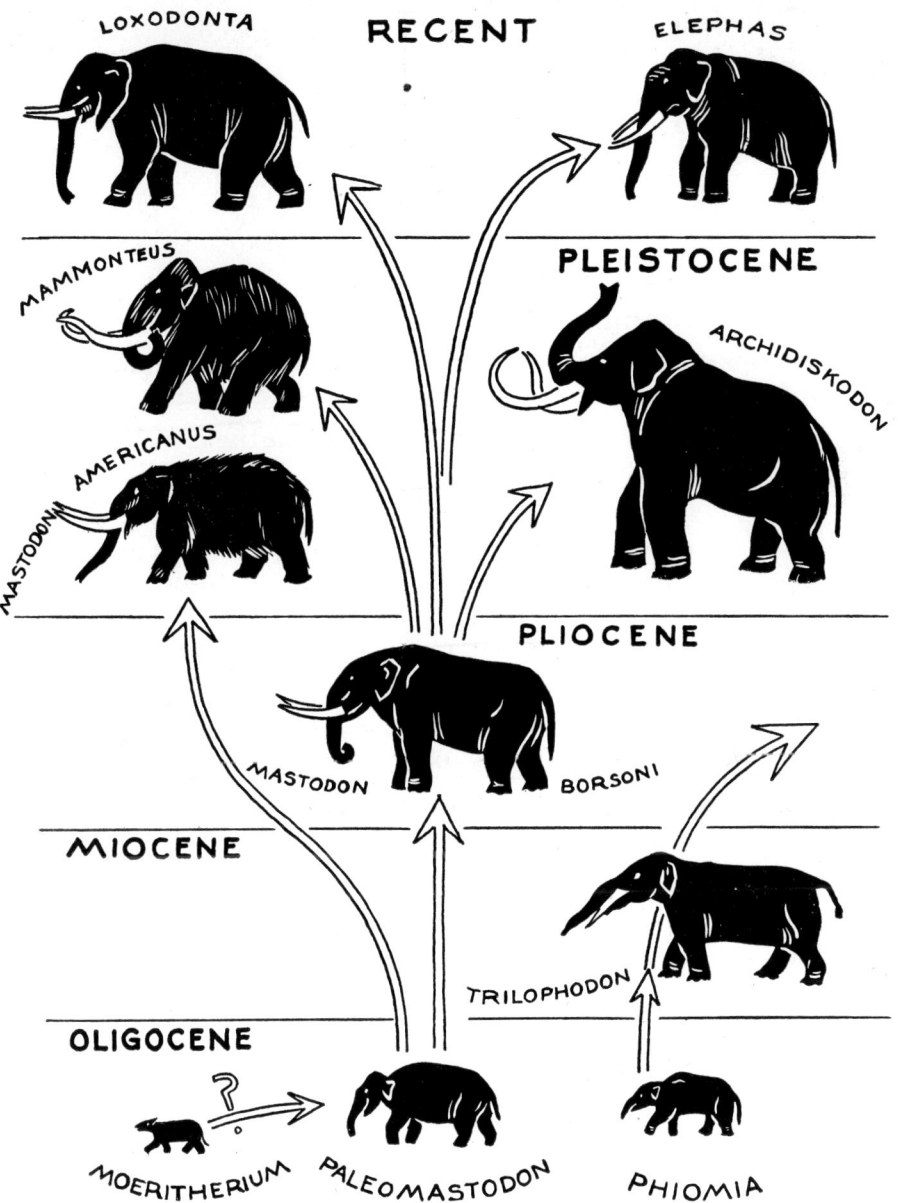

Evolution of the elephants. After H. F. Osborn

They make their first appearance in the early Cenozoic delta deposits of the Nile. *Moeritherium*, which is possibly the first of the line, possessed no real trunk or tusks and was no larger than a prize pig at a county fair. The much more elephant-like *mastodon* stock, characterized by conical cusps on the large molar teeth, arose only slightly later, and then migrated rapidly from the Egyptian center into Europe and Asia. During Miocene times the mastodons invaded North America by way of Siberia and Alaska. In their new home they persisted until the end of the Pleistocene.

The **true elephants are distinguished from the mastodons by the numerous folds in the enamel of their teeth.** These forms evolved out of the mastodon stock sometime in the late Miocene or early Pliocene. As they grew in size, they deployed into many habitats and became somewhat different in appearance. For instance, *Mammonteus*, or the "wooly mammoth," lived at or near the ice fronts during the glacial episodes, while *Archidiskodon*, the "imperial mammoth," roamed the dry plains of our southwestern states. In the advance from *Moeritherium*, or *Paleomastodon*, to the modern *Loxodonta* and *Elephas*, (1) **the molar teeth in the jaw at any one time were greatly reduced in number and the grinding surface perfected,** (2) **the trunk became elongated,** (3) **the tusks grew larger, and the** (4) **size of the body was greatly increased.**

South American ground sloth and glyptodonts

Space does not suffice to discuss a host of other interesting types of mammals such as the prehistoric armadillo-like *glyptodonts* and giant ground sloths of North and South America. Nor can we outline the development of the primates, except in so far as they are mentioned in the description of the rise of man in the following chapter. We must, however, refer briefly to the Californian tar pools at Rancho la Brea. These Pleistocene seeps entrapped many an unwary mammal and held them like so many flies on a gigantic

The California tar pits during the Pleistocene. Based on drawings by C. R. Knight

sheet of fly paper. As the doomed creatures struggled and cried out, they attracted carnivores and birds to the death scene. As the latter

Phororhacos

attempted to wrest a meal from the fatal grasp of the tar, they, too, commonly became entrapped. Little wonder, then, that as many as twenty skulls of saber-toothed tigers and wolves have been discovered in a cubic yard of the material.

The Pleistocene vultures that have thus been preserved are by no means the only Cenozoic birds of which we have a record. Modern toothless types appeared in the Eocene, during which time some of the flightless, ostrich-like forms were taller than a man. Most extraordinary of them all was a Miocene South American bird called *Phororhacos*, which stood some 8 ft. high and had a massive skull about 2 ft. long. *Dinornis* of New Zealand, which was slimmer but stood probably even taller still, has, like the famous Dodo, been exterminated by man. Eggs laid by another recently extinct bird, *Aepyornis*, are larger than a football; and some think that their early discovery may have given rise to Sinbad's story of the roc in the *Arabian Nights*.

We have now come to the point where we must once again say that these interesting animals were not the only creatures of the times. Obviously, the plants and the invertebrates have also marched rap-

Dinornis and *Didus*, the Dodo

idly ahead on their evolutionary paths which have brought them to their present positions but which lead ahead no one knows where. The reptiles and amphibians, however, are certainly no longer to be reckoned with in the age-old battle for organic supremacy. Nor are the fish, however numerous or meek, likely to inherit the earth. Even the mammals seem to have passed their climax in the Miocene. Is the future then dark for them also? Fortunately, no one definitely knows the answer. But you may as well amuse yourself by speculating on the subject in the light of what you have learned in the earlier chapters. Whatever your findings, you will not be embarrassed by anyone's checking up on you during your lifetime.

CHAPTER 49

HOMO DILUVII TESTIS

> Man is Nature's sole mistake.
> —WILLIAM GILBERT

JOHANN JACOB SCHEUCHZER, a contemporary of poor old Johannes Beringer, whose students so cruelly duped him, had even a more illustrious scientific reputation. Nor did he diminish it any when he announced, in 1726, that he had found, appropriately enough, in the gallows yard at Oeningen, Switzerland, "the sorrowful skeleton of an old sinner" who had actually witnessed the Noachian deluge. Scheuchzer described his find in a latin volume which he confidently dubbed *"Homo diluvii testis."* Years later Cuvier poured cold water on Scheuchzer's followers by showing that the "homo" was really nothing but the skeleton of a Miocene salamander; but by that time Scheuchzer had, as all scientists must, passed beyond ridicule.

.... beyond ridicule ....

Unfortunately for the modern paleontologists, not many ancient men who witnessed floods of any sort were foolish enough to get caught in them. Even their distant ancestors in the early Cenozoic were by that time sufficiently sagacious to avoid the tar pools, swamps, and quicksands which trapped their less knowing mammalian contemporaries. This is one reason that all primate skeletons, including those of lemuroids, apes, and monkeys, as well as those of man, are relatively rare as fossils. When man at last began to bury his dead, and his remains therefore become more common, the story has largely been inherited by the anthropologists. But there is also a growing store of geological data concerning our relatively recent ancestors, and we shall here summarize some of the outstanding features of this information.

In Scheuchzer's day if anyone had insisted that man and monkeys were even distantly related, someone surely would have made him eat his ill-considered words. Today the fact that man is merely another animal, related not only to the apes but to the snakes, fish, earthworms, and leeches as well, is such a commonplace that it requires little or no substantiation. But of all his relatives he is, of course, closest to those mammals which belong to his own order, namely, the **Primates**. Without getting technical we may define these creatures as mammals with specialized brains and generalized teeth and limbs. The Lemuroidea are primitive primates with foxlike faces, whereas the Anthropoidea are more or less manlike creatures like the monkeys and apes and man himself.

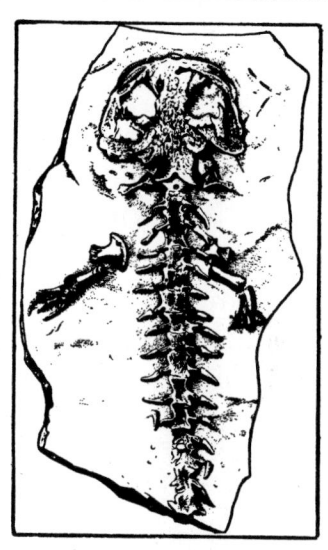

.... *Homo diluvii testis* ....

ONE of the best-known genera of the rare fossil primates is *Notharctus*, an Eocene mammal with a head about 2 in. in length. Since there is an argument as to whether *Notharctus* is a lemuroid or an anthropoid, we may well conclude that the two stocks sprang from a generalized ancestor at least very similar to *Notharctus*. This possible ancestor of man was a ratlike creature whose arboreal habits are indicated by the structure of the feet and the lightness of the bones. Its skull is long and low, and its teeth generalized. Eggs, nuts, fruits, and insects probably were its chief foods. But just when we seem to be getting somewhere in our story of man's ancient ancestors, we are blocked, at least in North America, by the fact that the Tertiary beds younger than Eocene have not yielded any primate fossils at all. In northern Africa, however, the Oligocene beds have produced *Propliopithecus*, a small generalized ape standing less than 2 ft. high. This form is sufficiently unspecialized that some scientists point to it as the ancestor of all the modern anthropoids. *Dryopithecus*, from the Miocene of Europe and India, is even more definitely a representative of the stock out of which the great apes and man deployed.

The Pliocene beds, which should be expected to tell us a great deal about early man, have proved a great disappointment. A good many anthropoid skeletons have been found in them, but they are nearly all fossils of ancient apes, not of ancient men. *Australopithecus*, however, is a primate nearer man than the apes. The specimens are a skull and lower jaw which were found imbedded in cave deposits near Taungs, South Africa. Authorities generally agree that this primate was intermediate between the apes and the as-yet-undiscovered primitive men of the time. The presence of the latter is indicated not only on a philosophical basis but by the crude flint implements of the late Pliocene, particularly those of Ipswich, England.

.... record of advancement ....
revealed in wall paintings ....

Now, as we come to the Pleistocene, we enter into the true "Age of Man" and find that a number of genera of fossil man have been discovered. These are known not only from their skeletons but from their *artifacts* as well. The latter are the weapons, tools, and other articles of culture which, of course, change with the passage of years. And in their changes we can read a part of the story of the progressive mental development in ancient man. The most primitive of the flint implements are called *eoliths*. They show wear, but little or no shaping. These are the only artifacts which occur in rocks older than Pleistocene. As the stones were shaped to serve definite purposes, the artifacts of the type known as *paleoliths* come into the picture. Finally, some 18,000 yr. ago, man began to make polished stone implements called *neoliths*. Naturally, as these flint or stone artifacts were improved, the same record of advancement was written in the art of the times, as revealed in bone carvings and wall paintings.

OF THE known ancient races of man, we can briefly mention only the outstanding. *Pithecanthropus*, or the "Java ape man," is the best publicized of all the fossil men. Discovered in 1891, by a Dutch army surgeon named Dubois, who regarded it as the true "missing link," *Pithecanthropus* turns out to be an apelike man probably not in the direct line of human ascent. The fossil specimens include a skull cap, some nasal bones, teeth, and a left femur. They indicate a fairly erect creature some $5\frac{1}{2}$ ft. high, with a cranial volume of slightly less than 1,000 cc., as compared, say, with that of a large gorilla with a volume of less than 600 cc. The Javanese river gravels, in which *Pithecanthropus* was found, contain other fossils which indicate its age as early Pleistocene.

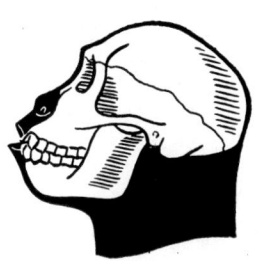
Java ape man

The "Dawn man," or *Eoanthropus*, is based on skull fragments found near Piltdown, England, in 1911; and additional finds have since come to light. Considered to have lived in the first interglacial stage, he is obviously not much younger than *Pithecanthropus*, but he has a brain volume estimated at 1,260 cc. The lower jaw was chinless, as in the apes.

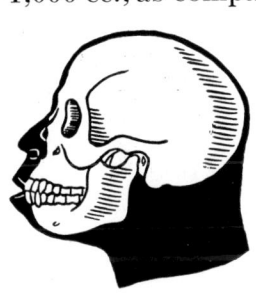
Piltdown man

The most recent important discovery of ancient man is that of *Sinanthropus*, or the "Peking man," whose remains were taken from a cave some 45 mi. from Peiping (Peking), China. Since the original find, in 1928, a great deal of excavation has yielded considerable additional information. The "Peking man" probably was a contemporary of *Eoanthropus*, and had a similar cranial volume. Charred animal bones and charcoal associated with his own bones indicate not only that he used fire but that he was a hunter of some skill.

THE modern genus, *Homo*, is first represented by *Homo heidelbergensis*, or the "Heidelberg man." Known only from a complete lower jaw found in a gravel pit near Heidelberg, Germany, this

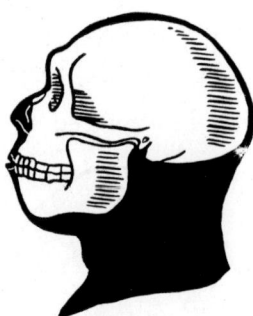
Heidelberg man

man apparently lived in the second interglacial stage. The "Neanderthal man," or *Homo neanderthalensis*, is the best understood of all the extinct species of man. He inhabited caves and rock shelters in many parts of Europe, and the lands bordering the Mediterranean, during the third interglacial stage. Although of short stature and stooped, as may be seen in Plate 60, he had a brain about the size of modern man. It is not surprising therefore to find that these men made good stone implements, used fire, hunted big game, and, judging from their burials, had some crude belief in a life after death. But during Würm glaciation they rapidly and rather unaccountably passed out of the Pleistocene picture and were replaced by the modern species, *Homo sapiens*.

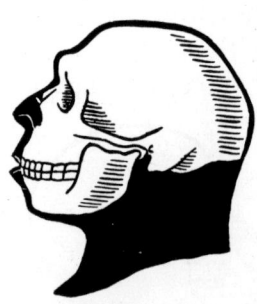
Neanderthal man

The race of modern men that supplanted the Neanderthals is known as the **Crô-Magnon**. These were the last of the paleolithic

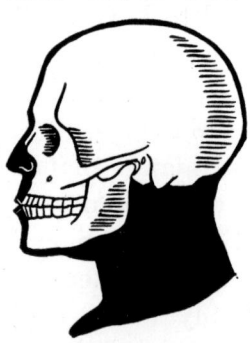
Crô-Magnon man. All skulls after McGregor

peoples of Europe. A tall, handsome type, with cranial volume fully equal to or greater than that of the average man of today, we might all well wish that some of the blood of this magnificent race coursed through our veins. Their art was especially noteworthy, for they decorated their cavern walls in a very realistic manner; and some of their drawings of contemporary animals compare favorably with the modern restorations. In addition, they originated a style of art which some of us, in our simplicity, are pleased to call "modern."

With the retreat of the Würm ice sheet in Europe, the neolithic peoples were introduced into Europe from the east, and they spread

HOMO DILUVII TESTIS

rapidly. Pottery was made, animals were domesticated, and agriculture began. But this is another story, and not ours to tell even if we could. We can, however, summarize in tabular form some of the things we have just said. And for those who are interested we have included the names of the more common cultural stages that are ordinarily recognized.

	Geological Ages		Man and Manlike Primates		Cultural Stages	
Quaternary	Pleistocene	Holocene or Recent	*Homo sapiens*	Modern man	Iron Bronze Neolithic	Paleolithic
		Würm glaciation		Crô-Magnon	Magdalenian Solutrean Aurignacian	
		Riss-Würm interglacial stage		*Homo neanderthalensis*	Mousterian	
		Riss glaciation				
		Mindel-Riss interglacial stage		*Homo heidelbergensis*	Acheulian	
		Mindel glaciation				
		Günz-Mindel interglacial stage		*Sinanthropus* Peking *Eoanthropus*	Chellean	
		Günz glaciation		*Pithecanthropus*	Pre-Chellean	
Tertiary	Pliocene			*Australopithecus*	Eolithic	
	Miocene			*Dryopithecus*		

TABLE SHOWING THE CHRONOLOGICAL DEVELOPMENT OF
ANCIENT MAN IN REFERENCE TO GEOLOGICAL TIME

SO FAR this synoptic account of ancient man has been largely an Old World story. What about the record in North America? It is poor at present, but new finds are being turned up in increasing abundance. There is general agreement that man entered North America by way of the Siberian-Alaskan bridge, but that he came much before the waning stages of the last glaciation has been vigorously denied. Discoveries of man, or evidences of his presence, in

 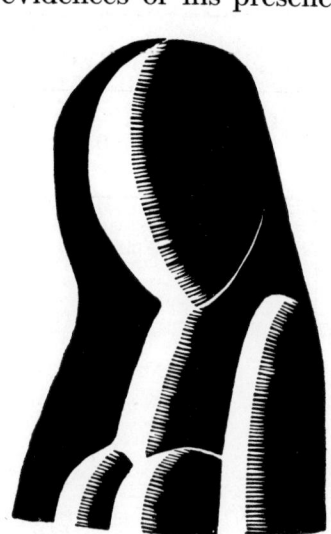

The Crô-Magnon men originated a style called modern

Left: Modern head by Brancusci   *Right:* Head by unknown caveman sculptor

geological settings indicating considerable antiquity have been reported at Mexico City, where bones and culture occur beneath a lava flow; at Folsom, New Mexico, where arrowheads are found with skeletons of extinct bisons; at Frederick, Oklahoma, where flint implements occur with the bones of many extinct mammals; and at Pelican Rapids, Minnesota, where a female skeleton was discovered in 1931 beneath about 10 ft. of laminated glacial-lake clays. Of course, doubt has been thrown on the validity of all these and many other finds, at least so far as their indicating great age for man is concerned. And then during the summer of 1935 artifacts and hearth pits were discovered in the White River districts of Nebraska and South Dakota. These relics of ancient man are pretty difficult to argue out of the picture, since they were found *beneath* undisturbed varved glacial clays made during the last glacial episode. Thus, although man may not be as old a settler in North America

Cave drawing of the woolly mammoth

Modern interpretation of woolly mammoth. After C. R. Knight

as in Europe, he probably has been an inhabitant of the continent for a good many thousand years. Future discoveries will doubtless tell us much concerning his activities here.

REGARDLESS how long man has lived either in Europe or North America, he obviously did not originate on either continent. For a number of reasons, most of which space does not permit us to mention here, many scientists consider the central Asiatic

Cave drawing of woolly rhinoceros

Modern interpretation of woolly rhinoceros. After C. R. Knight

area as man's ancestral home. It is, for instance, centrally located, not only with reference to the most primitive of living races, but also in relationship to the earliest civilizations. The ancient genera of men also are found in the area contiguous to this possible ancient center, and the stock out of which man evolved apparently was native to the area. More important still, so far as our story is concerned, is the fact that this area during the Tertiary was one of those in which uplift took place. With increased elevation, the area to the north of the growing Himalayas was rendered semi-arid, and consequently forests diminished. Two courses were then open to the arboreal primates of the region. They could gradually forsake their tree-living habits, or they could migrate to other forested areas. The latter course was not easy because of the growing mountains which lay between them and the forests. Under some such dire geological necessity as this, it is probable that man's direct ancestors in this region reluctantly gave up their arboreal habitat, largely because it was taken away from them. With hands freed from branches, these appendages became for the first time really useful servants to, as well as constant taskmasters of, the growing brain. Man, like many another lusty son of Earth, thus may well have established his organic success on a foundation of physical adversity.

CHAPTER 50

MONEY AND POLITICS

> There is many a rich stone laid up in the bowels
> of the earth.—BISHOP HALL

EVERYBODY talks about politics, and everybody wants money. Few there are who really know very much about either. But at least it is obvious that all money is based on something which the Earth has provided out of her vast storehouse of treasures. And all politics are founded on the desire of the supposedly most sagacious of the Earth's offspring to control that storehouse. *Ergo*, in the final analysis, money and politics are both geological, as well as sociological, subjects.

Money and ....

.... politics ....

OF COURSE, not only geologists, but chemists, physicists, and biologists as well, similarly like to enlarge their spheres of activity to include in their separate domains this poor planet of ours, lock, stock, and baggage. But do not be alarmed; we are not turning economists or social scientists. We are here interested only in pointing out that **the location, development, and utilization of natural resources have greatly influenced the course of civilization.** We have from time to time referred to these geological products in our discussion of the various periods, but there has not been space to do justice to the subject of economic geology, and even in this chapter we can only hint at its importance.

The soils which result from the decomposition of rocks are the basis of all agriculture. Our chief fuels, coal and oil, were formed through long and complicated geological processes. The very limestones, shales, and clays which we use as building-stones, or as the basis for the manufacture of cement and the ceramic products, are the stuffs out of which the outer crust of the earth is made. The metals are likely to occur in complex geological settings, and thus their discovery commonly requires detailed geological exploration. Most of the countries that possess considerable quantities of these geological products are great; those that lack them commonly are backward. The very balance of power in a complex modern world may hinge on the character and position of late Paleozoic terrestrial deposition! In fact, it may be said that the geological events of some ancient periods are often the ultimate basis for modern wars, generally the background of politics, and almost always the foundation of money.

EVERY mineral is a "one-crop resource." The United States has been singularly fortunate in having had bumper "crops" in most of the minerals. Nevertheless, we depend on foreign sources for all or part of a score of minerals. Furthermore, we know about where we stand in regard to our own mineral resources. We are just about scraping the bottom of what we once thought was an inexhaustible barrel, for no really *major* discovery has been made since 1907. Our gold production passed its peak in 1915; and oil, zinc, and lead are estimated to last only about 15 yr. more, at a maximum, with our present rate of production, and barring further large discoveries. Iron and copper reserves of high grade may be exhausted in some 40 yr., although there is enough low-grade iron ore to last for centuries. Coal, too, will last several thousand years; but the premium grades, in accessible positions, will not hold out a great deal longer.

All these serious situations have called for some national planning, and the National Resources Board recently has been trying to develop a long-term policy. But the difficulties obviously are manifold. The producers of any raw material cry for tariff protection against foreign competition. Since their cries commonly are heard, we develop and use our own reserves, whereas we could, by importing, be saving them. But if we save them, we do not have them developed at times when we need them most. And if we do not exploit our resources, our men are out of work. If we do protect our mineral producers, the cost of the minerals thus produced to our own people is raised. No wonder, then, that political troubles and financial problems go hand in hand with the exploitation of all geological resources.

The importance of exploiting geological resources was realized by Agricola, who wrote *De re metallica* about 1550. In the illustration above he showed how the "divining rod" was to be used in the search for ores

SINCE there is insufficient space to discuss even the outstanding group of these resources properly, let us choose a single one of them, oil, for a slightly more extended account. For many of the problems relating to oil are common to the mineral resources generally, and, in addition, its production involves some troubles peculiarly its own.

Oil and gas are mixtures of natural hydrocarbons which are relatively widely distributed through the post-Cambrian sedimentary rocks. Today we regard these fuels as indispensable. Nations that do not have adequate resources of oil and gas are willing to fight for them. Those that control them plan to keep competing nations from acquiring them. International incidents grow out of decisions to enforce or to abandon oil sanctions. Yet, less than a century ago not one drop of oil was produced in the modern sense of the word, although a little oil was skimmed from seeps for medicinal purposes. The pioneer wells of the late fifties were essentially hand drilled; those of today are put down with the most elaborate machinery. The early wells cost a few hundred dollars; many of those drilled today cost more than a hundred thousand; and some have required more than a quarter of a million dollars for their completion.

"Colonel" Drake's discovery well of 1859 yielded 25 bbl. of oil a day from a depth of less than 70 ft.; some of the modern producers have a potential capacity of more than 200,000 bbl. a day, and a few have produced from rocks more than 2 mi. beneath the surface! The total world-production in the early sixties was only a few thousand barrels a year, but by the end of 1935 it had mounted to the almost incredible total of twenty-seven billion barrels.

In these days in which ordinary conversations commonly involve astronomic figures, a billion is not the awe-inspiring figure that it once was. Perhaps, therefore, we did not make you gasp when we mentioned this great quantity of petroleum. But, of course, it represents a veritable Niagara of this precious single crop of the earth. And we are now producing "the black golconda" at the rate of a billion and a half barrels every year! The United States alone accounts for approximately 60 per cent of the total. This means that the United States has been able to produce the billion barrels of oil a year that it requires for domestic consumption. In chapter 1 we cited this great flood of petroleum and said that even this great amount could be increased. It could, indeed, for the moment, because most of our fields are "prorated." That is, they are, by law, limited to a certain percentage of what they might produce if all wells were run at full capacity or if more wells were drilled. But if the wells *were* run at full capacity, their *ultimate* total production would be markedly reduced.

INASMUCH as the rate of discovery of *important* new shallow fields is steadily declining, not only scientists but politicians have been wondering how long it will be before there is a shortage of oil in the United States. The proved oil reserves are generally given today as about twelve billion barrels. If no new fields are discovered, then we will soon be facing an oil famine. In fact, many experienced students of the subject are freely predicting a serious shortage of oil by 1945 at the latest. In this connection, however, it is interesting to note that the first estimates of this sort were made in 1908, at which time our reserves were calculated as eight billion barrels. Since the time the estimate was made, fortunately for our industrial

PLATE 49
EARLY MESOZOIC PHYSICAL FEATURES

Above: Petrified forest of Arizona (photo courtesy National Park Service). Silicified logs are shown weathering out of the reddish Triassic Chinle rocks, which are non-marine stream deposits. Many of these logs are of great length, and indicate that some of these ancient conifers were comparable in size to our modern redwoods.

Above: Many thick Mesozoic terrestial sandstones have been eroded into overhanging cliffs which made perfect sites for cliff-dwellers (National Park Service). *Below:* Great White Throne, Zion, formed of early Mesozoic terrestrial sandstone (Union Pacific Railroad). These thick sandstones are commonly eroded into sheer cliffs, which, like the Great White Throne, have never been scaled by man.

Above: Sierra Nevadas, as seen from the air (photo courtesy United Air Lines). This range originated near the close of the Jurassic period, and at about the time the granite, shown below, was intruded. Because mountain-building and igneous activity at the close of the Jurassic was centered in this area, this period of orogeny is often called the "Nevadian disturbance".

Yosemite Valley, a scenic wonder sculptured out of a Jurassic granite batholith (National Park Service). Note the hanging valley and the U-shaped glaciated main valley.

PLATE 50 END OF AN ERA

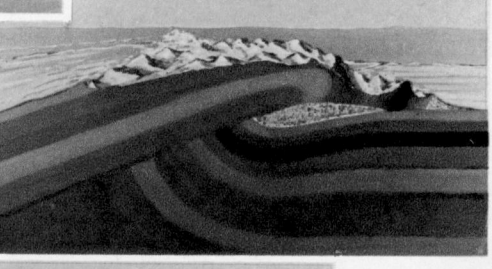

Mountains are built at the close of every geological era. At the end of the Mesozoic, the Laramide revolution formed the Rockies. *Above:* Mount Laramie, Wyoming, the type area of the Laramie formation, seen from a distance of 60 mi. (photo courtesy Union Pacific Railroad). At the right are a series of drawings from the moving picture "Mountain-building." These show the probable sequence of events in the formation of the Lewis overthrust (see text, pp. 396–97). The thrusting at the close of the Mesozoic was toward the east. Movements continued well into the Cenozoic era. The Rockies are not the only great mountain range formed at or near the end of the Mesozoic. Many other systems originated at this time, mountain-building being especially notable along the margins of the Pacific Ocean. The usual igneous intrusions also accompanied the later stages in the orogeny, and many other faults, similar to the Lewis overthrust, had their inception.

Below: Rocky Mountain front as seen from the air (photo courtesy United Air Lines). The eastern edge of the Rockies at places is marked by great faults (see also map on p. 396), but these structures, in some cases originated after the end of the Mesozoic era. With the birth of the Rockies the swamps lying to the east of them were drained, and the region doubtless became semi-arid. The western terrestrial Tertiary sediments were products of the erosion of the Rockies. Thus, always the rocks of one era are formed of débris from the older ones.

PLATE 51
MESOZOIC REPTILES OF AIR AND SEA

Pteranodon, a Cretaceous pterosaur, largest of all flying animals. From a restoration in the U.S. National Museum at Washington, D.C.

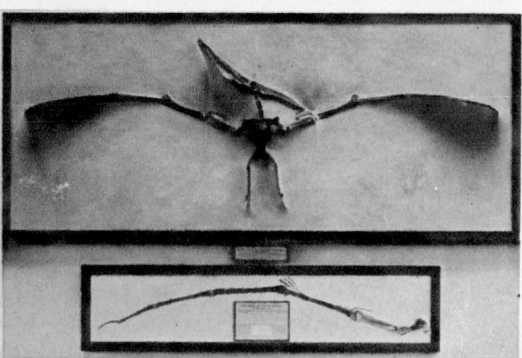

Above: A skeletal diagnosis of *Pteranodon;* and the bones of the arm and elongated digit of a large specimen (photo courtesy U.S. National Museum). Despite a wing spread of nearly 30 ft., *Pteranodon* skeletons weigh only a few pounds, since their bones were hollow as in the birds. Although there were a number of other relatively large flying reptiles, many of these creatures were as small as a pigeon. In rare instances the wing membranes are preserved.

Above: Excavating a plesiosaur skeleton from the Cretaceous chalk of Kansas. Note the large paddle, and the pile of stomach stones, or gastroliths. These apparently were used to help the creature digest its hurriedly gulped food, much in the manner the gizzard stones are employed by the common fowl.

Crocodile-like reptiles, such as the one at the left, were common but undistinguished Mesozoic reptiles; but since the close of that era they have remained relatively abundant, long after all the "ruling reptiles" became extinct.

Above: Skeleton of an ichthyosaur with the outline of the body preserved as a carbonized film. *Below:* C. R. Knight's restoration of plesiosaurs at left and ichthyosaurs at right. (Photos copyright by Field Museum of Natural History.) Both of these marine reptiles were well modified to fit their environment, the icthyosaur so much so that it gave birth to its young alive since it could no longer return to land to lay eggs.

PLATE 52
MESOZOIC MODERNS

During the Mesozoic era many animals and plants began to take on a modern appearance. *Right: Uintacrinus,* a Cretaceous free-swimming crinoid which lived in profusion in the area which is now Kansas (photo courtesy Field Museum of Natural History).

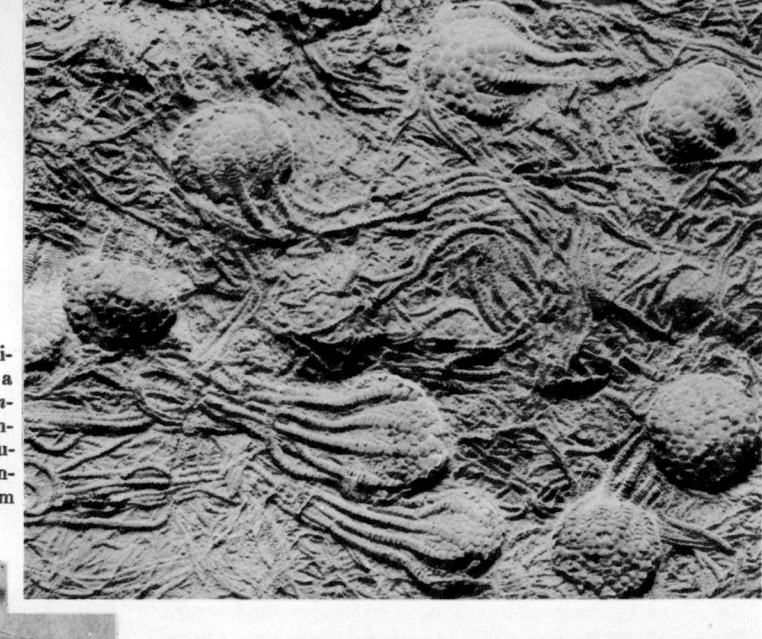

Left: A Jurassic (Solnhofen) lobster-like crustacean compared with a modern representative. *Lower left:* Restoration of a late Jurassic landscape, a mural by C. R. Knight in Field Museum of Natural History. Although pterosaurs are common, and small dinosaurs are shown in the left corner, the first birds (*middle foreground*) give a modern appearance to the scene. The cycads, too, are semi-modern in aspect. *Right, above:* A Jurassic horseshoe crab compared with a modern specimen (*below*). Although invertebrates, plants, and some vertebrates were progressing rapidly, the dominant reptiles, the dinosaurs, were still the archaic masters of the scene, in spite of the fact that the mammals have already appeared to challenge them. These early mammals were insignificant creatures no larger than a rat, and probably were not even noticed by the dinosaurs; but they, too, added a modern note to the Mesozoic scene.

PLATE 53
MEGALO-MANIACS

The most characteristic animal of the Mesozoic organic world was the dinosaur. At the left two bipedal carnivores are shown in mortal combat (courtesy Sinclair Oil Company). All carnivorous dinosaurs were bipedal; most, but by no means all, herbivorous dinosaurs were quadrupedal in habit.

The sauropod dinosaurs, like *Diplodocus* (*right*), were typical megalomaniacs. Largest creatures ever to walk the earth, they were too heavy for their limbs (restoration copyright Field Museum of Natural History). *Right:* Skeleton of *Diplodocus* in U.S. National Museum, a moderately large specimen about 65 ft. long. The largest mounted specimen is in the Carnegie Museum of Pittsburgh. This individual was about 88 ft. long and probably weighed nearly as much as a freight-car load of coal. But in spite of its great bulk, it had an extraordinarily small cranial capacity and probably was a creature largely motivated by reflex action.

Right: A real engineering problem is raised when a sauropod dinosaur skeleton is mounted. Erection of scaffolding for the job at the Field Museum of Natural History. Sauropods are actually relatively common as fossils, but great expense is involved in quarrying and preparing the bones for exhibition.

Above: Composite Jurassic landscape reconstructed under direction of Professor W. A. Parks (courtesy Royal Ontario Museum). Sauropod dinosaurs are seen in the rear, two bipedal carnivores at the left rear, and the armored *Stegosaurus* in left foreground. *Right:* Skeleton of *Stegosaurus* (U.S. National Museum). *Stegosaurus* is a typical representative of a group of quadrupedal herbivores characterized by a double row of great bony plates down the back. These animals were almost world wide in distribution during the Jurassic period.

PLATE 54
CRETACEOUS VERTEBRATES

Above: Battle between *Triceratops* at left, and *Tyrannosaurus*. From a painting by C. R. Knight (copyright Field Museum of Natural History). *Right:* Skeleton of *Triceratops* (courtesy U.S. National Museum), one of the most common of a large group of armor-headed Cretaceous herbivorous dinosaurs.

Left: Composite Cretaceous landscape prepared under direction of Professor W. A. Parks (courtesy Royal Ontario Museum). Carnivorous dinosaurs at left; horned and bipedal herbivores in middle background; flightless, toothed birds in middle foreground. Note modern appearance of the flora. *Below:* Skeleton of *Portheus*, giant bony fish of Cretaceous seas, which swarmed with the ancestors of the modern bony fish. These were the prey not only of the marine reptiles, such as the mosasaurs, plesiosaurs, and ichthyosaurs, but of *Hesperornis*, the toothed flightless bird shown in the restoration at the left.

Below: C. R. Knight's restoration of a Cretaceous seascape with *Pteranodon* in the air; mosasaur, and *Archelon*, a giant turtle, in the sea (copyright, Field Museum of Natural History). A skeleton of a small mosasaur is shown at the right. Many of these ancient sea serpents were 40 ft. long.

PLATE 55
CEPHALOPODS

Cephalopods constitute one of the most interesting and useful of the groups of invertebrate fossils. Some idea of their diversity may be gained from the photo on the left, which includes modern, as well as fossil, types. There are two main groups of cephalopods—the dibranchiate, or two-gilled, types; and the tetrabranchiates, with four gills. The dibranchiate type is geologically the younger.

Dibranchiate cephalopods, such as the squid in the jar above, have internal skeletons or none at all; tetrabranchiate cephalopods have external shells which are chambered. Mostly Paleozoic and Mesozoic in age, there is a modern survivor of the group, the pearly nautilus, whose shell is just in front of the jar. The diagonal sequence at the right shows increase in complexity of coiling from the Ordovician simple cone at the bottom to the Cretaceous gastropod-like form at the top. As might be expected, departures from the normal straight, curved, or coiled shells are uncommon except near the time of extinction of the group.

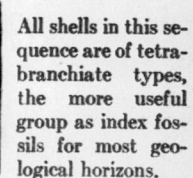

All shells in this sequence are of tetrabranchiate types, the more useful group as index fossils for most geological horizons.

Above: Many ammonoid cephalopods had suture lines more complex than those seen in this specimen. Since the sutures become more complicated throughout the history of the group, this feature is sometimes used as a crude index of geological time. The sutures are the lines of junction between the septa, or partitions separating the chambers, and the body wall. Since the latter must be removed before the sutures are exposed, most specimens in which the sutures can be seen are casts rather than actual remains.

The cuttle-fish bone at the lower right and the spiral shell adjacent are the last remnants of internal skeletons in the modern dibranchiate cephalopod group. The octopus, for instance is essentially without hard parts, though it has been fossilized in rare instances.

Above, right: Many Mesozoic ammonoids, in addition to possessing complicated suture lines, had heavy surface ornaments and some apertural modification. In some cases the body chamber was protected by a lid which closed over the aperture when the soft parts were withdrawn into the shell.

PLATE 56
CENOZOIC PHYSICAL FEATURES

Bryce Canyon as seen from Bryce Point. The extraordinary erosional features have been sculptured out of delicately tinted Eocene rocks (photo courtesy National Park Service). Many of the most unusual of our western scenic areas resulted from the erosion of Tertiary terrestrial sediments or from Tertiary volcanism.

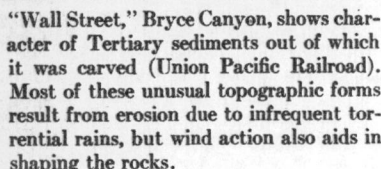

"Wall Street," Bryce Canyon, shows character of Tertiary sediments out of which it was carved (Union Pacific Railroad). Most of these unusual topographic forms result from erosion due to infrequent torrential rains, but wind action also aids in shaping the rocks.

The awe-inspiring lake-filled caldera of a relatively recent volcano, Crater Lake, Oregon, with Wizard Island, a small, younger cinder cone, rising out of its indigo waters (National Park Service).

Lower, right: Volcanic breccia, Crater Lake district (photo courtesy National Park Service). *Below:* Painted Butte, a Bad Land Tertiary erosional remnant near Medora, North Dakota (photo courtesy Northern Pacific Railroad).

PLATE 57
CENOZOIC VOLCANISM

Left: Mount Rainier from eastern side (photo copyright A. Curtis; courtesy National Park Service). *Below:* Mount Hood (photo courtesy Northern Pacific Railroad). *Lower right:* Popocatepetl, Mexico, seen from the air (photo courtesy Pan-American Airways).

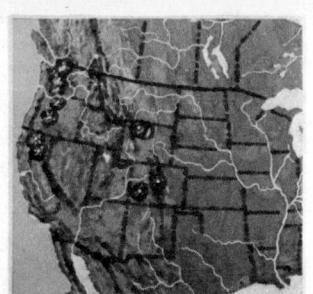

Rainier, Hood, Popocatepetl, and a number of other little-eroded volcanic peaks attest to the recency of volcanic action in western North America. *Left above:* Location of a few larger dormant cones. *Left:* Areas of Cenozoic lava flows in Columbia River district. (Photos from moving picture "Volcanoes in Action.") *Lower left:* Palouse Falls, Washington, plunges over layers of this lava (photo by McFaden, Walla Walla). At many places a dozen or more of the lava flows can be seen, one on top of the other. Several hundred thousand square miles in the Columbia and Snake River valleys were buried under these Tertiary lavas.

PLATE 58
THE WARM-BLOODED

Above: Bones of Tertiary mammals as they usually appear in rock. *Below:* Three oreodonts mounted in realistic positions. Specimens in Walker Museum, University of Chicago; collected and mounted by Mr. Paul C. Miller.

The Cenozoic era is the time of dominance of the warm-blooded mammals. By early Tertiary reptiles were reduced to a few forms like turtles and crocodiles (*above, left*), but mammals showed great diversification. Large archaic beasts such as the titanotheres (*above, right*) were abundant. Note the modern plants. Restoration of Cenozoic landscape under direction of Professor W. A. Parks (courtesy Royal Ontario Museum). *Below:* The Field Museum horse evolution series, showing changes in size, in skull, and in limb structure from the Eocene to the present. The American Eocene horse, *Eohippus*, was about the size of a fox, and had four toes on the front feet, three on the rear. Throughout the 60,000,000 yr. in their development, horses have not only grown progressively larger, but they have reduced their digits until their entire weight is carried on the middle "finger." Horses became extinct in North America during the ice age but were reintroduced to their native land by the Europeans.

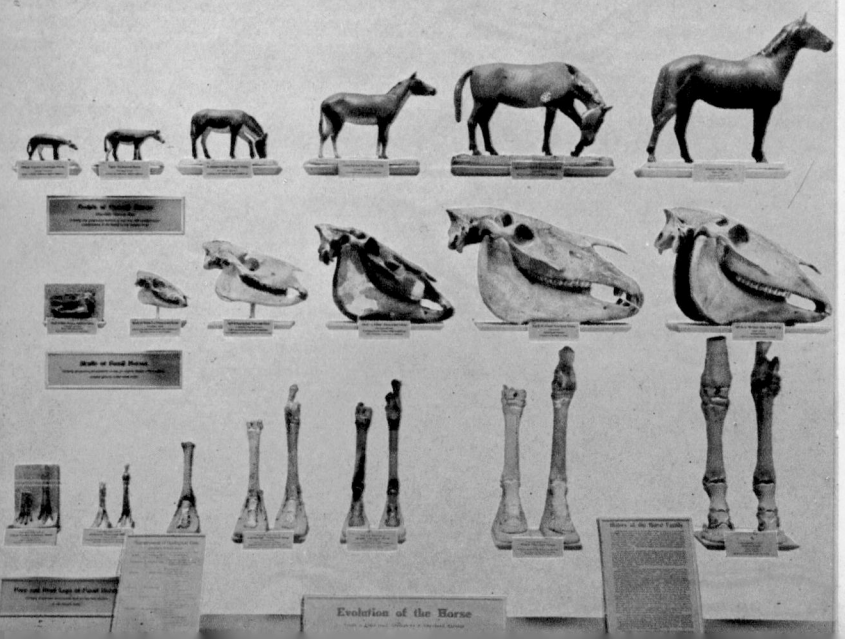

Above: Skull of a mastodon being excavated near Johnstown, Ohio. The tusks are broken, but teeth are intact. Many specimens of this type are discovered when swamps in the north central states are drained.

PLATE 59
THE PLEISTOCENE

Glacial striae on rock at the rim of Crater Lake, Oregon (photo courtesy National Park Service) demonstrate that glaciers scoured the sides of the much higher ancestral volcano whose destruction resulted in the present lake basin.

The relatively recent Pleistocene ice age has left evidences of its reality all about us, but today we must go to high mountains, or high latitudes, to simulate even approximately glacial conditions. *Above:* White Mantle Range, B.C. (Topographical Survey of Canada), a typical glaciated range in the Canadian Rockies. Glaciers of this type commonly are called "valley" or "alpine," because they flow in river-like valleys and because they are common in the Alps.

Above: An Illinois bedded glacial gravel deposit. *Below:* A well-developed soil profile showing depth to which leaching has taken place. (Photos courtesy Illinois Geological Survey.) To a certain extent it is possible to determine the relative amount of time involved in the leaching by measuring the thickness of the leached zone.

Yosemite Valley, a great glacial gouge showing characteristic U-shaped section and a hanging valley (photo courtesy National Park Service). Compare with the photograph of the same valley shown on Plate 49.

Left: Farm Creek glacial till section near Peoria, Illinois, showing deposits of several glacial stages (photo Illinois Geological Survey). *Right:* Varved, or annually layered, glacial lake clays typical in southern Canada and New England (photo by F. J. Pettijohn). Each light layer represents summer deposition, the dark layers having been put down during the winters. A light and dark layer taken together represents a year of sedimentation and are called a "varve." By counting varves a precise measurement of recent geological time is made possible.

Man is just another one of the many Quaternary mammals that survived the rigors of the last ice age. At the left is seen a typical Neanderthal family going about its daily tasks at the entrance of its cavern home (from a restoration in the Field Museum of Natural History). The Neanderthals were only one of several types of pre-historic man; but they are perhaps the best known, since a large number of complete skeletons have been discovered. Common in France, their remains also have been found, among other places, in northern Africa, and in Palestine. The Neanderthals belonged to the same species as modern man, but probably to a different species.

Above: Idealized chart to show the evolution of "the face from fish to man," and from Devonian to Recent (models by W. K. Gregory). Although the head of modern man is set apart, it belongs in the same line of ascent with the others, an ascent involving nearly a half-billion years.

PLATE 60
QUATERNARY MAMMALS

Above: Chart showing man's relationship to the other primates. Well differentiated by the Quaternary, the early Tertiary primates were all pretty much alike, and apparently similar to modern lemurs.

Below: Pleistocene mammoth and woolly rhinoceros only recently extinct in Eurasia (a mural by C. R. Knight, in Field Museum of Natural History).

Above: Restoration of Irish elk which became extinct seven centuries ago through man's activities (from a painting by C. R. Knight; copyright by Field Museum of Natural History).

Man in his search for oil has drilled wells out in the ocean (*right*) and has crowded derricks together until they look like a forest. *Above:* Torrence field, Los Angeles, California, representing an extremely wasteful type of development known as "town-lot" drilling.

PLATE 61
THE BLACK GOLCONDA

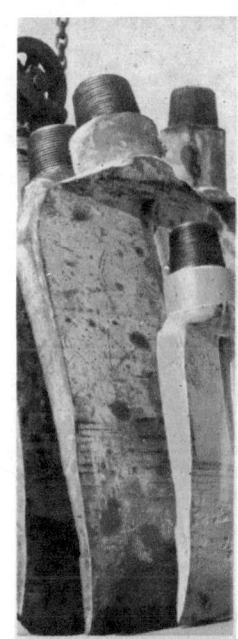

Above: Large fishtail drills.

Below: Frosted pipes in Calgary, Alberta oil field (photo by W. J. Oliver). *Lower right:* An oil-tank fire (courtesy of Illinois Geological Survey). Spectacular scenes such as these only add interest to the intrinsically romantic business of seeking, finding, and producing oil.

Above: Large Sullivan drill, contrasted with typical wildcat outfit, and crew (*below*). *Right:* Gusher in Crane Field, Texas (copyright by Jack Nolan). (All pictures except lower right courtesy Museum of Science and Industry.) In the early days of oil production most new fields were discovered by "wildcatting." Today most of the discoveries result from detailed geological exploration.

PLATE 62
THE SEARCH FOR OIL

Below: Model of a simple oil-field structure arranged to show the character of the folding at depth. (Models shown on this plate from geological exhibit of the American Petroleum Industries at A Century of Progress Exposition, Chicago).

Oil is where you find it, but it generally occurs in one of the structural conditions shown in the above model. Oil is always on top, water below; hence exhausted wells may "go to water."

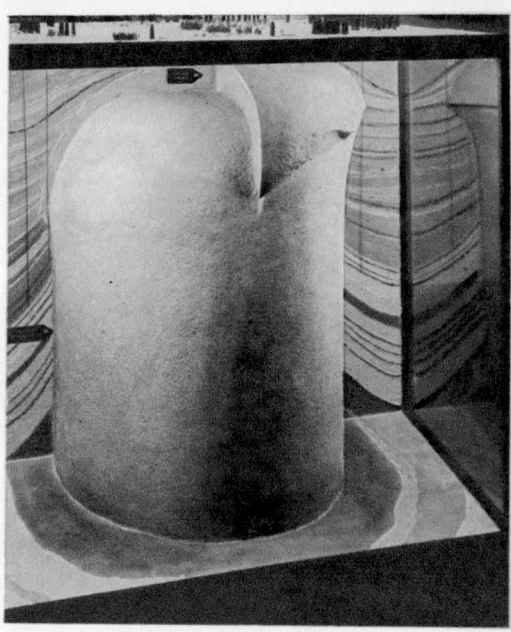

Below: An operating model to show the seismic, or artificial-earthquake, method of locating salt domes

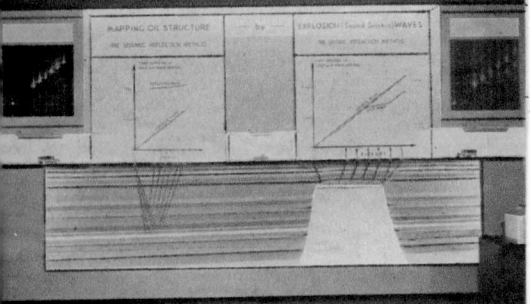

Above: Model salt dome showing how oil accumulates in upturned strata through which the salt was intruded. Such salt domes are located by seismological surveys as demonstrated by the model at the left. *Right:* Small slide with 100 species of radiolaria. *Below:* The same, highly magnified. These shells are used less commonly than the foraminifera in micropaleontology, since they are not common except in the relatively recent sediments, and are often with difficulty separated from the enclosing rock.

Left: Small slide with 100 species of foraminifera. ×½. *Below:* Same slide, ×2½. These tiny shells are commonly employed in subsurface correlation (slide prepared by Helen J. Plummer of the Texas Bureau of Geology).

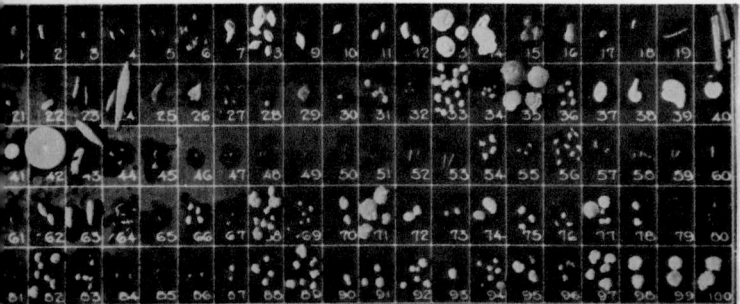

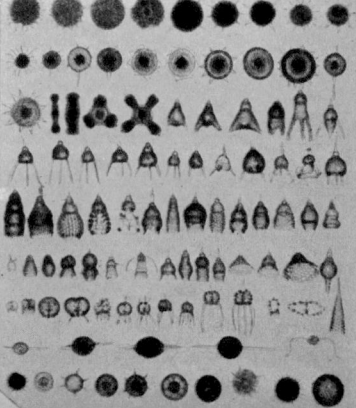

PLATE 63
GEOLOGICAL LOCATIONS

Cities, harbors, railroads, dam sites, all have had their locations determined by geological processes operating, in most cases, many million years ago. Enlarging on this subject, it might be said that every activity of man is geologically predetermined on the basis of happenings so ancient that in some cases they even antedate the appearance of life.

Above: The magnificent harbor of New York resulted from the drowning of the Hudson River valley (photo courtesy United Air Lines). *Right:* Black Canyon, with engineers' mark showing position of diversion tunnel for Boulder Dam (photo courtesy Union Pacific Railroad). Most harbors and many canyons are of recent origin, and they are changing rapidly today. In fact many harbors would not be usable without constant dredging, and sediments deposited in bodies of water ponded by dams eventually silt up the reservoirs completely.

Below: Boulder Dam, and the canyon across which it was constructed, as seen from the air (courtesy United Air Lines).

Dams such as the Boulder structure are examples of how man is now controlling the physical world. By thus harnessing nature he taps sources of tremenduous power and is able to make areas formerly desert bloom with luxuriant plant growths. It is not generally recognized that dams were invented by Nature long before man started to construct such structures. Some of the most interesting of the natural dams are those made by lava streams or by dikes, but glacial deposits or glacial ice itself are frequently responsible for ponding great bodies of water.

Right: Today a thorough geological investigation determines the suitability of every prospective dam site before a cubic yard of concrete for such structures is poured. Fault hazards expecially are evaluated. (Courtesy Museum of Science and Industry.)

PLATE 64
DE RE METALLICA

Above: Agricola, about 1550, wrote his famous treatise on metals, called *De re metallica*, which was translated into English by former President and Mrs. Herbert Hoover.

Man's preoccupation with geological resources is typified by ore docks such as those above, and by great batteries of furnaces like those shown at the right.

(Photos by Carroll Phelps; courtesy Museum of Science and Industry.) From Agricola to modern steel mill is a far cry, but the interest he fostered in metals grows greater yearly.

Above: "Giant Heat," another prize picture of the iron and steel industry by John P. Mudd (courtesy Museum of Science and Industry). *Left:* Man goes to the ends of the earth for gold. Alaska-Yukon gold mine, Juneau, Alaska (photo by courtesy the Canadian National Railways).

development, about twice that amount of oil actually has been produced in the United States. How did this great discrepancy arise?

In the first place, drilling has been carried to depths undreamed of a score of years ago. Production and refining methods have greatly increased in efficiency, and even abandoned fields have been made to yield again under the magic of modern engineering. But more important still, new methods of exploration have come into play. Of course, many a disillusioned prospector will tell you that "oil and gas, like gold, are where you find them." So they are—but through the intelligence of man in general and of engineers and geologists in particular, and, we must confess, through sad experience as well, we have learned that they are almost invariably found in *certain* places. Where are these certain places? A study of a map of the world's oil resources reveals the fact that the great oil-producing areas, although widely scattered, are limited to rather definite zones, and that very large sections of the face of the earth are entirely without oil reserves. Let us now examine a map of the world designed to show (1) the plateaus of ancient rocks, (2) the lowlands of ancient rocks, (3) the great folded mountain ranges, and (4) the lowlands of younger rocks. If we keep in mind the distribution of the world's great oil fields, we are struck by the fact that none of them is located either on

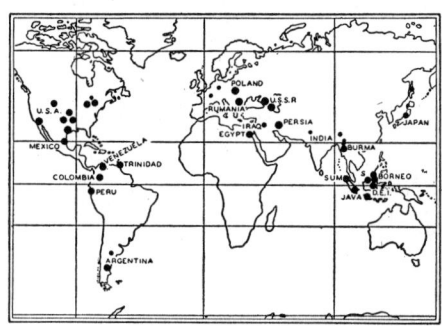

Map showing the principal oil fields of the world. After L. D. Stamp

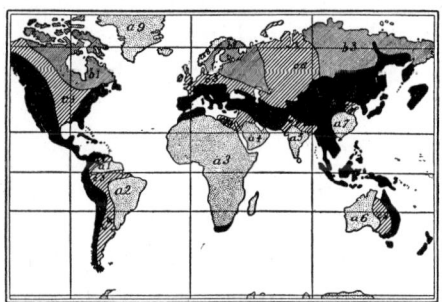

Map of world showing main morphological units. All areas labeled with prefix (*a*) are plateaus chiefly of ancient metamorphic rocks, those marked (*b*) are lowlands of ancient rocks. These areas are not oil-bearing. Lowlands of younger rock which *may* contain oil are labeled (*c*). The mountain regions are shown in black. The oil fields are concentrated on the flanks of these areas. After L. D. Stamp

the plateaus or the lowlands of ancient rocks; and equally apparent becomes the still more significant fact that oil and gas are found chiefly along the great mountain chains or in the lowlands of relatively recent rocks which border those chains. Such a world-wide alinement can hardly be coincidental. **Thus there seems to be a fundamental relationship between mountain-building and the distribution of oil and gas.** But in order to appreciate fully the real significance of this relationship, we must inquire somewhat further, if briefly, into the origin, migration, and accumulation of oil and gas.

THE question of the mode of origin has engaged the attention of scientists for more than a century, but it has not yet been surely answered. The theory that they have originated from the chemical combination of natural inorganic substances has been advocated mainly by chemists, and is based chiefly upon laboratory experiments. It is possible to form hydrocarbons through inorganic agencies, and the theory cannot summarily be dismissed; but geologists, for a number of reasons which cannot be entered into here, find the large quantities of petroleum and natural gas in sedimentary rocks incompatible with such a mode of origin.

Nearly all geologists are now convinced that **oil and gas are organic derivatives.** They believe that **natural hydrocarbons have been formed by the decomposition of organic material deposited with the sediments.** As to whether the organic material consisted of plant or of animal remains, or of both, there are still several different opinions. It is clear, however, that under certain conditions the natural decomposition of the remains of either or both animals and plants may supply the hydrocarbons found in oil and gas. It therefore seems reasonable to suppose that some oils are solely of animal origin, that others were derived from plant remains alone, and that many have originated from a combination of the two.

ALTHOUGH we now know that both oil and gas may migrate relatively great distances from their place of origin, nevertheless if they have formed in the manner geologists believe, then **the**

position of oil and gas pools today must in some measure be influenced by the sites of deposition of the animal and plant remains from which the oil and gas were derived. Marine life is most prolific along the strand and in shallow waters, that is, down to depths to which sunlight penetrates. It also happens that organisms living in such a habitat are at their death subject to burial with the sediments which are constantly being washed into the sea in such positions. In other words, the sites of abundant marine life and conspicuous marine deposition are in most cases the same. Hence, along ancient shores the muds that were washed out to sea buried either organic material, later to be converted to petroleum, or globules of oil which were already formed as a result of a special type of putrefaction carried on under the influence of marine waters. In either case—and here we are passing roughshod over many of the most difficult of the petroleum geologist's questions—the shales formed from the compacting of these muds are regarded as the most important "source beds" for oil and gas. Thus it becomes apparent that the geologist is supremely interested in ancient shore lines, for their position in large parts determines the location of the original source rocks.

BUT what, you may well ask, do ancient strands and sites of deposition have to do with the present location of mountain chains and the position of oil and gas pools? If we stop and consider, we find that we know the answer. We have already learned that great folded mountain ranges grow out of the very areas in which sediments rich in organic matter were formerly being laid down. And **all of the long and complicated steps by which these basins of deposition are formed, and by which mountains finally grow out of them, have also played their part in the migration and accumulation of oil and gas.** In the early stages of the formation of these products they are both widely disseminated through the sediments containing the animal and plant material from which they were derived. They are then gradually gathered together through the action of many agencies, among which may be mentioned capillary attraction, displacement, gravitation, gas pressure, differences in specific gravity, and the general circulation of water.

Capillarity, which moves a liquid in all directions through small openings in a solid, moves petroleum through minute pores in rocks much as kerosene rises in a lamp wick. Displacement, an effective cause of the migration of oil and gas, may result from the compacting of the beds in which those materials occur, owing to the weight of the overlying rocks. The muds in which oil and gas are assumed to originate are much more compacted than are the associated sands into which the hydrocarbons are driven. The progressive cementation of the pore spaces in these sandstones in turn may cause further migration of the oil and gas, and thus constitute another type of displacement.

Gravitation is rarely an important agent in the migration of oil and gas, and it is effective only where the sands are both porous and free from water. Gas pressure, however, has been effective in the movement of hydrocarbons. The expansion of the gas equally in all directions as it is formed not only pushes it through the openings in the rock but tends also to move any liquids present ahead of it. The differences in the specific gravities of oil, gas, and water is a further cause of migration of hydrocarbons, for, as the oil and gas are lighter than water, they are carried ahead of the water in its movement. The general circulation of water has also contributed to the migration of oil and gas, particularly in the early stages of their movement.

THE place of final accumulation of oil and gas after their migration depends largely upon the structure of the beds containing them. This structure, in turn, is in most cases the result of the mountain-building we have earlier described; in all cases it is the result of movements of one sort or another in the earth's crust. Oil and gas move along tilted and porous beds. In the comparatively rare dry porous beds, the oil moves downward and is concentrated in the synclines. If the beds are saturated with water, as they generally are, the oil and gas are driven upward by the water. This upward migration may continue to the surface, where the oil may emerge as "seeps," or it may be stopped in one or more of several ways. If the reservoir bed is lenticular and the lens is overlain by impervious strata, the oil and gas cannot move upward; if the reservoir is sealed by a fault, or by an intrusive mass, or by an asphaltic residue at the surface, these also prevent the escape of the oil and gas. But the most effective, and by far the most common, obstacle to the continued upward migration of oil and gas is a marked reversal or diminution of the dip of the reservoir bed. In such cases the hydrocarbons rise to the highest part of the anticline, but further movement is prevented by an impervious layer which overlies the reservoir.

The most productive of these structures are particularly common along the flanks of the great folded ranges where the movement has not been sufficiently severe to complexly fault the reservoir. In these anticlinal structures of large size considerable gas pressure is developed, and upon drilling such a structure the oil may be driven upward by the expanding gas to form a flowing well, or "gusher."

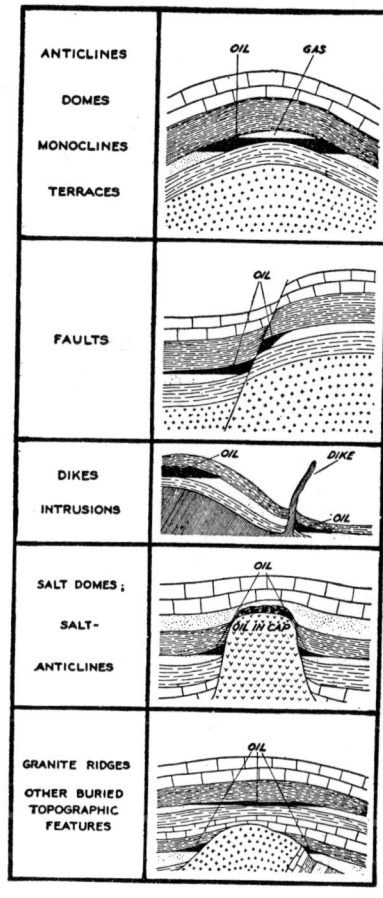

Some of the final places of accumulation of oil and gas. After C. A. Heiland. See also Plates 61 and 62

TODAY the geologist is interested not merely in anticlines but in any of a score of other structural conditions in which oil is now known to occur. Moreover, he now locates these structures by geophysical methods. The explanation of most of these procedures is beyond the scope of this volume. We may point out, however, that salt domes, for instance, are now discovered by seismic methods, as demonstrated on Plate 62. In this procedure an artificial earthquake is set up by discharging an appropriately placed explosive. The induced earth tremors can then be picked up on field seismographs. Either the reflection or refraction method makes it possible to determine the nature of the rock through which the waves passed or from which they were reflected. Since the salt mass reacts to the waves very differently than the enclosing sediments, these domes have been rather easily located in this fashion. One company, in a single year, located nine of the domes by this method; and each proved to be productive when drilled for oil and gas. When one stops to consider that all of these masses were far beneath the surface, and that they were located in the Louisiana lowlands, so that much of the seismic exploration was done from boats, then one sees that Agricola himself would surely approve of this modern type of water-, or better, oil-witching.

WE HAVE already proceeded far enough to see clearly that the oil industry, like many another, is inextricably linked with the science of geology. Not many of the millions of persons

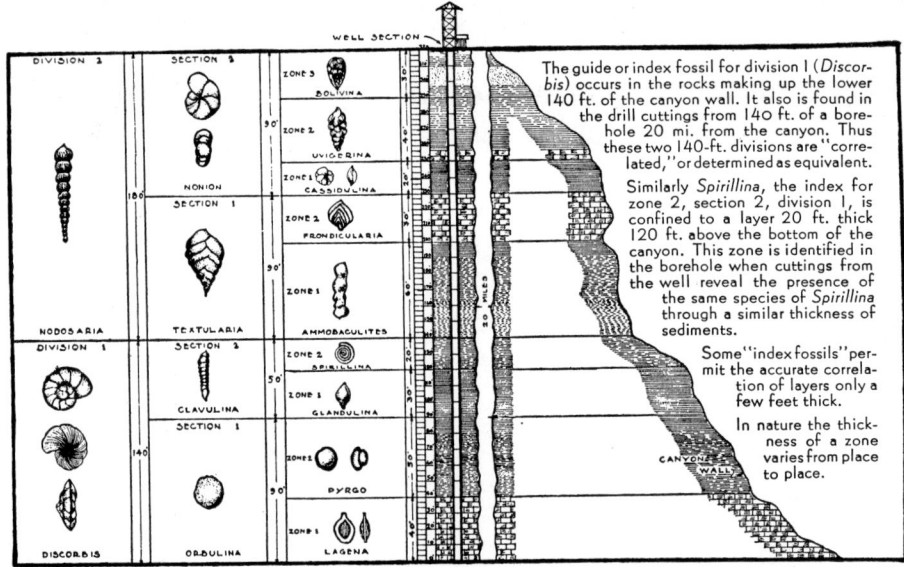

Generalized diagram to show how microscopic fossil foraminifera are used in oil field correlation. In actual practice the situation is far more complicated than shown here. In part modified after Cushman

who drive up to an oil and gas station and yell, "Fill 'er up!" have even a ghost of a notion of the importance of ancient geosynclines in the origin of the fuel they are burning. But you may be sure that the geological staff of each and every producing company is keenly aware of their significance. They also place great reliance in the information they can acquire from the microscopic fossils which come up undamaged in the bailings from the drilling wells. For after the new field has been located by geophysical methods, its plan of development is determined by "subsurface" studies. These involve micromineralogical studies of the sediments encountered by the drill, and micropaleontological investigations of the small fossils. Some of these are shown in Plate 62. Their use is predicated on the principles of correlation which we have commonly referred to in earlier chapters.

WE HAVE not had space to speak of one of the most troublesome of the features of oil and gas, namely, their mobility. For these fuels belong to those who can take them. The wells on our land will drain the oil out from your adjacent acres if you don't get your wells down to compete with ours. The troubles of overdrilling are tied up with this geological feature of oil migration through pervious rocks, and the modern trend toward "unit" operation of oil fields is a sensible move toward solving these difficulties. We have demonstrated that oil, from its origin to its production, presents a series of complicated geological problems. We need no additional words to tell you that it has, like other natural resources, such as dam sites, city locations, and harbor facilities, cooked up a seething financial and political brew as well.

In the final analysis, money and politics are geological, as well as sociological, problems.

CHAPTER 51

PROSPECTS OF AN END

> Last scene of all,
> That ends this strange eventful history.
> —SHAKESPEARE

WE HAVE now taken quite a ride on the Geological Express. Unfortunately it has been one of those modern trains which go too fast to enable one to see much of the landscape. Nevertheless, we stand on the observation platform and look back toward the illimitable horizon out of which we appear to have come. We can see but little except the great past stretching behind us. We become bewildered and are inclined to agree with Hutton that one can make out few real vestiges of our journey's beginning. Hopefully, we peer ahead. Here, also, the tracks seem to extend to the broad horizon, and we see no prospect of our journey's end. If we are honest, we confess that we are almost in the predicament of the absent-minded professor who, having lost his railroad ticket, didn't even know where he was going.

.... quite a ride ....

There are many other passengers riding along with us, and we know that most of them boarded our train long before we did. They do not speak our language; but by observing their behavior we judge that they neither know nor are concerned about when or where they got on, and that they do not know or care exactly where or when they are going to get off. We long ago came to the inescapable conclusion that each and every one of these fellow-sojourners might throw some light on the question as to our own destination. Accordingly, we have turned train detectives and gathered together all of the information about them that we could; and we have added it to what we have been able to learn about the ever changing landscape through which our train has been traveling. A synopsis of the results of our geological sleuthing has been presented in the preceding chapters. But, now that we have the information at hand, what can we do with it?

THE trouble is that, although we have learned many facts on our long journey, we do not yet know definitely how to interpret them. We have found that the physical world is changing as these very words are written. No hill, no stream, no continent, no ocean is ever precisely the same from one minute to the next. Countless living things die, and numberless new organisms are born, even as you read this sentence. But, although the geological record demonstrates that this has been the case for hundreds of millions of years, it shows just as conclusively that things were not ever thus. On the other hand, we have become aware of the fact that great geological revolutions have occurred as far back in our geological journey as we can read its records; and we have reason for believing that, since the time the first streams flowed on the earth, they have always carved their valleys in much the same manner as they do today. We can, therefore, make out a fairly strong case for the uniformitarianism of earth events.

But, in spite of the fact that we can trace geological cycles through millions of years, we see increasingly clearly that no two of them, after all, have been so very much alike. This is particularly true when one remembers that we love uniformity and attempt to match each newly discovered record with previously interpreted events. We are accustomed to say, for instance, that the marine inundation of the middle Ordovician was like that of the middle Silurian and that they both were similar to the Onondagan flood of middle Devonian times. This is a polite fiction maintained for the convenience of teachers and students. The farther one goes into the subject, the more one is convinced that the only compelling point of similarity is that they *are* three great periods of inundation. We are unable to compare the great Proterozoic iron-forming period with any later geological times, since iron deposition on a large scale has not been a recurring geological process. Coal has formed in abundance in many geological periods, but it could not form at all before the development of the primitive forests, and it has never again been formed so bounteously as in the Pennsylvanian.

Thus, it is obvious from our studies that there has been a general geological scenario which has been followed again and again. But throughout the long years many of the earth's actors have grown old and died. New players have taken their places. Scenes have worn out too, and new stage properties have been acquired. It therefore seems clear that, although we can interpret the future *reasonably well* in the light of the past, still we cannot place too much reliance in our own predictions.

IT MAY be worthy of note in this connection that the great geological cycles become somewhat shorter toward the present. Neglecting the apparently long pre-Cambrian cycles, whose records are rather poorly written, we may point out that the early Paleozoic cycle involved 200,000,000 yr.; the late Paleozoic, 150,000,000 yr.; and the Mesozoic cycle, 140,000,000 yr. If, however, we could regard the Mesozoic and Cenozoic eras as comprising a single cycle it would have a length almost precisely the same as that of the early Paleozoic. But, in addition to this unimportant coincidence, are there any *real* reasons for so extending the Mesozoic cycle? There are several which have been suggested.

THE continents stand higher today than during almost any of the periods of the past. The Cenozoic seas have all been marginal in character, such as one would expect during a time of mountain-building. Certainly they have not been typical of a period of submergence. The Cascadian revolution, although generally given as Miocene in age, really cannot be dated very accurately. Movements which might be referred to it are, in fact, still taking place. Our present mountain systems have all of the grandeur of any of the great ranges of the past, however liberal one is in making estimates of the height of the ancient mountains. There are large modern deserts and great salt lakes. Moreover, two continental ice sheets still exist, and we cannot be sure that others will not form in the near geological future. Terrestrial deposition is going on apace at many spots, but limestones are being formed only in small areas and at few places. There are some shallow seas, such as Hudson Bay and the Baltic, spreading out on the continents; and the lands are known to be sinking at a number of places. On many other coasts there is evidence of uplift. All of these features make one feel that **the earth is either just now coming to, or has recently passed, the climax of a great physical revolution.** But, of course, this is all relatively fruitless speculation. For, even if we could be reasonably certain of our exact present position in one of the great geological cycles, we could not definitely know how far the new cycle would depart from the general pattern followed by the older ones.

Another period of refrigeration?

TURNING now to the organic side of the picture, we find that, unless we are somewhat deceived by the freshness of the record, changes in the earth's living things have been more rapid during the Cenozoic than in any previous era. The major part of the evolution of the mammalian stock has taken place during this time; and, in fact, the climax of the group was probably reached as early as the Miocene, by the close of which many of the larger mammals had already become extinct. If we compare this mammalian record with the geological stories the amphibians and reptiles can tell, the future does not look too good for the warm-blooded. The amphibians evolved slowly during the cosmopolitan conditions of the late Paleozoic submergence; and they not only reached their climax, but went into their decline, during the late Paleozoic "depression" period. At the time the amphibians were near their climax, the reptile stock came into existence. It then deployed and survived the late Paleozoic mountain-building and rose to a climax during the prolonged period of Mesozoic physical quiet. The mammals, like the amphibians, originated during a period of quiescence and reached their climax during days of pronounced orogeny. If we could bank on the validity of the method of employing the record of the past in order to prognosticate the happenings of the future, we might well feel gloomy over the fact that the mammals chose a time of revolution for their big scene. Think how much more glorious the future mammalian history might be were the class still youthful enough to take advantage of the period of geological quiescence which surely is to come.

MAN, as such, has only been a passenger on the Geological Express for some five or six of the total hours in our Gargantuan geological year. But he has already terminated the journey of many of his fellow-travelers. This is just another way of saying that man and his activities have already notably altered the organic world. Fishing is poor where once it was excellent; our grandfathers hunted birds and beasts now nearly or quite extinct; and great dust storms originate in areas not long since covered with a lush growth of vegetation. Most of the responsibility for these and other notable changes can be laid at man's door. The changes he will effect in the future are limited only by his intelligence.

WHAT about man himself as a son of the Earth? New species of man do not seem to be originating as they characteristically should in a young and virile stock; but, on the other hand, **man is still a primitive primate.** Biologically speaking, he is still fairly generalized so far as his limbs, digits, and teeth are concerned. He is specialized chiefly in the great development of his brain. Paradoxi-

PROSPECTS OF AN END

cally, this seems to have resulted in overspecialization in many of his activities, to the extent that this new development is as serious a handicap to many men as biological specialization seemed to be to many a now extinct group among the lower animals.

If man really can come **DOWN TO EARTH** and at last actually plan his terrestrial course, he will be using his specialized organ to its best advantage, and his generalized body should carry him through all sorts of geological vicissitudes to further successes. But he probably will not live to achieve them in his present form. Sixty million years from now, when, barring some astronomic catastrophe of the type responsible for the earth's birth, the earth should again be well into the quiescent phase of the next great cycle, post-present man should be as different from his rather sorry modern ancestors as we ourselves are removed from our Eocene progenitors of the *Notharctus* type. His civilization also will be markedly different, for, unless he mends his ways, all his mineral resources long since will have been exhausted, and he will be using sources of energy of which, even now, we can only vaguely dream.

But in the final analysis, the matter is largely in man's own

LA MORT NOUS EGALE TOUS.

Will Death make equals of all animals, high and low, or ....

L'ETERNITÉ EST LE FRUIT DE NOS ESTUDES.

.... will immortality be the result of man's studies? From *Le Théâtre Moral*, 1672, a book illustrating some of the works of Horace

hands. We have already pointed out that no animal has as yet sentiently directed his own evolutionary course. This does not mean that it is beyond the bounds of reason for man ultimately to do so. If, like all the earlier rulers of the earth, man has his own sweet day of triumph, only to be succeeded on the throne by some menacing group, whether the challenge is to come from the arthropod strain or from another mammalian type, or even from representatives of a stock not yet in existence, then truly for man, as well as for the earlier animal dynasties, it will be just another case of "Death makes equals of us all." But it must be remembered that in only a minute of our Gargantuan year, man has achieved a mastery of the earth not even hinted at during the long months of the existence of the other animals. If he learns to control nature even reasonably completely, his mind and his studies will, indeed, have brought him immortality.

INDEX

INDEX

A

Acadian disturbance, 347, 348, 387
Acceleration of gravity, variations of, 135
Acorn worm, 368
Aepyornis, 455
Agassiz, Louis, 235, 236, 433, 434
Agents, geological, 27, 43
Ages, regional, 63
  valley, 57, 58
Agnatha, 368
Agricola, 222, 225, 422, 469, 476
Albertan epoch, 324
Alberti, von, 385
Alexandrian epoch, 328
Algae, 299, 310, 312, 358, 359
Algomian disturbance, 299, 303
Allegheny series, 350
Allosaurus, 411
Alluvial fans, 91, 110
Alluvial plains, 429
Alpha-particles, 296
Amber, burial in, 228, 229, 231
Amblypods, 450, 452
American Museum, 450
Ammonites, 419, 421
Ammonoids, 419, 420, 421, 422
Amoeba, 334
Amphibia, 299, 368, 371, 373, 387, 447, 481, 482
Anaximander, 223
Andesite, 193
Andrewsarchus, 450, 451
Angiosperms, 358, 363, 364, 407
Angle:
  of rest, 42
  of emergence, 150
Animal kingdom, 226

Animals:
  domestication of, 463
  one-celled, 299
Animikiean, 303
Annelid worms, 368
Annularia, 363, 365
Anorthite, 21
Anthozoa, 334
Anthracite field, 353
Anthropoids, 459
Anticlines, 140, 378, 475
Apes, Old World, 298
Appalachia, 325, 347, 355
Appalachian fault troughs, 386
Appalachian formations, 326, 328
Appalachian geosyncline, 248, 299, 346, 399
Appalachian region, 324, 348, 350
Appalachian revolution, 299, 345, 355
Appalachian trough, 324, 325, 327, 330, 331
Arachnids, 334, 343, 368, 384
Archaeocyathids, 335, 337
Archaeopteryx, 405
Archelon, 402
Archeozoic, 299, 304, 317
Archidiskodon, 453, 454
Archimedes, 375
Arduino, Giovanni, 424
Aristotle, 223
Armored fish, 299
Art, Crô-Magnon, 462
Arthropods, 310, 334, 335
Artifacts, 460, 465
Ash falls, 327, 429
Asteroidea, 334
Asthenoliths, 288
Atlantis, 256

Atmosphere, 14
   carbon dioxide in, 16, 264
   envelope of, 287
   gases of, 16
   water content, 264
Atmospheric circulation, 264
Atomic disintegration, 270, 292, 296
Atomic systems, 133
Australopithecus, 463
Axial plane, 140
Azoic era, 217, 317

B

Bacteria, 299, 311, 358, 359
Baculites, 421
Bad Lands, 429
Baltic, 481
Baluchitherium, 450
Barchan dunes, 113
Bars, 105
Basalt, 193, 298, 388, 398
Basic-eruption hypothesis, 288
Basins, geosynclinal, 330
Bat guano, 230
Batholiths, 179, 210, 347, 391
Bats, 448
Beach, 101
Beach sand, 117, 123
Bedding, 96
Belemnites, 422
Belt series, 311
Bennettitales, 358
Berea sandstone, 350
Beresovka mammoth, 229, 230
Beringer, J., 224, 225, 237, 458
Biosphere, 27
Birds, 298, 368
Bisons, skeletons of, 464
Black River limestone, 326
Blastoids, 334, 374, 375, 383
Block faulting, 388
Bogs, peat, 361
Borderlands, 247, 248
Bottom land, 63
Boulder batholith, 250, 397
Boulder clay, 122
Boulders, 122
   drift, 432, 433
Brachioles, 374
Brachiopods, 299, 334, 335, 336, 337, 338, 339 375, 383

Brachiosaurus, 412
Bradford sandstone, 346, 347
Brassfield limestone, 328, 329
Breadfruit tree, 398
Bromoform, 119
Brongniart, 389
Brontosaurus, 411, 412
Bryophytes, 358, 360, 361
Bryozoans, 334, 339, 342, 375
Buch, Leopold von, 389
Buffon, 237, 275
Bunter series, 385
Burgess shale, 338, 406
Burrows, 233

C

Calamites, 363, 365
Calcification, 231
Calcite, 37, 77, 107, 122
Calcium bicarbonate, 78
Caledonian disturbance, 315, 330, 345
Caledonian mountains, 345
Calyx, 374
Cambrian period, 299, 309, 322-24, 327, 334
Camels, 452
Canadian epoch, 326, 327
Canadian shield, 248, 249, 304, 305, 307, 325
Canyons, 62
Carbon cycle, 213
Carboniferous, 348, 350, 352
Carbonization, 231, 232
Carvings, bone, 461
Cascades, 257
Cascadia, 324, 325, 349
Cascadian revolution, 298, 430, 480
Cats, 451
Cave, 78
Cave men, 230
Cave paintings, 230
Cavendish experiment, 8
Cavern, 78
Cayugan epoch, 328, 329
Cenozoic era, 298, 424-31, 447-50, 451, 458, 480
Centipedes, 334
Centrifugal force, 277
Cephalopods, 298, 299, 334, 335, 342, 376, 383, 406, 416-22
Ceramic products, 468
Ceratites, 419
Ceratosaurus, 411

INDEX 489

Chamberlin, T. C., 277, 278, 283, 287, 317, 318, 434
Chamberlin-Moulton hypothesis, 282
Chambers, Robert, 268
Champlainian epoch, 326, 327
Chazy Limestones, 326
Chemical reactions, endothermic, 204
Chemung beds, 346
Chester series, 348, 375
Chico series, 393, 395
Chimney rocks, 104
Chitin, 232, 314
Chlorophyl, 320
Chondrichthyes, 368
Chordates, 335, 368
Chromosphere, 270
Cimarron series, 354
Cincinnati Arch, 326, 327
Cincinnatian epoch, 326, 327, 342
Circulations, atmospheric, 26
Cirque, 88
Clams, 334, 339, 367, 375
Clay minerals, formation of, 38
Clay, 37, 108, 121, 122, 228
Climate, 330
Climates, 257–67
Climatic controls, 264
Climatic zones, 330
Clinton rocks, 328, 329
Club mosses, 384
Coal, 124, 266, 267, 299, 351, 353, 356, 357, 359, 365, 387, 388, 394, 395, 399, 480
Cobalt series, 308
Cobbles, 122
Cockroaches, 367
Coelenterates, 334, 335, 368
Collini, 404
Colloids, 107, 108, 299
Colorado beds, 393, 395
Comanchian series, 394
Compsognathus, 410
Concretions, 80
Condylarths, 450
Conemaugh series, 350
Conglomerate, 122, 244, 307, 308
Conifers, 358, 364, 384, 407
Connecticut Valley, 388,
Continental-migration hypothesis, 262, 431
Continental platforms, 437
Continental shelves, 260

Continental submergence, 247
Continental uplift, 299
Continents, origin of, 247
Convection currents, 285
Copper, 307, 356, 468
Coprolites, 229, 234
Corals, 263, 299, 330, 339, 342, 343, 346, 374, 383
Cordaites, 358, 363, 365, 384
Cordilleran geosyncline, 248, 298, 322, 324, 326, 327, 346, 348, 355, 390, 436
Cordilleran glacial center, 435–37
Correlation, principles of, 235–45, 253, 296
Cosmic evolution, 279–81
Cosmogony, 268, 275
Crayfish, 334
Creation, 316, 368
Creodonts, 450
Cretaceous, 298, 335, 358, 392, 393, 394, 396, 397, 398, 399, 401, 402, 404, 413, 414, 447
Crinoids, 299, 334, 339, 341, 342, 374, 375
Crocodiles, 409
Croixan, 325
Crô-Magnon man, 298, 462, 463
Crustaceans, 242, 243, 298, 406
Crustal instability, 355
Crustal movements, 131, 349
Crustal shortening, 131, 132, 134, 258
Crustal stresses, 249
Crystal, 19
Culmide disturbance, 350
Culture:
　Acheulian, 463
　Aurignacian, 463
　Bronze Age, 463
　Chellean, 463
　Eolithic, 463
　Iron Age, 463
　Magdalenian, 463
　Mousterian, 463
　Neolithic, 463
　Paleolithic, 463
　pre-Chellean, 463
　Solutrean, 463
Currents, longshore, 102
Cuvier, 404, 458
Cycadeoids, 358, 364
Cycadofilicales, 358, 362, 363
Cycads, 298, 358, 364, 365, 387
Cycle:
　of erosion, 67, 103
　of stability, 345
　of unrest, 345
Cycles and matter, 214

Cynodonts, 447
Cystoids, 334, 341, 342, 374, 383

D

Dakota series, 393, 394, 395
Dana, J. D., 288
Darwin, Charles, 227, 275
Darwin, Erasmus, 275
David, Sir T. E., 310
Da Vinci, Leonardo, 225, 235
Dawn man, 461
Dawson, Sir William, 310, 312
Deciduous trees, 407
Deep-seated environments, 17, 138, 140, 187, 193, 201, 210
De Geer, Gerard, 443
Degradation, 64
Dekkan district, 398
Delta, 93, 110
   Catskill, 347
De Moines series, 350
Deposition, 261, 318, 329–30
   geosynclinal, 258
   glaciofluvial, 99
   Nile, 454
   saline, 331
   varved, 441–43
   volcanic, 429
Depth-velocity curves, 161
De re metallica, 469
Devonian period, 299, 329, 330, 334, 337, 338, 345, 346, 348, 357, 363, 369, 370, 371, 372, 375, 392, 480
D'Halloy, O., 393
Diastrophism, 29, 130, 138, 201, 245, 260, 274, 343, 396
Diatomaceous ooze, 360
Diatoms, 358, 360
Dibranchiates, 422
Dicotyledons, 358
Didus, 455
Dikes, 178, 199, 218
Dinichthys, 369
Dinornis, 455
Dinosaurs, 298, 387, 390, 395, 401, 407, 408, 409, 410, 411–14, 415, 449
Diorite, 193
Dip of strata, 142
Diplodocus, 411, 412
Disconformity, 218, 351
Divide, stream, 53
Dogger, series, 389

Dogtooth spar, 38
Dogs, 451
Dolomites, 325, 326
Dolphins, 402
Dome, Adirondack, 326
   Ozark, 326, 328
Dragon flies, 367
Drake, "Colonel," 470
Drawings, cave, 464, 465
Drift boulders, 432, 433
Driller's log, 241–42
Drowned-valleys, 130
Dryopithecus, 460, 463
Dubois, 461
Dune sand, 123, 382
Dunkard rocks, 354
Dust storms, 49

E

Earth, 280, 281
   axis, inclination of, 10
   center, 262
   central core of, 161, 164
   circumference, 7
   compression of interior, 133
   crust, 175, 274
   density stratification, 282
   determination of size of, 6, 7
   diameter, 270
   energy, 25, 30, 32
   layers, 284
   mass, 8
   materials, density of, 160
   movements, 322
   nucleus, 285
   revolution, 9, 10
   rotation, 8
   shells, 161, 163
   shrinkage, 131, 132, 134, 135, 258
Earthquake, San Francisco, 156
Earthquake zones, 146
Earthquakes, 29, 145
   body waves, 159
   center of, 154
   damage by, 150, 151
   distances from seismographs, 154
   elastic rebound theory of, 147, 148
   epicenter of, 151
   focus of, 150
   frequency of, 145
   relation to faults, 146
   waves of, 153, 157, 158, 159, 161, 162
   and volcanoes, 174
Echinoderma, 334, 335, 342, 343, 374, 375, 406
Echinoidea, 334
Elasmosaurus, 401

INDEX

Elephants, 298, 453, 454
Elephas, 453
Elms, 358
Embayments, 312, 322, 329, 425
Energy, 23
    flow, 31, 128, 214
    kinetic, 56, 101, 269
    potential chemical, 204
    sources of, 24
    versus matter, 208
Environments:
    deep-seated, 17, 138, 140, 187, 193, 201, 210
    factors in, 209
    surface, 17, 187, 193, 210
Eoanthropus, 226, 461, 463
Eocene epoch, 298, 426, 428, 430, 431, 449, 451
Eohippus, 451
Eoliths, 460
Eozoan, 312
Epicontinental seas, 249, 298
Epoch:
    definition of, 253, 254
    names, 326, 328
Equilibrium, 17, 38, 207, 208
    complexities of, 209
Equisitales, 358
Equus, 451
Era:
    Archeozoic, 299, 304
    Azoic, 299
    Cenozoic, 298, 447–51, 458, 480
    definition of, 254
    Mesozoic, 298, 447–49, 480
    Paleozoic, 254, 299, 333, 480
    Proterozoic, 299, 304
Erian group, 346
Erosion, 53, 218, 261, 318, 323, 428
    cycle of, 67, 103
    forms, 111
    processes, 269
    rates, 292, 294
    regional, 61
    by shore agents, 103–5
Erosional breaks, 252
Eurypterids, 299, 343, 382
Eurypterus, 342
Eutaw series, 393, 395
Evidences of pre-Cambrian life, 308–10
Exponential curves, 296, 297
Extrusions, 170, 217, 302

F

Facies, 240, 327
Fault:
    formation, 397
    normal, 144
    plane, 143
    thrust, 144
Faulting, 143, 249, 256
Fauna:
    Cambrian, 335
    Hamilton, 346
Feldspar, 21, 38, 189, 192
    formation of, 190
    weathering of, 37
Fenestelloids, 375
Ferns, 358, 361, 362, 387
Ferromagnesian minerals, 21, 38, 189, 192
    formation of, 190
    weathering of, 37
Field, British Columbia, 338
Filicales, 358
Fish, 368, 402
    armored, 299
    bony, 370
    lobe-finned, 372
    Mesozoic, 406
    ray-finned, 370
Floating continents, 288–89
Flood plain, 51
Floods, 246–54
Flora, Devonian, 363
Florissant beds, 429
Flow, laminar, 45, 75, 86, 103
    turbulent, 45, 47, 102, 107
Flying dragons, 406, 415
Flying reptiles, 298
Folds, 217, 248
Folsom, New Mexico, 464
Foot wall, 143
Foraminifera, 243, 376, 476
Forests:
    Paleozoic, 365
    primeval, 357–65
Formation, definition of, 254
Fossils, 121, 236–39, 242–45
    ancient history of, 223, 224
    bacterial decay of, 228
    classification of, 229
    definition of, 222
    derivation of word, 222
    evolutionary history, 227
    index, 238, 420
    microscopic, 242
    pre-Cambrian, 310–12
    pseudomorphic, 232
    requisites for preservation, 228
Foucault pendulum, 9
Fracastoro, 225, 235
Frederick, Oklahoma, 464
Fredericksburg group, 393, 394

Fronds, reproductive, 364
Fuels, 468
Fundamental processes, 27, 29, 30
Fungi, 358–60
Fusulina, 376

G

Gabbro, 193
Galena dolomite, 326
Ganoid fish, 370, 387
Gargantuan calendar, 291–300, 385, 408, 424, 447, 482, 484
Gas, natural, 348, 350, 356, 469, 472, 473–77
Gaseous-Tidal hypothesis, 282
Gastroliths, 229, 233, 234, 401
Gastropods, 299, 334, 335, 337, 376
Geological ages, 463
Geological chronology, 252, 253
Geological evidence, 119, 135
Geological periods, 254, 255
Geological thermometers, 198
Geological time, 291–300
Geology:
    defined, 2
    dynamical, 4
    historical, 258
    and industrialism, 2
    time factor in, 61
    versus other physical sciences, 121
Geosyncline, 248, 298, 306, 322–25, 329, 390, 394, 477
    Appalachian, 248, 299, 324, 326, 327, 346, 378, 387
    belts, 249, 260
    Cordilleran, 248, 298, 299, 322, 324, 326, 327, 348
    Ontarian, 299
    Ouachita, 299, 326
    Pacific, 387, 395
    Paleozoic, 355
Geothermal gradient, 24, 133
Gesner, Conrad, 222, 225
Gilboa, forest of, 364
Ginkgos, 358, 364, 365
Glacial deposits, 97–99
Glacial epochs, 435–40
Glacial episodes, 438–40
Glacial ice, 43, 84, 86, 87
    distribution of load, 87
    erosion by, 87
Glacial Lake Agassiz,
Glacial movement, 86, 87
Glacial scour, 87, 440
Glacial till, 97, 118

Glacial varves, 441, 443
Glaciation, 264, 299, 362, 433–46
    Cobalt, 303
    Permian, 378, 379
    Proterozoic, 308, 309
    South American, 381
Glacier, 83, 84, 85, 298
Glaciers:
    Alpine, 85, 433
    centers of radiation, 434, 437
    continental, 85, 433
    deposition by, 98
    movement of, 86, 118
    structure of, 84
    valley, 84
Glossopteris, 362, 381, 382, 384
Glyptodonts, 454
Gneiss, 205
Gold, 468
Gold-quartz veins, 391
Gondwana, 381, 382, 398
Gradation, 27, 126, 127, 128
Grains, shifting of, in transportation, 47
Granite, 188, 189, 193, 205, 297, 306
Granite batholith, 397
Granite crust, 286
Graptolites, 227, 232, 239, 339, 342
Grasses, 358
Gravel, 107, 121
Green River shales, 429
Greenland, 85
Grenville limestone, 306, 310
Ground water, 43, 70, 75, 77
    features of, 113
    lower limit of, 73
    movement of, 76
    solution and deposition, 80
    source of, 71
    table, 72
    transportation by, 77
Growth rings in trees, 266
Guano, 230
Gullies, 52
Gumbotil, 444
Gunz glaciation, 463
Gunz-Mindel interglacial period, 463
Gymnosperms, 358, 363, 364
Gypsum, 124, 267, 329, 354, 382, 388
Geysers, 81, 173–74

H

Hagfish, 369
Hamilton shales, 346, 347
Hanging wall, 144

INDEX

Haüinite, 183
Heat, radiation of, 30, 214, 269
Heavy minerals, 119
Heidelberg man, 226, 298, 462
Helderbergian group, 346
Helmholtz, 269
Henry Mountains, 177
Hepaticae, 358
Herodotus, 294
Hesperornis, 406
Himalayas, 465
Historia naturalis, 224
Holocene, 298, 426, 428, 463
Hominidae, 226
Homo diluvii testis, 458
Homo heidelbergensis, 226, 462
Homo neanderthalensis, 226, 462
Homo sapiens, 226, 462
Horses, 298, 451
Horsetails, 358
Hudson Bay, 481
Humboldt River, 92
Huronian period, 299
Hutton, James, 55, 292, 478
Hydnoceras, 266
Hydrocarbons, 469–75
Hydrosphere, 14, 285
Hydrozoa, 232, 239, 334
Hyolithids, 337
Hypothyridina cuboides, 238
Hypsograph, 16

I

Ice age in America, 130
Ice as a burial medium, 228, 229, 230
Ice sheets:
    continental, 85, 380, 435
    Permian, 381
Icebergs, 85
Icecaps, 85, 433, 435, 437
Ichthyornis, 405, 406
Ichthyosaurs, 402, 422
Idaho batholith, 397
Igneous dikes, 178, 199, 218
Igneous injections, 179
Igneous intrusion, 170, 244, 302
Igneous rocks, 182, 243, 247, 263, 277, 297, 299
    acidic, 191
    basic, 191
    composition of, 191
    crystals of, 185
    differentiation of, 194
    effect of environment on, 187
    extrusive, 181, 217, 302
    intermediate, 191
    intrusive, 181, 217, 219, 302
    minerals in, 192, 193
    normal series, 190, 191, 193, 195
    order of crystallization, 187, 188, 189, 192
    origin of, 187
    Tertiary, 429
Igneous sills, 178, 199, 218
Igneous transgressions, 180
Iguanodon, 413
Illinois River, 115
Implements:
    flint, 460, 464
    stone, 462
Index fossils, 238
Ink sac, 422
Insectivores, 448
Insects, 334, 384, 429
Interglacial periods, 298, 391
Intrusion:
    granite, 299
    heating effect of, 180, 181
    igneous, 170, 204, 244, 302
    minor, 178
Invertebrates:
    land, 384
    marine, 238, 239, 298
    Mesozoic, 406
    Paleozoic, 299
    pre-Cambrian, 310, 311
Iron, 468
Iron oxide, a cement, 122
Isoseismal lines, 150
Isostasy, theory of, 261
Isostatic equilibrium, 261, 262

J

Jatulian series, 311
Jeans, Sir James, 282
Jeffreys, Harold, 282
Jellyfish, 334
Juniata sandstone, 326
Jupiter, 280
Jurassic, 298, 361, 385, 389, 404
Jurassic coals, 391

K

Kames, 98, 434
Kanimbla disturbance, 347
Kant, E., 275
Kaolin, 38

Katmai, eruption of, 265
Keewatin center of ice, 435–37
Keewatin period, 299, 303
Kelvin, Lord, 269, 292
Keuper series, 385
Keweenawan period, 299, 302, 303, 306, 307, 311
Key beds, 241
Killarney range, 306
Killarney revolution, 299, 303
Kimmswick limestone, 244
Kinderhook series, 348
Kinetic energy, 56, 101, 269
Kircher, A., 69
Kootenai beds, 394
Krakatoa, eruption of, 168, 265

L

Labradorian center of ice, 435–37
Labyrinthine tooth structure, 370, 372
Laccolith, 177
Lag sediment, 46
Lake Bonneville, 444
Lake Superior, 130
Lakes, Cambrian, 361
Lamarck, J. B., 227, 237
Laminar motion, 45, 75, 86, 103
Lamp shells, 334
Lamprey eels, 369
Land features, table of, 111
Land forms, complexity of, 114
Landslides, 42
Lapilli, 172
Laplace, 275–77
Laplacian hypothesis, 282
Laramide revolution, 298, 397, 426, 429
Laramie group, 393–95
Laurentian Mountains, 257
Laurentian revolution, 299–303
Lava, 22, 168, 185
  Keweenawan, 307
  rapid cooling of, 171, 186
  temperature of, 171
Lava flows, 171, 184, 244, 429, 430
Lawson, A. C., 302
Lead, atomic weight of, 296
Lemuroidea, 459
Lepidodendron, 362, 365, 368
Levee, natural, 92
Lewis overthrust, 396–97
Lias series, 389

Life:
  birthplace of, 319
  borderlands, 316
  creation of, 268
  definition, 317
  origin, 315–20
  requisites for, 318, 319
  synthesis of, 316–20
Light micas, 202
Lime mud, 109
  as a burial medium, 228
Limestone, 77, 113, 123, 239, 241, 244, 266, 325, 328, 329, 390, 481
  cherty, 349
  deposition of, 123, 329
  fresh-water, 351
  marine, 351
  precipitation of, 123, 264
  Tertiary, 427
Lingula, 337
Lion, 451
Lithographica würceburgensis, 224
Lithosphere, 14
  chemical composition of, 18
  oxides in, 20
Liverworts, 358–61
Load, suspension, 46, 47
Lockport limestone, 328
Loess, 95, 123, 228, 439, 440
Lorraine sandstone, 326
Louisiana lowlands, 476
Louisville limestone, 328
Loxodonta, 453, 454
Lyell, Sir Charles, 425

M

Magma, 185, 194, 197, 198, 288
Maidenhair trees, 365
Mallet, Robert, 149, 151
Malm series, 389
Mammals, 226, 298, 368, 406, 415, 447, 448, 449, 452, 459, 481, 482
Mammonteus, 453, 454
Mammoth, 229, 230, 464, 465
Mantle rock, 34, 41
Maples, 358
Marble, 198, 205, 331
Mastodon americanus, 453
Mastodon borsoni, 453
Mauch chunk rocks, 348, 349
Mauna Loa, 169
Media for burial of fossils, 228–34
Megalomania, 408–13

INDEX

495

Meltwater, 83
Mercury, 271, 280, 281, 282
Mesabi Range, 307
Mesozoic era, 298, 363, 370, 371, 380, 384, 397, 398, 399, 400, 401, 402, 406, 413, 415, 417, 419, 420, 447, 448
Metamorphic rocks, 197, 217, 243, 299
    table of, 205
Metamorphism, 197, 199, 201, 202, 203, 204, 310, 331
    contact, 199
Meteor Crater, 273
Meteor scars, 285
Meteorites, 163, 272, 273, 319
Meteors, 162
Mexico City, 464
Mica schist, 203, 205
Micas, 202
Micropaleontology, 242
Miller, Hugh, 344
Millipeds, 343, 366
Mindel glaciation, 463
Mindel-Riss interglacial period, 463
Minerals, 19
    accessory, 189
    grains of, 47
    primary, 20, 38
    secondary, 38
Miocene epoch, 298, 426, 428, 429, 430, 454, 463, 481
"Missing link," 461
Mississippi River, 392
Mississippi Valley, 348
Mississippian period, 299, 326, 328, 345, 348, 350, 362, 369, 370, 374–76
Missouri series, 350
Mobile belts, 260
Moeritherium, 453, 454
Molds, 229, 233
Molluscoidea, 334
Molluscs, 334, 337, 367, 375
Molten-earth hypotheses, 285–87
Moment of momentum, 272
Monocotyledons, 358
Monograptus, 342
Monongahela series, 350
Monotremes, 447
Montana beds, 393, 394, 395
Monte Nuovo, 166
Moon, 282, 283, 285
Moons of Mars, 372
Moraines, 98, 434, 435, 438, 440, 441

Morrison beds, 390, 411
Morrow series, 350
Mosasaurs, 402
Moss animals, 334
Mosses, 358, 360
    club, 360, 362
Mother lode, 391
Moulton F. R., 277, 278
Mount Vesuvius, 167, 168
Mountain-building, 248, 251, 255, 259, 262, 263, 267, 274, 305, 309, 328, 330, 345, 347, 350, 351, 352, 355, 356, 378, 431, 472
Mountain chains, 258
Mountain systems, 305
Mountains, 16, 131, 257–67
Mu, 256
Mud, 107
Multituberculates, 448
Murchison, Sir Roderick, 255, 321, 344, 345
Muschelkalk, 386
Musci, 358
Muscovite, 202
Myriopods, 334

N

National Resources Board, 468
Natural laws, 316
Nature of earth's interior, 160
Nautiloids, 334, 417, 419, 420, 421
Nautilus pompilius, 417, 418
Neanderthal man, 226, 298, 462
Nebular hypothesis, 275
Neolithic man, 298
Neoliths, 460
Neptune, 280–81
Nevadian disturbance, 391, 394
New Red Sandstone, 385
Newark series, 387, 388
Niagara Falls, 293
Niagara epoch, 329
Niagara series, 254, 255
Niedermendig, Germany, 183
Niobrara chalk, 394
Notharctus, 459, 483
Nothosaurus, 403

O

Oaks, 358
Obelia, 339
Ocean basins, 15, 106, 136

Ocean deposits, 109
Octopods, 416, 422
Ogygopsis, 335
Oil, 348, 354, 399, 468, 469, 470, 471, 472, 473, 474, 475, 476, 477
    as a burial medium, 228–30
    Devonian, 348
    Mississippian, 350
    Permian, 356
    Tertiary, 427, 430
Oil shales, 429
Old Man of Hoy, 344
Old Red Sandstone, 344, 385
Olenellus, 335
Oligocene epoch, 298, 426, 428, 429, 430, 449
Onondagan, 346, 374, 379
Oölite, 385, 389
Operculum, 370
Orbits:
    circular, 271
    planetesimal, 271
Order, definition of, 358
Ordovician period, 299, 322, 325, 326, 327, 328, 331, 332, 334, 338, 369, 416, 479
Oreodonts, 452
Ores, Proterozoic, 331
Organic revolution, 354
Origin of species, 227
Oriskany sandstone, 346
Ornithiscians, 409, 414
Ornitholestes, 410
Ornithopods, 409, 413
Orogenies, 349–50, 327, 346, 378, 379, 396, 431
Orthoclase, 21
Osgood limestone, 255
Osteichthyes, 368
Ostracoderms, 299, 369
Ouachita geosyncline, 299, 326, 351
Outcrops, 141
Outwash plains, 98
Overthrusts, 262, 397
Oysters, 334
Ozark Dome, 326, 328

P

Painted Desert, 386
Paintings, wall, 460
Paleocene epoch, 426, 428
Paleogeography, 255
Paleolithic culture, 298
Paleoliths, 460
Paleomastodon, 453, 454

Paleontology, definition of, 226
Paleophonus, 366
Paleoscincus, 414
Paleozoic era, 254, 299, 340, 357, 362, 367, 370, 375, 399, 417
Paleozoic:
    climate, 330
    cycle, 333, 345, 348
    early, 321–32, 416
    floras, 363
    geosynclines, 355
    late, 344–56, 373, 374, 376–77, 379, 380, 419, 420, 480
Paleozoic Alps, 379
Palisade disturbance, 388
Palisades, 388
Pallial line, 375
Palms, 398
    ancestral, 384
Paris Basin, 425
Passing star, 280–81
Pearly nautilus, 417
Peat, 124, 228, 361
Pebbles, 87, 89, 122
Peking man, 461
Pelecypods, 298, 299, 334, 375, 406
Pelican Rapids, Minnesota, 464
Pelycosaurs, 373
Peneplain, 67, 134, 218–19, 262
Pennsylvanian coals, 399
Pennsylvanian flora, 384
Pennsylvanian period, 299, 362
Period, definition of, 253–54
Periodic melting, 262, 263
Permian fauna, 383
Permian glaciation, 362, 384
Permian period, 299, 345, 355, 356, 376, 380, 409, 448
Petrification, 231–32
Petroleum, 3, 359, 470–72
Phiomia, 453
Phororhacos, 455
Photosphere, 270
Phylum, 334
Physiographic cycle, 213
Physiography, 68
Pigs, 452
Piltdown, England, 461
Pines, 358
    Oligocene, 231
Pisces, 368
Pithecanthropus, 226, 461, 463

INDEX

497

Placer deposits, 39
Placodermi, 368
Plains, 16
Planetesimal hypothesis, 273, 281–84, 313, 319
Planetoids, 277, 280–81
Plankton, 360
Plants:
    classification of, 358
    geological range, 358
    primitive, 284
Plateau, 16
Plateau, Colorado, 62
Platform, areas, 247–53
Platteville limestone, 326
Pleistocene, 298, 426, 428, 438, 451–52, 454, 463
Plesiosaurs, 233, 401, 403
Pliny, 224
Pliocene, 298, 426, 428, 454, 463
Pluto, 271, 280, 281
Pocono group, 347, 348, 349
Poles, position of, 264
Polyps, 374
Pompeii, 167, 233
Porifera, 334
Porphyry, 194
Portage rocks, 346, 347
Potash salts, 356
Potomac series, 393
Pottsville, series, 350
Pozzuoli, Italy, 166
Pre-Cambrian, 301–14
    cycles, 480
    fossils, 311
    granites, 302
    orogenies, 397
    metals, 307
    submergences, 311
Pressure of earth's interior, 133
Primary minerals, 20, 38
Primary rocks, 301, 302, 424, 425
Primates, 226, 459, 466, 482
Proboscideans, 452
Productids, 375
Propliopithecus, 459
Prospecting, geophysical, 164, 475
Proterozoic era, 299, 304, 322, 334, 397, 480
Protoceratops, 414
Protozoa, 299, 334, 335, 406
Psilophyta, 361
Psychozoic era, 217
Pteranodon, 404, 405
Pterichthys, 369

Pteridophytes, 358, 361
Pterodactyl, 404
Pteropods, 337
Pterosaurs, 404, 406, 415
Pterygotus, 342, 343
Pumice, 171
Pyritization, 231–32
Pyroxene, 21
Pythagoras, 223

Q

Quarries, 331
Quartz, 21, 22, 119, 189, 190, 192
    formation of, 190
    stability of, 37
Quartz sand, 119, 331
Quartzite, 205, 324
Quaternary period, 298, 381, 425, 432, 463

R

Radioactive decay, 28, 259, 269, 270, 284, 292, 296, 300
Radiolaria, 312
Radium, 270
Raindrop impressions, 121
Rancho la Brea, 454
Ranges, serrate, 257
Ravines, 53
Recent epoch, 370, 428
Region:
    mature, 64
    old, 66
    youthful, 63
Regional age:
    elements of, 63
    versus valley age, 65
Reptiles, 298, 368, 371, 373, 384, 401, 405, 481, 482
    armored, 409
    bipedal, 409, 410
    flying, 298, 403, 404, 409
    horned, 409
    mammal-like, 387
    marine, 298, 403
    Permian, 373
Residual concentration, 307
Residual sands, 327
Resin, 228, 229
Resisting medium, 282
Rhinoceros, 229, 230, 465
Rhyolite, 193
Richardson, Rev. B., 235
Richmond basin, 387
Ripley series, 393, 395

Ripple marks, 120
Riss glaciation, 463
Riss-Würm interglacial period, 463
River gravel, 391
River sand, 123
Rochester formation, 255
Rock:
  changes:
    constructive, 206
    destructive, 206
  cleavage, 203
  cycle, 207, 210, 211, 212
  débris, 35, 42, 44, 90, 102
  defined, 19, 20
  fabric, 204
  flow, 138, 200
  folds, 132, 134, 140, 141
  granite, 288, 289
  igneous, 119, 172, 243, 247, 263, 297
  metamorphic, 144; 242, 247
  permeability, 71
  phosphate, 348, 355
  platforms, 247–53
  porosity, 70
  silicate, 284
  stratification, causes of, 97
  structures, 138, 141
  weathering, 33–40
Rocks:
  crystalline, 20, 206
  elasticity of, 147
  fragmental, 20, 97, 206
  heat, conductivity of, 259
  indurated, 122, 197
  melting of, 262, 263
  methods of study of, 39, 218, 220
  Primary, 301
  sedimentary, 119, 142, 242
  stratified, 97
  thin sections of, 39
Rocky Mountains, 257–58
Roundness of rock particles, 118
Running water, 43
Rushes, 361

S

St. Hilaire, G., 223
St. Peter sandstone, 244, 326, 327, 331
Salamander, Miocene, 458
Salina beds, 331
Saline basins, 383
Saline deposits, age of earth by, 291, 295
Salt, 38, 124, 267, 331, 354
  Permian, 356
Salt domes, 428, 475
Salt Range of India, 383
Salts, potash, 356
  soluble, 38, 124

San Andreas fault, 147
Sand, 94, 101, 107, 121
  angle of rest of, 94
  glass, 348
Sand dunes, 94, 112
Sand grains, 117
Sand ripples, 95
Satellites, 282
  of Neptune, 271
  retrograde motion of, 282
  of Uranus, 271
Saturn, 280, 281
Saurischians, 409, 411
Sauropods, 409, 411, 412, 413
Scavengers, pre-Cambrian, 311
Scheuchzer, J. J., 458, 459
Schuchert, C., 294
Scorpions, 343
Sculpture, cave, 464
Scyphozoa, 334
Sea bladders, 334
Sea buds, 334
Sea cliff, 104
Sea lilies, 334
Sea scorpions, 343
Sea urchins, 334, 342, 375
Seaweeds, 299, 359
Sedgwick, Adam, 321, 344
Sediments:
  analysis of, 116
  cross-bedding in, 120
  cyclical deposition of, 351
  genetic classification of, 123
  mineralogical analysis, 119
  shape analysis, 118
  size analysis, 117
  table of, 124
Segregation hypotheses, 287–88
Seismograms, 153, 157
Seismographs, 152, 476
Seismology, 148, 162
Selective transportation, 46, 96
Selma series, 393, 395
Senecan group, 346
Septa, 374
Sequoias, 365
Series, definition of, 254
Shale, 123
  facies, 327
Shale oil, 429
Shape of rock particles, 46, 118
Sharks, 299, 369
Shastan series, 393, 394

INDEX

Shields, definition of, 305
Shore:
    of emergence, 106
    of submergence, 105
Shore agents, 43, 101–109
Sigillaria, 362, 365
Silica, 122, 190, 191
Silicate, 19
Silicate minerals, instability of, 34
Silicification, 231, 232
Silicon, 18
Silicon oxide, 19, 21, 190
Sills, 178, 199, 218
Silt, 121
Silurian period, 299, 328, 329, 334, 341, 343, 359, 367, 479
Simiidae, 226
Sinanthropus, 461, 463
Sinkholes, 79
Size frequency curves, 117
Slate, 203, 205, 331, 332
Slope creep, 42
Sloth, 230
Smith, William, 226, 227, 235, 236, 316, 385, 389
Snails, 367
Snow line, 83
Solar constant, 23
Solar equator, 279
Solar eruptions, 271
Solar radiation, 269, 279
Solfatara, 173
Solnhofen beds, 404, 406
Sorting of sediments, 96, 97, 107, 118
Specific heat, 26
Spermatophyta, 358, 363
Sphenophyllales, 358, 361, 363
Sphericity of rock particles, 118
Spiders, 334
Spits, 105
Sponges, 266, 299, 310, 312, 334, 335
Springs, 73
    hot, 173
Squantum, Massachusetts, 380
Squids, 314, 416, 422
Stacks, rock, 104
Stalactite, 79
Stalagmite, 79
Starfish, 334, 339, 342, 374, 375
Stegocephalians, 372, 373, 384
Stegosaurus, 414

Steno, 225, 235
Strabo, 223, 224
Stratification, 97, 124, 286
Stratified rocks, 97
Stratigraphy, definition of, 226
Stream:
    bed, 55, 56
    deposition, 91
    discharge, 56
    flow, 56
    gradient, 56
    levees, 110
    meanders, 57
    velocity, 55, 90, 91
    volume, 56
Streams, loaded, 90
Strike of rocks, 142
Struthiomimus, 410
Stylonurus, 343
Sulphur, 427
Sun, 264, 269, 271, 281
    gravitational pull of, 282
    oxygen in, 284
    spots, 264
    surface of, 270
Sundance formation, 390
Suture lines, 421
    Ammonite, 419, 420
    Ceratite, 419, 420
    Goniatite, 419
Sylvania sandstone, 331
Synclines, 140, 474
Synthesis, organic, 316
System, definition of, 254
Systema naturae, 225

T

Taconic disturbance, 328
Talchir tillite, 380
Tar beds, 454–56
Taungs, South Africa, 460
Temiskaming period, 299, 303
Temple of Jupiter Serapis, 129
Terebratulina, 336
Terminal velocity, 45
Terrace, 105, 111
Terrestrial deposition, 427
Terrestrial heat, escape of, 264
Tertiary, 298, 396, 424–31, 452, 463, 465
    igneous rocks, 429
    insects, 429
    marine embayments, 425
    orogenic movements, 431
    plants, 429, 430

Tetraseptate corals, 374
Thallophytes, 359
Thecodonts, 409, 410
Theophrastus, 223
Therapsids, 384, 447
Theriodonts, 373
Theropods, 411
Thinopus, 371, 372
Thorium, 270, 296
Thrust faults, 143, 396, 397
Tidal filament, 282
Tiger, 451
    saber-tooth, 455
Till plains, 98
Tillite, 123, 308, 309, 434
Titanotheres, 449, 450
Trachodon, 413
Traction loads, 94, 96
Transportation of rock débris
    agents of, sorting by, 96
    effect of particle shape on, 46
    by gravitative pull, 42
    selective, 46
    by wind, 47
Transverse waves, and central core of earth, 162
Trenton limestone, 326
Triassic, 298, 383, 385, 387, 389, 393, 403, 420, 447, 448
Trilobites, 299, 334–36, 338, 339, 340, 375, 383
Trilophodon, 453
Trinity group, 393, 394
Trinucleus, 340
Trough:
    Appalachian, 330, 331
    Cordilleran, 351
    late Paleozoic, 260
    Ouachita, 351
Turbulent flow, 45, 47, 102, 107
Turrilites, 420
Turtles, 402
    Tuscaloosa series, 393, 395
Tuscarora deep, 261
Type area, 254, 302, 346
Tyrannosaurus, 411, 414

U

Uintatherium, 450
Ulsterian group, 346
Umbo, 375
Unconformity, 127, 218, 244, 302, 428
    angular, 218, 219, 328, 330
Undertow, 101
Ungulates, 298

Upland, 63
Uranium, 270, 296, 306
Uranus, 280, 281
Ussher, Archbishop, 291

V

Valley glaciers, 88, 380
Valleys, 16, 51, 54, 57, 59, 62, 65
    hanging, 88
Valmeyer series, 350, 375
Variscan Mountains, 349
Varves, 295, 296, 442, 443
Veins, 81
"Vermes," 334
Venus, 280, 281
Vermont disturbance, 325
Vertebrates, ancestral, 367–68
Vestiges of creation, 268–74
Virgil series, 300
Viruses, 316–17
Viscosity, 75
Vishnu schists, 304, 305
Volcanic activity, 264, 355, 427
Volcanic ash, 172
Volcanic bombs, 172
Volcanic cones, 172
Volcanic dust, 172, 265
Volcanic glass, 185
Volcanism, 22, 30, 169, 287, 298, 299, 391
    causes of, 175, 176
    and diastrophism, 176
    early, 284
    explosive, 168
Volcanoes, map of, 174
Vultures, Pleistocene, 455

W

Walchia, 364, 365
Washita group, 393, 394
Water cycle, 213
Water limes, 328, 329
Waucobian series, 323, 324
Wave of translation, 101
Wave-cut terrace, 104
Waves:
    elastic, in rocks, 157
    of translation, 101
    longitudinal, 157
    transverse, 157
    water:
        "breaking of," 101
        pulsating effect of, 48, 102
        velocity of, 101
Wegener hypothesis, 288–89

INDEX

Wells, artesian, 74
Whales, 448
White Jura series, 389
White River, 465
White River beds, 429
Willis, Bailey, 288
Wind, 43, 45
Wind action, 48, 49
Wind deposits, 112
Wolves, 455
Worms, 334, 335
    pre-Cambrian, 310
Würm glaciation, 462, 463

X

Xenophanes of Colophon, 223

Y

Yellowstone National Park, 81
Yosemite Valley, 88

Z

Zinc, 350
Zion National Park, 386
Zone of flow, 261

Eau Claire State Teachers College
LIBRARY RULES

No book should be taken from the library until it has been properly charged by the librarian.

Books may be kept one week but are not to be renewed without special permission of the librarian.

A fine of two cents a day will be charged for books kept over time.

In case of loss or injury the person borrowing this book will be held responsible for a part or the whole of the value of a new book.

DUE	DUE	DUE	DUE
My 27 '37	Jan 25 '44		
My 31 '37	Je '47		
Je 7 '37	No 9 '49		
Ap 21 '39			
De 10 '43			
De 17 '43			
Ja 6 '44			
Ja 7 '44			
Ja 14 '44			